P9-ANZ-744

HISTORY AND FICTION IN GALDÓS'S NARRATIVES

History and Fiction in Galdós's Narratives

GEOFFREY RIBBANS

CLARENDON PRESS · OXFORD
1993

Oxford University Press, Walton Street, Oxford OX2 6DP
Oxford New York Toronto
Delhi Bombay Calcutta Madras Karachi
Kuala Lumpur Singapore Hong Kong Tokyo
Nairobi Dar es Salaam Cape Town
Melbourne Auckland Madrid
and associated companies in
Berlin Ibadan

Oxford is a trade mark of Oxford University Press

Published in the United States
by Oxford University Press Inc., New York

British Library Cataloguing in Publication Data
Data available

Library of Congress Cataloging in Publication Data
Ribbans, Geoffrey.
History and fiction in Galdós's narratives/Geoffrey Ribbans.
Includes bibliographical references and index.
1. Pérez Galdós, Benito, 1843–1920—Criticism and interpretation.
2. Historical fiction, Spanish—History and criticism.
3. Literature and history. I. Title.
PQ6555.Z5R46 1993 92–41137 863'.5—dc20
ISBN 0-19-815881-5

Typeset by Best-set Typesetter Ltd., Hong Kong

Printed in Great Britain
on acid-free paper by
Bookcraft Ltd
Midsomer Norton, Bath

To my wife Madeleine,
for her renowned patience

Preface

THIS book has had a long and laborious gestation. It began with an interest in how specific incidents from recent Spanish history were utilized and integrated, in different ways, into Galdós's novels and the *Episodios nacionales*. From this start grew a desire to trace on a broader scale the manner in which Galdós treated historical material in each of these forms and to distinguish between them in as precise a fashion as possible. I further set out to draw some overall conclusions about his attitude (or attitudes) to history and about his narrative technique against the general backcloth of nineteenth-century history and the genres of the realist and historical novels. It was necessary to impose certain time-limits. I have therefore confined myself to the period of the novelist's immediate past. The beginning of Isabella II's effective reign (1843) thus constitutes the starting-point of my investigations, with occasional throw-backs to earlier years when justified; the examination grows in intensity during the troubled period from the 1860s onwards and continues until the consolidation of the Restoration in the late 1870s.

Critical fashion has evolved considerably in the ten years or so in which I have been working on themes which led eventually to the present work. It will be clear from my text that I do not share many of the theoretical attitudes which are currently prevalent in literary scholarship, both in Europe and in the United States. David Lodge, expressing his regret at the 'loss of common language of literary discourse', remarks that post-structuralist 'discourse is so opaque and technical' that a cultivated reader outside the academic Establishment can no longer fail to be 'totally baffled and bewildered, unable to make any sense at all at what purports to be literary criticism'.[1] I share these regrets and, without going all the way with the rather simplistic arguments of the anti-theory reaction provoked by Steven Knapp and Walter Benn Michaels,[2] I am convinced that theoretical considerations have become dangerously top-heavy.

My disagreements are on three counts. The most important concerns history itself. Distrust of history—all historical considerations, not just an outdated deterministic historicism—has been endemic in literary criticism and creative writing since the 1920s; it has characterized both the New Criticism and structuralist and post-structuralist theory, as well as fiction itself. More

[1] *After Bakhtin: Essays on Fiction and Criticism* (London: Routledge, 1990), 4.

[2] W. J. T. Mitchell (ed.), *Against Theory: Literary Studies and the New Pragmatism* (Chicago UP, 1985).

recently, I have been gratified to see this distrust—which I have never shared—gradually recede and new forms of historical consciousness begin to emerge; Dominick LaCapra's essay 'On the Line: Between History and Criticism', in the MLA publication *Profession 89* (4–9), and the critical discussions concerning Paul de Man's work are cases in point. A second disagreement relates to the advisability, often deliberately scorned today, of maintaining a clear distinction between literature and criticism, just as one needs to distinguish between literary theory and literary criticism, as Tzvetan Todorov, among others, has indicated.[3] Third, I am conscious of the desirability of focusing on a strict division between genres as a starting-point for the evaluation of past literature; the fact that there has been a constant and fertile infiltration of one genre by another, more than a sign of the artificiality of the division, is an indication of its initial importance. In formulating these judgements, I am not seeking to supplant modern critical tendencies—particularly the subtle insights linguistic-based criticism has brought to literary studies—which have clearly proved their value, but I am anxious not to forgo methods which have been, are still, and will be in the future appropriate for the time and place of the texts I am studying.

Critical opinion has ebbs and flows like other casts of thought—history, philosophy, linguistics. Each generation tends to emphasize some facet of reality to the point of exaggeration, and in so doing both adds a new modicum of wisdom and leads by its excesses to a different, equally unbalanced emphasis; discrimination is required to assess which theoretical fads are valid or not. The present critical trend is still—or has been until very recently—at a phase of overreacting to the diachronic rigidity of nineteenth-century positivism; without returning to this inflexible position, I suggest that it is necessary to maintain a proper distance from linguistic deconstruction while accepting its undoubted contributions to a more cogent formulation of the nature of narrative. Moreover, the tendency of much modern criticism, especially that of French origin, is towards a pretentious self-assertion, an implicit claim to absolute certainty, the creation of deliberately opaque jargon, and an attempt to provide a substitute or rival for creative literature: all of which I find irritating and distasteful as well as misguided. I have little sympathy either with the constant punning and plays on words prevalent in semiotic criticism, especially in its deconstructive variety; at once capricious and frivolous, it is often based on a single language and has no more solid foundation than casual phonic identity. My personal conviction, which makes no claims to originality, is that only by a pluralistic use of appropriate criteria, both extrinsic and intrinsic, can a reasonably balanced and productive view of the literature of the past, including nineteenth-century narrative, be achieved.

[3] Quoted in Seymour Chatman, *Story and Discourse: Narrative Structure in Fiction and Film* (Ithaca: Cornell UP, 1978), 17.

My debts are many and diffuse. I am grateful, especially, to Rudi Cardona, who has done so much to stimulate Galdós studies and who first encouraged my interest in the fourth series of *Episodios nacionales*; to my colleague from Liverpool, the late Harold Hall, so unobtrusively knowledgeable about Galdós; to Alison Sinclair, who has generously shared with me her knowledge of nineteenth-century illustrations; to the stimulus provided by many friends on both sides of the Atlantic—Eamonn Rodgers, James Whiston, Pedro Ortiz Armengol, John Kronik, Peter Bly, Harriet Turner, Peter Goldman, Diane Urey, Germán Gullón, and many more; to my colleagues at Brown, especially the current Chairman of Hispanic Studies, Antonio Carreño, but also David Kossoff, Alan Trueblood, Frank Durand, José Amor y Vázquez, Stephanie Merrim, Mercedes Vaquero, for their collegiality and encouragement; to many graduate students—too many to list—who have shared my concerns and enthusiasms.

I am grateful for the assistance afforded by many academic institutions and libraries: the University of Liverpool (my home base for many years) and its Sidney Jones Library; Brown University, which has made my research possible by generous special assignments and travel grants; Brown's Rockefeller Library; the Providence Athenaeum; the Houghton Library, Harvard University; the Casa-Museo de Galdós, Las Palmas, and its director Alfonso Armas Ayala; the Biblioteca Nacional, Madrid; the British Library and the Colindale Newspaper Library; Cambridge University Library; Sheffield University Library.

Special appreciation is due to the University of Cambridge, especially Professor Colin Smith, now Emeritus, of St Catherine's College, for the flattering invitation to deliver the Norman Maccoll Lectures there in 1985. These four lectures gave rise to the first draft of a text which has been extended and revised many times since. My book also incorporates material from the following previous publications on the subject, but in every case the text has been very substantially revised and amplified, often beyond recognition: '*Historia novelada* and *novela histórica*: The Use of Historical Incidents from the Reign of Isabella II in Galdós' *Episodios* and *Novelas contemporáneas*', in John England (ed.), *Hispanic Studies in Honour of Frank Pierce* (Sheffield: Department of Hispanic Studies, University of Sheffield, 1980), 133–47 (Chapters 2, 3, and 4); 'The portrayal of Queen Isabella II in Galdós's *Episodios* and *Novelas contemporáneas*', *LA CHISPA, '81: Second Louisiana Conference on Hispanic Languages and Literature* (New Orleans: Tulane UP, 1981), 277–86 (Chapters 2, 3, and 4); '"La historia como debiera ser": Galdós's Speculations on Nineteenth-Century Spanish History', *BHS* [Memorial number for H. B. Hall], 59 (1982), 267–74 (Chapter 7); 'Galdós' Literary Presentations of the Interregnum, Reign of Amadeo and the First Republic (1868–1874)', *BHS* 63 (1986), 1–17 (Chapter 5); 'The *Restauración* in the novels and *episodios* of Galdós', *AG* 21 (1986), 79–93 (Chapter 6);

'¿Historia novelada o novela histórica? Las diversas estrategias en el tratamiento de la historia en las *Novelas contemporáneas* y los *Episodios nacionales*', in Peter Bly (ed.), *Galdós y la historia* (Ottawa Hispanic Studies, 1; Ottawa: Dovehouse, 1988), 167–86 (Chapter 1); 'Galdós's View of the Bourbon Restoration in *Cánovas*', in Dian Fox, Harry Sieber, and Robert TerHorst (eds.), *Studies in Honor of Bruce W. Wardropper* (Newark, Del.: Juan de la Cuesta, 1989), 221–36 (Chapter 6). Grateful acknowledgement is made to the copyright-holders for permission to republish this material.

G. R.

Brown University, Providence
April 1992

Contents

List of Figures

Permission from the authors to reproduce these diagrams is gratefully acknowledged.

Abbreviations

Act 1 [*2*, *3*, etc.] *AIH*	*Actas del* 1^{r} [2°, 3^{r}, etc.] *Congreso de la Asociación Internacional de Hispanistas* (1962–)
Act I	*Actas del Primer Congreso Internacional de Estudios Galdosianos* (Las Palmas: CIGC, 1977)
Act II	*Actas del Segundo Congreso Internacional de Estudios Galdosianos* (2 vols.; Las Palmas: CIGC, 1978–80)
AG	*Anales galdosianos* (Pittsburgh, Austin, and Boston, 1966–)
BAE	Biblioteca de Autores Españoles (Madrid)
BBMP	*Boletín de la Biblioteca Menéndez Relayo* (Santander)
BH	*Bulletin hispanique* (Bordeaux)
BHS	*Bulletin of Hispanic Studies* (Liverpool)
BRABL	*Boletín de la Real Academia de Buenas Letras* (Barcelona)
Cent.	*Galdós: Centenario de 'Fortunata y Jacinta' (1887–1987). Actas* (Congreso Internacional, 23–28 de noviembre, 1987; Madrid: Universidad Complutense. Departamento de Ciencias de la Información, 1989)
CG	Critical Guides to Spanish Texts (London: Grant & Cutler)
CHA	*Cuadernos hispanoamericanos* (Madrid)
CIGC	Cabildo Insular de Gran Canaria
CSIC	Consejo Superior de Investigaciones Científicas
EC: *FyJ*	*Fortunata y Jacinta*, ed. Germán Gullón (El Escritor y la Crítica; Madrid: Taurus, 1986)
EC: *G*	Douglass M. Rogers (ed.), *Benito Pérez Galdós* (El Escritor y la Crítica; Madrid: Taurus, 1973)
GS I	J. E. Varey (ed.), *Galdós Studies* (London: Tamesis, 1970)
GS II	Robert J. Weber (ed.), *Galdós Studies II* (London: Tamesis, 1974)
GyH	Peter Bly (ed.), *Galdós y la historia* (Ottawa Hispanic Studies, 1; Dovehouse, 1988)
HFCE	Harry Furness Centenary Edition (Thackeray)
Hisp.	*Hispania* (Mississippi)
HR	*Hispanic Review* (Philadelphia)
KRQ	*Kentucky Romance Quarterly* (Lexington, Ky.) (see *RQ*)
LD	*Letras de Deusto* (Bilbao)
LT	*La torre* (San Juan de Puerto Rico)
MLN	*Modern Language Notes* (Baltimore)
MLR	*Modern Language Review* (Cambridge)
MP	*Modern Philology* (Chicago)
Neo.	*Neophilologus* (Amsterdam)
NRFH	*Nueva revista de filología hispánica* (Mexico)

OC	*Obras completas* [omitted in the case of the Galdós Aguilar edition, where the volume-numbers are given in bold type]
OI	Benito Pérez Galdós, *Obras inéditas*, ed. Alberto Ghiraldo (4 vols.; Madrid: Renacimiento, 1923–30)
PMLA	*Publications of the Modern Languages Association* (New York)
PQ	*Philological Quarterly* (Iowa City)
PSA	*Papeles de Son Armadans* (Palma de Mallorca)
PUF	Presses Universitaires de la France (Paris)
RCEH	*Revista canadiense de estudios hispánicos* (Toronto)
REH-PR	*Revista de estudios hispánicos-Puerto Rico*
RF	*Romanische Forschungen* (Tübingen)
RFE	*Revista de filología española* (Madrid)
RHM	*Revista hispánica moderna* (New York)
RN	*Romance Notes* (Chapel Hill)
RNC	*Revista nacional de cultura* (Caracas)
RO	*Revista de Occidente* (Madrid)
RQ	*Romance Quarterly* (Lexington, Ky.) (formerly *KRQ*)
Symp.	*Symposium* (Syracuse)
WLE	Waverley Library Edition (Scott)

Treatment of Spanish Proper Names

It is difficult to strike a satisfactory balance between common English usage and consistency in dealing with foreign historical figures. My solution has been to use English forms for reigning monarchs in all countries, including Spain: hence Charles IV, Ferdinand VII, Isabella II; and for all other Spaniards to use the Spanish originals, even when the same names are involved: thus Don Carlos Isidro, María Isabel, Jaime, Juan, Francisco de Asís, etc.

Bibliographical References

Bracketed references in the text and notes are to the various sections of the Bibliography. Thus *1a* refers to General History; *3c* to Galdós: Novels; *4b* to Other Literary and Critical Works: other literatures. Where there is more than one item under an individual author, the date (with *a*, *b*, *c*, as necessary) is given. Volume-numbers are listed immediately before the page-numbers. References to Galdós's texts are to *Obras completas* (Aguilar); the volume-number is given in bold type. Since the various editions have different pagination and to facilitate the consultation of alternative editions, reference is also made, in the case of the three lengthy novels, *La desheredada*, *Fortunata y Jacinta*, and *Ángel Guerra*, to part, chapter, and subchapter divisions, according to the following matrix: **V**: I. 2. iii, 44 (Aguilar volume-number; part; chapter; subchapter; page). References are always to volume **V** of the *Obras completas* where a text is duplicated in volume **VI** (see the footnote to section *3a* of the Bibliography).

Introduction
The Novel and History

THE relations between history and creative literature (poetry) have always been a matter of critical debate.[1] For generations since the Renaissance the distinction culled from Aristotle's *Poetics* held sway:

The poet's function is to describe not the thing that has happened, but the kind of thing that might happen, i.e., what is possible as being probable or necessary. The distinction between historian and poet is not in the one writing prose and the other verse—you might put the work of Herodotus into verse and it would still be a species of history; it consists really in this, that the one describes the thing that has been, and the other a kind of thing that might be. Hence poetry is something more philosophic and of graver import than history, since its statements are of the nature rather of universals, whereas those of history are singulars (*2*, 25).

History and narrative literature—epic poetry, later the novel—were thought of as parallel narrative forms, with certain similarities implied by the common concern with story but with a clear superiority accorded to literature and marked differences of content and treatment.

By the nineteenth century this parallel scheme no longer holds. The rise of empirical knowledge based on observation and experiment—what we can loosely call 'science'—during the eighteenth century introduced a new factor which steadily enhanced history's status and pulled it away from literature. The new historical approach, exemplified by Leopold von Ranke, was founded, not on literary language or imagination, but on chronology, documentation, and a claim to impartiality:

[Ranke] wanted the past to speak for itself, by recording exactly what had happened, '*wie es eigentlich gewesen*.' Ranke wanted to picture the past in words as it offered itself to the observer's view, in exactly the same manner as the natural historian of his time began to prefer exact illustrations of the phenomena of nature over the traditional literary descriptions.[2]

[1] The following account is not intended to be a full discussion of the enormous issue of history and the novel. Its purpose is limited to providing an adequate background for the subsequent treatment of Galdós, to which it is expressly directed.

[2] Ilsa N. Bulhof (*1a*, 5). Jauss has shown decisively that Ranke's claim to treat events as they happen is illusory: 'the historian Ranke continually reveals himself through *a posteriori* viewpoints and aesthetic classifications that could have played no part in the lives of those who actually experienced the historical event' (*2*, 55). Brook Thomas comments ironically, 'World history confirmed God's will to have a higher state of civilization progressively emerge in Europe' (in *2*, Veeser 189).

History in this new guise became, thanks to the astounding achievement of the German historical tradition, the predominant humanistic form running in tandem with the physical and biological sciences and increasingly influenced by them. It thus enjoyed an unchallenged position of privilege.[3] Under the auspices of Hegel, every intellectual and cultural activity was permeated with historical values,[4] as part of what Hayden White calls 'a rage for a realistic apprehension of the world' (*1a*, 1973: 45). History in turn, in common with other branches of knowledge and culture, was seized by the urge to become more and more scientific, to develop inescapable laws, to adopt a posture of God-like irrefutability.

The representative metaphor of the nineteenth century, for history as well as life and literature, is perhaps the tree: a steady organic growth, constructive, solid, predictable, uninterrupted.[5] Of not infrequent occurrence is a faster and more dynamic image which has found even more favour in modern times: the stream. In both concepts there is a progression representing a diachronic orientation. Temporal rather than structural, the stream denotes both a continuum and a sign of unperceived change: 'like those who drift down the stream of a deep and smooth river, we are not aware of the progress we have made until we fix our eye on the now distant point from which we have been drifted' (*2*, Scott, quoted in Fleishman 25). Used in this way, the flow is a controlled one conditioned by chronological considerations.[6] The image lends itself, however, to further diversification, becoming more varied, less predictable, capable of ebbs and flows. This is rather the sense of Thomas Hardy's intriguing note in his diary (1885), which also uses the tree metaphor only to cast it aside:

History is rather a stream than a tree. There is nothing organic in its shape, nothing systematic in its development. It flows on like a thunderstorm-rill by a road side; now a

[3] See Geoffrey Barraclough: 'For a century and a half, from the time of the French Revolution, historical principles and historical conceptions dominated, shaped and determined the character of European thought' (*1a*, 1). See also Lukacs (*1a*, 18). The schematic categories Lukacs establishes—'Eighteenth century: History as *literature*; the *narrated* past. Nineteenth century: History as *science*; the *recorded past*. Twentieth century: . . . history as *a form of thought*; the *remembered* past' (22)—though over-facile, are useful as a rough guide.

[4] As Stephen Bann has recently observed, '19th-Century man did not simply discover history; he needed to discover history, or, as it were, to remake history on his own terms' (*2*, 103). John Lukacs comments, '*Wie es eigentlich gewesen* could have been Tolstoy's or Zola's motto besides Ranke's' (*1a*, 119).

[5] See Carlyle, *The Hero as Poet*: 'It is all a Tree: circulation of sap and influences, mutual communication of every minutest leaf with the lowest talon of a root with every other greatest and minutest portion of the whole' (*4b*, 102). For organic images of growth in literary invention from the 18th cent. onwards see Abrams (*2*, 184–225).

[6] Compare Carlos Clavería: 'Siglo de la Historia ha sido llamado el siglo XIX. . . . los hombres de la época adquieren ya definitivamente entonces la conciencia de su conexión con el mundo circundante, la conciencia de estar insertos en el fluir de los tiempos, en la evolución de una Humanidad en constante progreso hacia el futuro, la conciencia de que la Historia es la esencia del vivir humano' (*3b*, 1957: 170).

straw turns it this way, now a tiny barrier of sand that. The offhand decision of some commonplace mind high in office at a critical moment influences the course of events for a hundred years. (*4b*, F. E. Hardy 172)

History conceived of as a stream rather than a tree, its directions not determined by any law but casually, by mediocre individuals: the idea of organic evolution is beginning to break down. The water image is also amenable for development into the tide of unconscious association we associate with William James, Bergson, Proust, Joyce, or Antonio Machado.

The scientific or mechanistic criterion did not have it all its own way. Though its perspective is no less historicist,[7] the powerful Romantic impulse, entranced by the sweep of human endeavour, moves history towards grandiose concepts and transcendental issues. Schopenhauer had already cast serious doubts upon the significance of history, which he saw as inferior to philosophy and science, and akin to journalism, concerned only with individual things and incomplete in its findings (*4b*, 221–2).[8] By the late nineteenth century, a stronger reaction has begun to set in.[9] The full blast came, of course, with Nietzsche, who rails against the Hegelian concept of history, against history as a science, and emphasizes the dangers of an inert cult of history not harnessed to 'life': 'When the historical sense reigns *without restraint*, and all its consequences are realized, it uproots the future because it destroys illusions and robs the things that exist of the atmosphere in which alone they can live' (*4b*, 95).

Typical twentieth-century metaphors for life, now that it can no longer be automatically identified with history, break with the coherent linear development seen earlier; they are static, reiterative, circular, or ambiguous, synchronic rather than diachronic. A few random examples might be Proust's preoccupation with ocular devices,[10] Virginia Woolf's Nietzschean metaphor of the waves,[11] distorting mirrors in Valle-Inclán and Machado,[12] the library in Borges.[13]

[7] Lehan (*2*, 533–6) gives a useful survey of historical attitudes, but telescopes Romantic and positivist concepts of history.

[8] Schopenhauer's most vehement remark is that 'Clio, the muse of History, is as thoroughly infected with lies as a streetwalker is with syphilis' (*4b*, 22).

[9] The strongest—but not the most subtle—attack on scientific history comes in Popper's *The Poverty of Historicism* (*1a*).

[10] 'L'ouvrage de l'écrivain n'est qu'une espèce d'instrument optique qu'il offre au lecteur afin de lui permettre de discerner ce que, sans ce livre, il n'eût peut-être pas vu en soi-même' (*4b*, quoted in Shattuck 106).

[11] See the title and the last paragraph of the novel: 'the eternal renewal, the incessant rise and fall and fall and rise again' (*4b*, 297). Compare Azorín's image of the clouds: 'Vivir es ver pasar: ver pasar allá en lo alto las nubes. Mejor diríamos: vivir es *ver volver*' (*4a*, 137).

[12] Valle-Inclán: 'Los héroes clásicos reflejados en los espejos cóncavos dan el Esperpento. El sentido clásico de la vida española sólo puede darse con una estilística sistemáticamente deformada' (*4a*, *Luces de bohemia*, XIII, p. 168). Machado: 'allí te vi vagando en un borroso | laberinto de espejos' (*4a*, XXXVII, p. 133); 'El alma que no sueña, | el enemigo espejo, | proyecta nuestra imagen | con un perfil grotesco' ('Introducción', LXV, p. 180).

[13] 'El Universo (que otros llaman la Biblioteca)', 'La Biblioteca de Babel' (*4a*, 85).

As confidence in the privileged role of history sagged, there arose a cautious distrust in universal theories, and with it all attempts—such as Spengler's, Arnold Toynbee's, or Ortega y Gasset's—to discern an all-embracing pattern in history. In this respect W. H. Walsh makes a useful distinction between the meaning *of* history, a subject more appropriate for philosophers and theologians, and meaning *in* history. The latter consists of an enquiry into 'a stretch of events restricted in time and to some extent also in space of which historians are trying to make sense' (*1a*, Gardiner 302): a much more manageable task.

Despite all hostile incursions, a concept of historical-mindedness subsists: what Huizinga called 'the intellectual form... in which a civilization renders account of its past' (*1a*, quoted in Kelley 14). A workable modern definition which I shall utilize in what follows is Donald R. Kelley's:

> 'historicism' refers to that cast of mind which, consciously or unconsciously, turns not to nature but to the world of man's making; which seeks out not the typical but the unique; which emphasizes the variety rather than the uniformity of human nature; which is interested less in similarities than in differences; and which is impressed not with permanence but with change. (*1a*, 4–5)

Meanwhile, the influential *Annales* school has been conspicuous in continuing the scientific tendency, rejecting narrative history as literary and accusing it of dramatizing or novelizing its material. Its leading light, Fernand Braudel, declared that,

> In fact, though, in its own covert way, narrative history consists of an interpretation, an authentic philosophy of history. To the narrative historians, the life of men is dominated by dramatic accidents, by the actions of those exceptional beings who occasionally emerge, and who are often masters of their own fate and even more of ours... A delusive fallacy, as we all know. (*1a*, 11)[14]

Other historians have shown more affinity with literature. Herbert Butterfield, for example, in a pioneer study of 1924, treated the historical novel with great respect. He seems to me partially right, if rather simplistic, in the contrast he then draws between the historian and the novelist:

> With a set of facts about the social conditions of England in the Middle Ages, the historian will seek to make a generalisation, to find a formula; the novelist will seek a different sort of synthesis and will try to reconstruct a world, to particularise, to catch a glimpse of human nature. Each will notice different things and follow different clues; for to the historian the past is the whole process of development that leads up to the present; to the novelist it is a strange world to tell tales about. (*1a*, 1924: 113)

The last phrase, in particular, indicates a naïve view of the discipline inherent in imaginative creation. In a later study ('History as a Branch of Literature',

[14] Hayden White does not hesitate to judge the reasoning behind this approach as 'jejune' (*1a*, 1987: 32–3).

1951), though still very aware of the achievements of narrative (i.e. literary) historians, Butterfield draws away from them by admitting that they 'are in danger of being too intent on scenic display ... to be too merely pictorial in their attempted resurrections of the past' (*1a*, 1951: 250).

John Lukacs also argues for a close relationship between history and fiction, though he tilts the balance in favour of the former, opting for the historicity of fiction rather than the fictional nature of history (*1a*, 115). As he rightly says, 'To illustrate moral issues, therefore, the novelist may invent plausible potentialities; this the historian cannot do' (127). It does not follow, however, as he goes on to claim, with understandable pride in his craft, that the historian's is the more difficult task. Finally, a recent essay by Stephen Bann has urged an eclectic approach to all historical matters: 'What I am therefore proposing ... is the need for a historical view which will include both the amateur and the professional, Carlyle and Ranke, the historical novel and the text edited by the Public Record office, rather than insisting that they differ like chalk and cheese' (*2*, Bann 103).

Philosophers of history have likewise often shown a concern with its relations with literature. A major example is Benedetto Croce, who took an individual idealistic stance divergent from the main European tradition. His all-pervading, 'absolute' historicism is anti-scientific and anti-deterministic, essentially concerned with the present and seeking discrete individual progress in a constant movement towards freedom. 'In Croce's historicist world, the fundamental reason for studying the past is to understand the historically specific reality of the present' (*4b*, Roberts 271). It is, moreover, inextricably related to humanism: 'Poetry and history are, then, two wings of the same breathing creature, the two linked moments of the knowing mind' (*1a*, Croce 1921: 313–14).

A comparable case is R. G. Collingwood, influenced as he was by Croce. He emphasizes that both history and fiction share the quality he calls the *a priori* imagination:

> Each of them—the historian and the novelist—makes it his business to construct a picture which is partly a narrative of events, partly a description of situations, exhibition of motives, analysis of characters. Each aims at making his picture a coherent whole, where every character and every situation is so bound up with the rest that his character in this situation cannot but act in this way, and we cannot imagine him as acting otherwise. The novel and the history must both of them make sense; nothing is admissible in either except what is necessary, and the judge of this necessity is in both cases the imagination. Both the novel and the history are self-explanatory, self-justifying, the product of autonomous or self-authorizing activity; and in both cases this activity is the *a priori* imagination. (*1a*, 245–6)

Collingwood's approximation of history and fiction is constructive and tempting, but I have none the less some reservations. The novelist does, it is true, to use Collingwood's own image, fill in the gaps of the web of unascertainable facts with plausible data spun from his imagination; and to do this he needs a special degree of historical discernment which George Eliot calls the 'veracious imagination'.[15] Avrom Fleishman comments pertinently: 'We might compare the historical novelist to the restorer of a damaged tapestry, who weaves in whole scenes or figures to fill empty places which a more austere museum curator might leave bare' (*2*, 6–7). As with Butterfield, however, Fleishman is a little uneasy about Collingwood's possible attribution of too wayward a concept of 'fancy' to the novel, with all that the term has implied since Coleridge.[16] It is evident that imagination, in Coleridge's sense, not fancy, is the more applicable term. And the novelist does much more in addition, for he has to operate on a different plane of reality. Hans Robert Jauss puts it very well: 'Literary works differ from purely historical documents precisely because they do more than simply document a particular time, and remain "speaking" to the extent that they attempt to solve problems of form or content, and so extend far beyond the silent relics of the past' (*2*, 69).

Robert Scholes makes another important distinction, which applies not only to historical texts but to a lesser degree to historical fiction:

> History is a narrative discourse with different rules than those that govern fiction. The producer of a historical text affirms that the events entextualized did indeed occur prior to the entextualization. Thus it is quite proper to bring extra-textual information to bear on those events when interpreting and evaluating a historical narrative. Any important event which is ignored or slighted by a historical narrative may properly be offered as a weakness in that narrative. It is certainly otherwise with fiction, for in fiction the events may be said to be created by and with the text. They have no prior temporal existence, even thought they are presented as *if* they did. (*2*, 1981: 207)

Even if the novelist does not tamper with the historical evidence, he does inevitably select it, modify it, rearrange it, by fitting it into the fictional structure he has chosen. Moreover, if he wishes to recast the historical material, he may do so, and justifiably. Alfred de Vigny, in the 'Reflexions sur la Vérité dans l'Art', which serves as a preface to *Cinq-Mars* (1827), proclaims the novelist's right, in the name of artistic truth and moral example, to subordinate factual reality to the idea that a historical character was meant to present. Moreover, the images he employs are relevant to our purpose:

> Ne voyez-vous pas de vos yeux la chrysalide du FAIT prendre par degré les ailes de la FICTION? — Formé à demi par les nécessités du temps, un FAIT est enfoui tout obscur

[15] 'By veracious imagination, I mean the working out in detail of the various steps by which a political or social change was reached, using all extant evidence and supplying deficiencies by careful analogical creation . . . all these grand elements of history require the illumination of special imaginative treatment' (*2*, quoted in Fleishman 158).

[16] In contrast with Imagination, which 'dissolves, diffuses, dissipates, in order to recreate' and is 'essentially *vital*', Fancy is 'indeed no other than a mode of Memory emancipated from the order of time and space' (*4b*, 159–60). For a discussion of these concepts see Richards, especially ch. IV, 'Imagination and Fancy' (*4b*, 72–99).

et embarrassé, tout naïf, tout rude, quelquefois mal construit, comme un bloc de marbre non dégrossi; les premiers qui le déterrent et le prennent en main le voudraient autrement tourné, et le passent à d'autres mains déjà un peu arrondi; d'autres le polissent en le faisant circuler; en moins de rien il arrive au grand jour transformé en statue impérissable. Nous nous recrions . . . (*4b*, 26)

While this criterion holds for real historical figures, it applies *a fortiori* to the invented characters on whom the construction of the fiction without doubt depends. In other words, the liberty any novelist takes with his material, the manner in which and the degree to which his discriminating selectivity is applied, is and must be so much greater than that permitted to even the most imaginative historian in constructing his necessarily provisional and incomplete vision of the past as to constitute a difference of kind rather than of degree.[17]

At the same time, we must be careful not to build up historical novelists into pseudo-historians.[18] One should not look in a historical novel for a documentary contribution to historical knowledge. Flaubert himself got himself off on the wrong foot in his reply to Sainte-Beuve by seeking to defend the accuracy of *Salammbô*.[19] It is not just that a novelist is unlikely to have the same access to primary sources as a professional historian; even if he had, it would be liable to get in the way of his fictional narration. It is not the novelist's function to provide information for the social historian; this is purely marginal to his activities, even though what he does carry out—the assimilation of known material into a wider context—may, and in practice does, elucidate in valuable ways the social environment of concern to the historian. The novelist's purpose may or may not need any factual underpinning, so that his imagination has a freer range without being exempt from intellectual restraint, a freedom which has been progressively more and more exploited since the decline of realism.[20] For future reference, finally, we may note how very close Collingwood comes to saying that both historian and novelist are completely determined by their material.

The historical imagination is also stressed in Hayden White's challenging approach to historiography. White reverses the hitherto normal tendency to impose historical methodology on literature, such as we have just seen with John Lukacs, by applying distinctively rhetorical criteria to the main forms of

[17] Contemporary writing clearly accepts no limit on the imaginative distortion of historical material; one example, among many, is Juan Goytisolo's treatment of the legend of Count Julian.

[18] '. . . as soon as the past is converted into the language of the novel, into the form of an imaginative literary work, the sacred essence of history has been adulterated' (*1a*, Huizinga 1926: 49).

[19] As Amado Alonso says, it is a question of having 'triunfado en la minucia pero aceptando la derrota en lo fundamental' (*2*, 76). A much better-founded criticism is Flaubert's own: 'la faute énorme . . . le piédestal est trop grand pour la statue' (*4b*, 175; see also 'Introduction', vol. i, pp. civ–cx).

[20] The fact that contemporary writers—novelists, chroniclers, critics—often choose deliberately to blur the old distinctions of genre does not mean that these distinctions do not continue to apply, in a practical rather than theoretical sense, to 19th-cent. practitioners.

historical writing in the nineteenth century.[21] If 'what was at issue throughout the nineteenth century, in history as in both art and the social sciences, was the form that a genuinely "realistic representation of historical reality" ought to take' (*1a*, 1973: 432), White claims that, for historians as well as novelists, there is no certain scientific or objective proof of superiority of one response over another. The crisis of modern historiography lies in the fact that it finds itself squeezed into an uneasy middle position between science and literature (*Tropics of Discourse*: *1a*, White 1978). The term *history* produces an 'inevitable' confusion as a result of its diversity of meanings, and approaches literature in its use of the imagination:

> The difficulty with the notion of a truth of past experience is that it can no longer be experienced, and this throws a specifically historical knowledge open to the charge that it is a construction as much of imagination as of thought and that its authority is no greater than the power of the historian to persuade his readers that his account is true. This puts historical discourse on the same level as any rhetorical performance and consigns it to the status of a textualization neither more nor less authoritative than literature itself can lay claim to. (*1a*, 1987: 147)[22]

Such a view virtually equates past events which can no longer be captured with purely imaginative experiences. This is of course precisely the illusion a realist novelist is endeavouring to create, but it is important to recognize that it is an illusion. Furthermore, I would suggest that this *rapprochement* to literature may introduce a perilous imprecision into historical writing, to such an extent that it gets White into hot water with his fellow historians; a representative figure complains that by his criterion, 'the formalist literary and linguistic analysis of the end product of the historian's labor, the narrative, obscures the interpretation of the meaning of the signifying evidence that lies at its basis' (*1a*, Bulhof, 16).[23] White also makes a valuable addition to our critical terminology, which he calls 'emplotment' and defines as follows: 'Emplotment is the way by which a sequence of events fashioned into a story is gradually revealed to be a story of a particular kind' (*1a*, 1973: 7).

The fact is that the novelist does not attempt at all, in a sizeable if variable part of his endeavour, to make as close an approximation to the concrete

[21] In *Metahistory* (*1a*) White defines four key historians of 19th-cent. realism according to a rigid rhetorical pattern derived from Northrop Frye: Michelet, as Romance; Ranke, as Comedy; Tocqueville, as Tragedy; Burckhardt, as Satire.

[22] *The Content of the Form*, which has the highly pertinent subtitle *Narrative Discourse and Historical Representation* (*1a*), traces the complex relation between historiography and narration in various authors, among whom Foucault, Jameson, and Ricœur are included; some use is made of this material in the course of this chapter.

[23] See also Fox-Genovese, 'Literary Criticism and the Politics of the New Historicism': 'Contemporary critics tend to insist disproportionately on history as the ways in which authors have written about the past at the expense of what might actually have happened, insist that history consists primarily of a body of texts and a strategy of reading or interpreting them' (*2*, in Veeser 216). Hayden White is cited in a footnote to this passage.

historical reality as he can; on the contrary, he deliberately constructs fictional tales which by definition did not happen, though they may have as much plausibility, or perhaps even more, as what actually happened. The Aristotelian distinction—to which I shall have occasion to return—continues to be useful, not as indicating an incompatibility, much less a superiority, but a divergence of perspective, descriptive rather than normative.

Given the nineteenth century's essentially historicist vision, it is inevitable that the novel of this period will likewise be imbued with historicity. It is what is loosely called the realist novel, characterized, in J. P. Stern's words, by 'an unabating *interest* in this world and in this society as a thing real and, as to its reality, wholly unproblematic' (*4b*, 5). The two forms, novel and historiography, share many features, such as a sequential narrative technique and a concern with society, character, and chronology. Such characteristics make for a notably biographical approach,[24] which in its extreme form embraces the cult of the great man, as advocated by Carlyle,[25] while recurring characters, such as we find in Balzac, serve to reinforce, at least externally, the illusion of life (*2*, Robert 253 ff.).

Inevitably, in literature as in historiography, there occurred a parallel powerful reaction against all-embracing nineteenth-century historical values. Thus the Symbolist aesthetic, in an assertion of literary identity and autonomy, showed a revulsion against history as well as against naturalism. Not only was traditional narrative in crisis, but a work of fiction which contained explicit and recognizable historical figures, milieux, or references was seen as guilty of a sort of literary lese-majesty, an adulteration of a pure literary aesthetic. Such an attitude characterizes the largely aesthetic concept of the novel still represented, for example, by E. M. Forster,[26] who proclaims the rough-and-ready truth that 'History develops, Art stands still' (*2*, 28), and resolves to ignore chronology. History, he declares, is a 'great tedious onrush' which 'only carries people on, it is just a train full of passengers' (173). His is a view essentially determined by an intrinsic criterion of the novel as an art form; it thus has little patience with history, precisely because historical elements contained within the novel seem to him to lack the two qualities he prizes most: selectivity and discrimination. Percy Lubbock's well-known strictures

[24] As Stephen Gilman notes: 'The notion of biography as such . . . depends . . . on perception of the historicity of individual lives' (*3c*, 11).

[25] 'Universal History, the history of what man has accomplished in this world, is at bottom the History of the Great Men who have worked here' (*4b*, 1). The shadow of Napoleon looms large, as Marthe Robert (*2*, 237–47) has shown. Julien Sorel, in Stendhal's *Le Rouge et le noir*, consciously models himself on Napoleon. For another example, from the work of the Brazilian novelist Machado de Assís, see Gledson (*4b*, 97–8). It is noteworthy that Carlyle treats Napoleon as the 'second modern king' (*4b*, 237), but inferior to Cromwell: he is 'a great implement too soon wasted, till it was useless: our last Great Man' (243).

[26] In his famous Clark Lectures at Cambridge in 1927, published as *Aspects of the Novel* (*2*).

on the historical dimension of *War and Peace* (*2*, 26–58), more or less contemporary with Forster's lectures, have a similar motivation, though they derive from a different source: the elaborate and fastidious theory and practice of Henry James.[27] James is, naturally, equally intolerant of the historical novel.[28]

The concern with intrusion within a closed narrative structure is an ancient one. For our purposes we may take it back to Stendhal, in whom so many of the problems and options of narrative fiction are found. In the famous passage in *Le Rouge et le noir*, the author resists the publisher's demands to include an explicit political allusion:

> — La politique, reprend l'auteur, est une pierre attachée au cou de la littérature, et qui, en moins de six mois, la submerge. La politique au milieu des intérêts d'imagination, c'est un coup de pistolet au milieu d'un concert. Ce bruit est déchirant sans être énergique. Il ne s'accorde avec le son d'aucun instrument. Cette politique va offenser mortellement une moitié des lecteurs, et ennuyer l'autre qui l'a trouvée bien autrement spéciale et énergique dans le journal du matin ...
>
> — Si vos personnages ne parlent pas politique, reprend l'éditeur, ce ne sont pas des Français de 1830, et votre livre n'est plus un miroir, comme vous en avez la prétention ... (*4b*, II. xxii, pp. 575–6).

Thus, in a brilliant if untidy piece of self-reflexivity,[29] the essential question of aesthetic integrity or mimetic realism is posed without any solution being offered.

The crisis of the novel form, then, parallels the renewed uncertainty about history and science. It causes an inward-looking direction which pulls fiction away from positivism, history, and referentiality and towards the more imaginative established genres of poetry and drama, whose aesthetic superiority is clearly asserted. Thus the New Criticism, with its insistence on the inviolate autonomy of the literary work, radically separated the text—the verbal icon, the well-wrought urn—from its circumstances[30] and condemned history and politics, together with genres like narrative contaminated by them, to outer darkness.

Concern over the inherent difficulties of merging history and fiction had of course been voiced far earlier. The classic case is Manzoni's well-pondered

[27] 'Life being all inclusion and confusion, and art being all discrimination and selection' (Preface to *The Spoils of Paynton*: *2*, James 529); 'I delight in a deep-breathing economy and an organic form' (Preface to *The Tragic Muse*: *2*, 515). For representative samples of novelists' critical views consult Allott (*2*).

[28] 'The "historical" novel is, for me, condemned ... to a fatal *cheapness* ... *the* real thing is almost impossible to do ... you have to simplify back by an amazing *tour de force*—and even then it's all humbug' (5 Oct. 1901: *2*, 348).

[29] Alter (*2*) does not make use of the potentialities of Stendhal's passage for his purposes. Victor Brombert notes, without reference to this passage, that 'The real action of the novel is thus indistinguishably associated with the very process of writing the novel' (*4b*, 75).

[30] 'Rescuing the text from author and reader went hand in hand with disentangling it from any social or historical context' (*2*, Eagleton 1983: 48).

acceptance of Goethe's rather incidental criticism, on Aristotelian grounds, of *I promessi sposi*:

In picking up a 'historical novel,' the reader knows well enough that he will find there *facta atque infecta*—things that occurred and things that have been invented, two different objects of two different, fully contrary, sorts of belief... [B]oth critics are right: both those who want historical reality always to be represented as such and those who want a narrative to produce in its reader a unified belief. But both are wrong in wanting both effects from the historical novel, when the first effect is incompatible with its form, which is narrative, and the second is incompatible with its materials, which are heterogeneous.[31]

Amado Alonso, in his profound commentary on Manzoni's arguments, concludes that he commits 'la noble herejía' of interpreting verisimilitude historically and not universally: 'de las dos parejas aristotélicas de conceptos, lo verosímil-universal y lo cierto-particular, Manzoni eliminó de su razonamiento el segundo de cada una' (*2*, 67–8). This results in an imbalance due to excess of documentation which, from a position perhaps unduly close to the ahistorical criteria of the New Criticism, Alonso finds inherent in the historical novelist.

Modern linguistically orientated theory—semiotics, structuralism, and post-structuralism—accentuates the ill-concealed scorn felt for historicist criteria[32] or for its cognate form, the realist novel.[33] To an enduring influence of the Symbolist aesthetic[34] is added a powerful impulse drawn from Saussurean linguistics. Following Saussure's reaction against the diachronic tradition of philological study—in itself parallel to the scientific school of historians—in favour of synchronic linguistics, structuralists and post-structuralists have revealed an acute awareness of the shifting nature of discourse and have centred their attention on the analysis of discourse as such: semiosis replaces mimesis. This criterion, though well justified in itself, leads them not only to

[31] I quote from Sandra Bermann's translation: *4b*, 70–2. The edition contains a thorough introduction, but Bermann kills off the historical novel far too early: 'By 1850... the genre was already in decline' (44). Ortega's view is not dissimilar; he speaks of 'la enorme dificultad—tal vez imposibilidad—aneja a la llamada "novela histórica". La pretensión de que el cosmos imaginado posea a la vez autenticidad histórica mantiene en aquélla una permanente colisión entre dos horizontes' (*2*, 1925: 135).

[32] Paul Ricœur complains about the disregard literary critics show concerning time and hence history (*1a*, quoted in White 1987: 167). He registers a similar complaint about historians who have severed their ties with storytelling (169–84).

[33] As David Lodge observes, 'It will be obvious how this view of language [that of the "Tel Quel" critics] militates against any mimetic theory of literature and especially against that status of the most mimetic of all the genres, the realist novel' (*2*, 1979: 62). Compare also Frank Lentricchia's comment: 'The traces of the New Criticism are found in yet another way: in the repeated and often extremely subtle denial of history by a variety of contemporary theorists' (*2*, Preface, p. xiii).

[34] The influence of Mallarmé is very evident, for instance, in both Barthes (*2*, 1977: 143) and Derrida (*2*, 1974: xiii).

challenge the nineteenth-century privileged status of history, but to reduce radically its overall importance: 'historical knowledge has no claim to be opposed to other forms of knowledge as a supremely privileged one', declares Claude Lévi-Strauss (*2*, quoted in De George 229), in his polemic with Sartre, who himself is described as having 'an almost mystical conception of history' (221). In similar fashion, Lévi-Strauss denies the totalizing effect of history: 'historical facts are no more *given* than any other . . . History is therefore never history, but history-for' (222–3).

Roland Barthes, for his part, equates 'historical' and 'fictional' discourse in a common self-referentiality at the same time as he discards the presuppositions of realism and objectivity. As Hayden White points out, his 'principal aim was to attack the vaunted objectivity of traditional historiography' (*1a*, 1987: 35). In a famous essay, 'The Discourse of History', Barthes insists that narrative offers, not mimesis, but merely *l'effet du réel*: 'in "objective" history, the "real" is never more than an unformulated signified, sheltering behind the apparently all-powerful referent. This situation characterizes what might be called the *realistic effect*' (*2*, 1981: 17). The point that it is impossible to transfer reality directly into literature, so often uncritically assumed by realist writers as their goal,[35] is well taken; but it needs to be emphasized once more that the fact that historical and fictional narrative share a common rhetorical mode does not mean that they can be identified as far as aim and content are concerned.

Together with the relegation of history goes Barthes's other essential and controversial contention, central to post-structuralist theory, that the literary text is completely autonomous and can be entirely separated from the author, whose 'death', in a famous phrase, is accordingly celebrated (*2*, 1977: 142–8). Jacques Derrida's doctrine of deconstruction takes this concept to its limits. It not only dislodges history, as it does all other referential concerns, from its privileged status, but by claiming that the signified and the signifier are inseparable, he 'decentres' the text and arrives at his famous dictum: 'il n'y a pas de hors-texte' (*2*, 1974: 158; see *2*, Harvey 102–3). Thus 'In the undecidable textual moment called the *aporia*, all historically locatable thematic categories are "torn apart"' (the words are de Man's) (*2*, Lentricchia 181). Derrida's American followers of the Yale group—Paul de Man, Geoffrey Hartman, J. Hillis Miller—in this respect still not entirely free from the influence of the New Criticism, are even less concerned with history. As Frank Lentricchia declares, '*Mise en abyme*, thanks to Miller and company, has

[35] See Robert's severe judgement on Balzac (*2*, 237–91). Galdós's 'Imagen de la vida es la novela' (*3a*, Menéndez y Pelayo—Pereda—Pérez Galdós, *Discursos* . . . 11) is another case in point; see Ribbans (*3c*, 1977: 14). On the whole question of realism, consult Stern's perceptive remarks on the two contexts (life and literature) in which the term is used: 'a way of thinking (in the one case [real life]) and a way of writing (in the other [literature]), each of which is positively related to the world' (*4b*, 42).

become a battle cry, along with two corollary themes: that reading is no mimesis but a violence of mastery and substitution, and that history, in the models of Lovejoy and Abrams, is a chimera' (*2*, 179). In Edward Said's words, 'Textuality has therefore become the exact antithesis and displacement of what might be called history' (*2*, 1983: 3–4).

This is not the place to discuss at any length the complicated ramifications of the divorce between author and text, exemplified by critics like Tzvetan Todorov.[36] My own concern is to stress the validity within literary works of outside referents like historical data, a criterion which also involves making a pragmatic separation of the signifier and the signified and giving the author due credit for his creation; I readily embrace the sort of pluralism which (in his typically leisurely humanistic fashion) Wayne Booth elucidates from the practice of M. H. Abrams.[37] Derrida's free-ranging textuality I find so fluid as to be incoherent[38] and its expression so intricate as to lose all application to actual texts—'an endless labyrinth without an outlet', as Frye once called it (*2*, 118; quoted in Lentricchia 166).

Writing from a semiotic perspective in his influential book *Partial Magic*, Robert Alter explains the 'eclipse' of self-reflexivity in the nineteenth-century novel by its concern with history: 'The imaginative involvement with history . . . is the main cause for an almost complete eclipse of the self-conscious novel during the nineteenth century' (*2*, 89). The negative perspective is symptomatic: history, by its intrusive presence, has the deleterious effect of 'eclipsing' the self-reflexivity which is assumed to be the mainstream of the novel. But is reference to external circumstance—history, politics, society—superfluous and indeed harmful in narrative fiction? Can any judgement one makes on the subject be absolute? It should be evident by now that my view is firmly eclectic: there is no pure literary form which demands the exclusion of extraneous elements. Historical facts or any other component are therefore neither good nor bad, neither required nor redundant. If historical material is present, as it clearly is in this heyday of the novel, as an intrinsic and central component of nineteenth-century narrative, its function must be examined, free from any preconceived prejudice. In much the same way, self-consciousness or metafiction is widely regarded as marking the transition to

[36] For instance: 'Toute œuvre, tout roman raconte, à travers la trame événementielle l'histoire de sa propre création, sa propre histoire' (*2*, 1967: 49). It is the subject of a celebrated polemic between Hillis Miller and M. H. Abrams, organized by Wayne Booth (*2*, 1979).

[37] Abrams notes that 'If one takes seriously Miller's deconstructionist principles of interpretation, any history which relies on written texts becomes an impossibility' (*2*, quoted in Booth 1979: 187). See also E. D. Hirsch, who by making a sharp distinction between fixed 'meaning' and modifiable 'significance' argues for an intentionalist concept of hermeneutics (*2*, 1976: 79–82), and Scholes (*2*, 1985: 86–110), who voices cogent reservations about deconstruction.

[38] LaCapra comments, 'there is something ideologically misleading in the "rewriting" or reprocessing of all texts in the same universalizing terms, even when those terms stress the impasses of universalization or theorize the aporias of theory' (*2*, 1985: 207).

the twentieth century and, possibly, associated with the decline of the novel;[39] it, too, is subject to the same considerations.

Michel Foucault gives a different twist to the debate.[40] For him, traditional history, like traditional authorship, is equated with the exercise of power. Getting away from purely chronological categories organized in causal sequences, he seeks a more permanent spatial concept which he terms 'archaeology', with an 'archive' of discursive experience which acquires a 'monumental' persistence: 'In our time history is that which transforms *documents* into *monuments*' (*2*, 1972: 7). The organizing and constraining principle is called the episteme: 'a constantly moving set of articulations, shifts, and coincidences that are established only to give rise to others' (192).

Foucault's repeated emphasis on discontinuity separates him radically from nineteenth-century historicism, and he has come under criticism on historical grounds in recent years.[41] 'To a great extent, Foucault's flawed attitude to power derives from his insufficiently developed attention to the problem of historical change' (*2*, Said 1983: 222). Margaret A. Rose comes to the complementary conclusion that, though it is highly desirable 'to work toward a *real* communication between historical and literary research, which will make our knowledge of the *particular facts* of history more *universal*, but thereby also more open to a self-critical understanding . . . Foucault and others have in fact made [this] more difficult by their insistence on the reading of history as "text"' (*1a*, Schulze and Wetzels 35). Yet Foucault's may be, of all modern systems, the most likely to prove adaptable to the revision of historicist criteria which is now developing.[42]

At the same time, of course, another profound current of thought continues a nineteenth-century trend: the socio-economic interpretation of literature, essentially grounded in the historical process, which has found its strongest, though not only, impulse in Marxist thought. In the nineteenth-century novel it is clear that both the period and the genre partake of Auerbach's famous concept of 'historism' in *Mimesis*, drawn from the work of the historian Friedrich Meinecke (*1a*). According to this criterion, it is important to be aware of the dynamics of history, of the uniqueness of historical phenomena

[39] Ortega y Gasset, as is well known, considers the subject-matter of the novel to be exhausted to such an extent that it can be redeemed only by delicacies of form: 'creo que el género novela si no está irremediablemente agotado se halla de cierto en su período último y padece una tal penuria de temas posibles que el escritor necesita compensarla con la exquisita calidad de los demás ingredientes necesarios para integrar un cuerpo de novela' (*2*, 1925: 89). Bakhtinian criticism is a striking rebuttal of that view.

[40] I am grateful to my research student Pilar Tirado for drawing my attention to these aspects of Foucault.

[41] Lentricchia (*2*, 231–42), in particular, finds Foucault's position politically inadequate and aesthetically self-protecting.

[42] As a result of their famous confrontation over Descartes, Foucault and Derrida have been increasingly presented as the two alternative approaches: Said (*2*, 1983: 183–5) and Ann Wordsworth (*2*, Attridge *et al.* 116–25).

and of their constant fluidity. No less important is the vital unity of individual periods, so that the character of each age is reflected in each of its manifestations, be these artistic, economic, or intellectual:

> The meaning of events cannot be grasped in abstract and general forms of cognition and ... the material needed to understand it must not be sought exclusively in the upper strata of society and in major political events but also in art, economy, material and intellectual culture, in the depths of the workaday world and its men and women ... [the resultant] insights will be also transferred to the present and ... in consequence, the present too will be seen as incomparable and unique, as animated by inner forces and in a constant state of development; in other words, as a piece of history whose everyday depths and total inner structure lay claim to our interest both in their origins and in the direction taken by their development. (*2*, Auerbach 444)

It is not difficult to see in these criteria features, such as a concern with the relation beween past and present and with an undercurrent of unrevealed continuity (related to what Unamuno calls 'intrahistoria'),[43] which will be directly relevant to Galdós's attitude to history. The literary manifestation of 'historism' is realism, defined by Auerbach as follows:

> The serious treatment of everyday reality, the rise of more extensive and socially inferior groups to the position of subject matter for problematic-existential representation, on the one hand; on the other, *the embedding of random persons and events in the general course of contemporary history*, the fluid historical background—these, we believe, are the foundations of modern realism ... (*2*, 491; my italics)

I shall refer with some frequency to the concept of 'embedding' in the italicized phrase in the course of this study; it is the historical complement of Hayden White's structural 'emplotment'.

We owe some of our most valuable insights to the influential study of the historical novel by Georg Lukács. As is well known, Lukács traces a direct trajectory from the historical 'romances' of Sir Walter Scott to the realist tradition established by Balzac. He has provided a powerful revindication of Scott—towards whom, incidentally, E. M. Forster displayed a disdainful attitude[44]—as the founder of the classical form of the historical novel:

> Scott's greatness lies in his capacity to give living human embodiment to historical-social types. The typically human terms in which great historical trends become tangible had never before been so superbly, straightforwardly and pregnantly portrayed. And above all, never before had this kind of portrayal been consciously set at the centre of the representation of reality. (*2*, 34–5)

[43] 'Esa vida intra-histórica, silenciosa y continua como el fondo mismo del mar, es la sustancia del progreso, la verdadera tradición, la tradición eterna, no la tradición mentida que se suele buscar al pasado enterrado en libros y papeles y monumentos y piedras': *En torno al casticismo*, first published in 1895 (*4a*, i. 793). Note once more the image of water. As Eamonn Rodgers has pointed out, the concept has its origin in Herder (*3b*, 45).

[44] 'He is seen to have a trivial mind and a heavy style. He cannot construct. He had neither artistic detachment nor passion' (*2*, 38). Forster grants Scott only the one important ability of telling a story.

In this way Scott carries out a 'conquest of Romanticism', avoiding the falsely heroic and picturesque: 'what in Scott has been called very superficially "authenticity of local colour" is in actual fact this artistic demonstration of historical reality' (45). Similarly, Duncan Forbes praises Scott's 'unique blend of sociology and romance, of "philosophical" history and the novelist's living world of individuals, of the general and the particular' (*2*, quoted in Fleishman 40).[45]

As the direct and ongoing heir of Walter Scott, Balzac sought to remedy the lack of cohesion he found in his predecessor's novels by providing the detailed systematic classification characteristic of *La Comédie humaine*. Within Balzac's grandiose plan[46] one aspect, the *Études de mœurs*, is disproportionately favoured over the other divisions; as he said in the *Avant-Propos* to the Furne edition of 1842: 'peut-être pouvais-je arriver à écrire l'histoire oubliée par tant d'historiens, celle des mœurs' (*4b*, 15). From this Lukács proceeds to the all but complete identification of the historical novel with the near contemporary historico-social setting of *La Comédie humaine* (*2*, 177).

While these aspects of Lukács's thought show great perception, his approach is essentially external and ideological, and he has nothing to say about the structural and linguistic techniques involved in the act of writing. He shows an excessive attachment to the realist novel, which he tends to see as the model against which to judge all other literary forms. As Laurence Lerner has shown, in his praise of Balzac he undervalues Stendhal (*2*, 150–2); and his use of the year 1848 as a historical turning-point is arbitrary and ideologically motivated (153).

The spatial and temporal conditions appropriate for the historical novel are severely circumscribed. By identifying the realist and the historical novels, Lukács fails to tackle the question of when 'history' merges with contemporary society and ceases to be historical. He shows a special penchant for novels of the recent past; and when he remarks of Balzac and Tolstoy that 'the powerful pressure of contemporary problems was too great for either of them to dwell for long on the prehistory of these questions' (*2*, 97), he clearly reveals how peremptory his priorities are.[47] His harsh criticism of the model of a profoundly documented novel set in the remote past, Flaubert's *Salammbô*, as

[45] Note Scott's reflections in his *Journal*: 'in my better efforts, while I conducted my story through the agency of historical personages and by connecting it with historical incidents I have endeavoured to weave them pretty closely together, and in future I will study this more. Must not let the background eclipse the principal figures—the frame overpower the picture' (18 Oct. 1826: *4b*, 215; see also *4b*, Green 9).

[46] Harry Levin comments: 'Organization, for Balzac, meant all-embracing inclusion rather than discriminating selection' (*4b*, 184).

[47] Lukács's attitude to Manzoni is typical of his limitations. He sees *I promessi sposi* as a worthy successor to Scott, surpassing him in many ways, but since Italian history was incapable of providing a variety of themes, it remained an isolated masterpiece lacking 'the world-historical atmosphere which can be felt in Scott' (*2*, 79).

escapist[48] shows marked intolerance of a novel set in a distant civilization; are the problems posed by the fate of a civilization such as Carthage's of no potential concern to us?

What does remain, however, is a coherent vision of the modern historical novel, which 'has to *demonstrate* by *artistic* means that historical circumstances and characters existed in precisely such and such a way . . . it is the portrayal of the broad living basis of historical events in their intricacy and complexity, in their manifold interaction with acting individuals' (*2*, 45). Despite the tendency revealed in this passage to speak in terms of types, the close correlation it establishes between historical events and representative individuals is apposite.

We have now discussed two apparently incompatible critical attitudes to prose fiction, anti-historical textual theory and socio-historical analysis. For some years there have been uneasy and inconclusive attempts to find common ground between them, but it has proved, in Scholes's words, a 'treacherous no-man's land' (*2*, 1985: 8). As early as 1972, Fredric Jameson ended his stimulatingly entitled *Prison-House of Language* with the seemingly remote hope (hardly a 'confident' assertion, as Christopher Norris contends [*2*, 79]) that a new and genuine hermeneutics might emerge which will bring together the two lines of theory:

> [This hermeneutics] would, by disclosing the presence of existing codes and models and by reemphasizing the place of the analyst himself, reopen text and analytic process alike to all the winds of history . . . it is only, it seems to me, at the price of such a development, or of something like it, that the twin, apparently incommensurable, demands of synchronic analysis and historical awareness, of structure and self-consciousness, language and history, can be reconciled. (*2*, 1972: 216)

And in a later book, *The Political Unconscious* (1981), Jameson sets out to make a sustained attempt at such reconciliation: its starting slogan is 'Always historicize!', but as Scholes notes, although Jameson is seeking the middle ground, 'he leaves vague or blurred the whole matter of reference' (*2*, 1985: 80–1). Michael Ryan, too, makes a valiant if in my view ultimately unsuccessful effort to harness deconstruction to an open-ended and activist conception of Marxism (*2*, 21, 57). As a consequence, Norris, in 1982, finds little ground for this approximation, concluding that 'it is difficult to square deconstruction in this radical, Nietzschean guise with any workable Marxist account of text and ideology' (*2*, 80).

Recent years have witnessed a much sharper change of emphasis towards a

[48] Its intention, according to Lukács, is 'to reawaken a vanished world of no concern to us' (*2*, 220). Seen in this way, 'history becomes a large, imposing scene for purely private, intimate and subjective happenings' (237). This seems to me a deplorably limiting view of the potentiality of the novel. For a sound defence of *Salammbô* see Green (*4b*, 58–72).

renewed concern with history.[49] One of the clearest milestones of the change came in J. Hillis Miller's Presidential Address to the MLA in 1986. Miller declares, regretfully, that 'literary study in the past few years has undergone a sudden, almost universal turn away from theory in the sense of an orientation toward language as such and has made a corresponding turn toward history, culture, society, politics, institutions, class and gender conditions, the social context, the material base'.[50] As Louis Montrose indicates, Miller establishes an opposition which works 'not only to oversimplify both sets of terms but also to suppress their points of contact and compatibility.' Much of the discussion centres around the latest fashionable term, the as yet ill-defined 'new historicism'.[51] How far is this movement, if it merits such a title, merely a continuation, under a new guise, of an essentially formalist orthodoxy, as Vincent Pecora (243–76), from a Marxist perspective, and Elizabeth Fox-Genovese (213–24), from a historical one, suggest? Or how far, on the contrary, is it politically orientated, seeking to regalvanize an ailing Marxist inspiration, as H. Aram Veeser implies in his rather over-heated Introduction to *The New Historicism*, with the object of incorporating into academic discussion a series of contemporary issues, seemingly urgent but essentially short-term? Certainly, robust and welcome defences of historical involvement from an overtly Marxist standpoint, such as Barbara Foley's, are not lacking:

> Despite these theoretical advances, however, many recent discussions of mimesis (=*fiction*) have been flawed by the tendency . . . to minimize its embeddedness in material history. The text is fetishized, the author and reader are dehistorized and mimesis is divested of the capacity to make assertions. We must demonstrate that the mimetic contract involves a commitment to take seriously the text's propositional claims. (*2*, 63)

The important debate is no doubt destined to continue.

What interests me in these discussions is not so much the theoretical arguments, much less the particular political stances which to a large extent determine the arguments, but the possibility of attaining a vitally necessary pragmatic *rapprochement* between literature and history.[52] A critic who sets out (albeit in forbidding substantivized terminology) to avoid the two extremes of

[49] Fox-Genovese is quite categorical: 'In recent years, literary critics, surfeited with the increasingly recognized excesses of post-structuralist criticism in its various guises, have discovered history' (in *2*, Veeser 213).

[50] Quoted in Montrose (in *2*, Veeser 15; all the essays cited in this paragraph are taken from *The New Historicism*).

[51] It is significant that the propagator of the expression, Stephen Greenblatt, is ill at ease with the term, preferring to substitute 'Cultural Poetics' (15–17). Catherine Gallagher notes that it is 'charged on the one hand with being a crude version of Marxism and on the other with being a formalist equivalent of colonialism' (37). There are also links with the British 'cultural materialism' movement associated with Raymond Williams. See also Lehan (*2*).

[52] While all such rethinking is for me most welcome, I cannot fail to see in it a great deal of *déjà vu* for those of my generation who without in any way forgoing a close critical reading of texts have never abandoned an essentially historicist approach.

'overcontextualization, historicization, and documentary reductionism, on the one hand, and formal hypostatization—including the universalizing fixation (or compulsive "fetishization") of deconstructive techniques—on the other' is Dominick LaCapra (*2*, 1987: 209–10), with the result that he is able to claim that 'I have tried to keep a distance from both historicism and related formalisms by attempting to articulate ways in which a critical concern with history implies an interest in form that need not eventuate in formalism' (211).[53] Among those commenting from the 'new historicism' perspective in Veeser's compilation, Montrose seems to come closest to offering such a convergence. Using Greenblatt's term 'Cultural Poetics', he declares that its 'interests and analytical techniques are at once historicist and formalist; implicit in its project, though perhaps not yet adequately articulated or theorized, is a conviction that formal and historical concerns are not opposed but rather are inseparable' (*2*, Veeser 17).

In any coming rearticulation several specific objectives are to be hoped for: that the blanket term 'historicism' will not be confined to an outmoded Tainian determinism, as it seems to be in Lentricchia's view (*2*, Veeser 231–4), but that a more adequate definition, such as Kelley's, quoted on p. 4, will be employed; that it will incorporate certain aspects of Foucault's analysis of power; that it will subsume much reader-response criticism, as it has done in the case of Jane Tompkins (*2*, Thomas in Veeser 184); and that the very real achievements of feminist criticism (e.g. *2*, Newton in Veeser 152–67; Montrose in Veeser 26, 356), hardly sustainable without a firm historical point of reference, will be fully integrated into the new scheme.

Moreover, it is to my mind of crucial importance to beware of too dogmatic or exclusivist an approach. The pursuit of absolute truth is illusory, and the negative arguments in any metaphysical discussion invariably outweigh the affirmative. We should remind ourselves constantly, therefore, that neither of the two opposing factors—inner structure and outside referentiality—can possibly attain perfect logical coherence. Nor do they in practice, even in writings deliberately pushed to a theoretical extreme (a noteworthy example is *le nouveau roman*), exclude each other totally; individual authors may contain considerable doses of what appears to be the unorthodox quality at the time.[54] Deconstructionist theory itself has found its exemplary (though anomalous) cases in nineteenth-century narrative: *Sarrasine, The Purloined Letter, Billy Budd*.[55] On the other side, it is not difficult to find substantial traces of

[53] Lodge also argues judiciously the need for a synthesis: 'since art is supremely the province of forms and since literature is an art of language, I believe such a synthesis can only be found in linguistic form. But the synthesis must be catholic: it must account for and be responsive to the kind of writing normally approached via content, via the concept of imitation, as well as to the kind of writing normally approached via form, via the concept of autonomy' (*2*, 1979: 70).

[54] Witness the persuasive arguments of John Kronik on Feijoo (*3c*, 1984: 40) and Robert Spires (*4a*, 36) regarding self-reflexivity in Galdós.

[55] For all these, consult Johnson (*2*, 5–12, 110–46, and 79–109 respectively).

historical awareness in largely aestheticizing texts; Proust, Joyce, and Valle-Inclán all provide more than sufficient evidence of this.

Norris puts his finger on a fundamental defect of much recent confrontational criticism:

> Structuralist theory was clear enough about the basic distinction between 'story' and 'plot', the one an implied (and imaginably real-life) sequence of events, the other a pattern imposed by the requirements of narrative form. They represent two different kinds of reading: the latter is attentive to structure and device, while the former rests on a willing—but not necessarily naïve—suspension of disbelief. To see them locked in conflict or paradox is to mistake the conventions of narrative for the rigours of logical discourse. The tactics of 'double-reading' automatically generate the kind of paradoxical impasse they set out to find. Like the structuralist attack on that ubiquitous 'classic realist text', such strategies ignore the variety of possible relations between language, text and reality. (*2*, 134–5)

It is my firm conviction that the coexistence—not the predominance of one over the other—of both directions (objective reality and formal structure) must be recognized to produce balanced critical work on nineteenth-century narrative. This book is dedicated, in a modest way, to that task.

Recent narratological theory establishes a clear demarcation between history (external referentiality) and the text. A typical and influential exponent of narratology like Gérard Genette rightly argues (*2*, 27–8) that while in studying a text such as Michelet's *Histoire de la France* it is legitimate to take into account outside documents concerned with French history and external biographical data, such outside documentation is not pertinent to *A la recherche du temps perdu*. This seems to me correct with one important proviso: if and when extrinsic items of reality—history, in the first instance, but other aspects such as place also—intrude in the text, these too demand special consideration. The extensive references to Dreyfus or, for that matter, the Champs-Élysées, in Proust's novel, for example, bring in such an outside dimension in a way that references to Combray do not: whatever relation Combray has to Illiers is suppressed as far as the text is concerned by its being called the one and not the other. The novel must be judged on its own terms as an artistic construct, with all that it contains, the chunks of external reality included. These cannot be ignored or passed over; they may be essential building-blocks in the fictional edifice, but they must be considered from inside outwards, not from outside in. Their ultimate reality depends on the use made of them in the structure of the novel, though bearing in mind the distinctive resonances, and the special limitations, their historical context gives them.

Just as I see no reason why historical incidents and characters should not have a place in fiction, it seems to me clear that a novel with historical content does not on that account forgo the demands which all narrative fiction has to meet: certain common artistic principles must be applied to it as to any other narrative. Genette states this clearly: 'Story and narrating thus exist for me only by means of the intermediary of the narrative . . . As narrative, it lives by its relationship to the story that it recounts; as discourse, it lives by its relationship to the narrating that utters it' (*2*, 29). The fact that various historical incidents occur within the narration has nothing to do with this essential factor; because they are present in a work, historical materials are turned into components of the narration, just as, if a writer puts himself (or rather his name) into the action, he himself becomes a character.[56] The narrative and discursive structure is not thereby destroyed: on the contrary, this structure is fully subject to analysis and assessment just as it stands, according to the norms that proceed from the structure itself.[57] A deep concern for history, therefore, does not exclude the application of narratological criteria, some, but not all, of which are shared with non-fictional writing; those which are not shared, like self-referentiality, tend to be those most prized by semioticians. All writing is interpretation, organization, selection, and this applies to historiography as much as it does to creative literature, but fiction allows a much wider scope for variation of the narrative mood or voice, for intertextuality, for the utilization of symbolic or mythological material, for metonymic or metaphorical patterns, for ironic presentation, for self-reflexive analysis, and so forth.

Substantial advances have been made in recent years in the understanding of narrative techniques. First, the distinction between 'telling' and 'showing', what Percy Lubbock called 'picture' and 'scene' (*2*, 267–72) or, in more modern terminology, 'summary' and 'scene', has been perceptively linked with Plato's diegetic and mimetic modes (*2*, Genette 162–6). Although the former is more characteristic of the traditional realist novel and the latter of the modern novel, both are essential techniques of all fiction. Summary is fast-moving, apparently objective, a narrative account of the past from one who claims to know what is happening; scene is slow (its discourse time is the same as story time), reveals the subjective feelings of the characters, is unmediated and simultaneous with the action. Genette has made a distinctive contribution in at least two directions. One is to the elucidation of time-scale, in his systematic defining of discrepancies between 'story time' and 'discourse time'

[56] Compare Lodge: 'And need one say that the more nakedly the author appears to reveal himself in such texts, the more inescapable it becomes, paradoxically, that the author as a *voice* is only a function of his own fiction, a rhetorical construct, not a privileged authority, but an object of interpretation?' (*2*, 1990: 43) As Spires comments with reference to Unamuno's intervention in *Niebla*: 'Real people cannot exist within a fictional world; signs merely masquerade as people' (*4a*, 36).

[57] Again, Genette puts it well: 'Story and narrating thus exist for me only by means of the intermediary of the narrative . . . As narrative, it lives by its relationship to the story that it recounts; as discourse, it lives by its relationship to the narrating that utters it' (*2*, 29). For a lucid synthesis of Genette's views consult Rimmon-Kenan (*2*, 46–58).

(called 'anachronies'), like analepsis (flashback) and prolepsis (flash forward), and of the iterative type within the frequency mode. The second concerns narrative technique: what used to be called 'point of view' and Genette terms 'focalization'. Non-focalized narrative is at one with an omniscient narrator. Internal focalization may be *fixed*, with a 'reflector' like Strether in Henry James's *The Ambassadors*, *variable*, as in *Madame Bovary*, which shifts its focus from Charles to Emma and then to Charles again, or *multiple*, as in epistolary novels. Full internal focalization is found only in interior monologues or in the sustained rigidity of the *nouveau roman*, but in a less rigorous sense it can be determined by the test devised by Barthes: an internally focalized text can be translated without loss or incongruity into the first person (*2*, 1977: 189–94).

As increased attention has been paid to scene, direct speech (which can also cover thoughts and speaking to oneself) has gained in importance. As a spoken utterance to others, it is traditionally reproduced between quotation marks and the speaker indicated. When used without quotation marks and tags ('direct free thought': *2*, Chatman 182), it is even more immediate. An extended form is the 'interior monologue'; with certain differences[58] it is related to the 'stream of consciousness', which tends to be less coherent and conceptualized.

It has been demonstrated, too, that diegesis and mimesis are not entirely separable, since a form of narrative, now generally called 'free indirect style' (or 'speech' or 'discourse'), partakes of the subjective utterances of a particular character while retaining the narrative mode; it is what Stephen Ullman, one of the pioneers in discerning this style, called 'reported speech masquerading as narrative' (*4b*, 113). Roy Pascal has carefully traced the origin of the concept and the phrase 'style indirect libre' to the Swiss linguist Charles Bally and has explored its relationship with the cognate phrase 'erlebte Rede' in German. Free indirect discourse has been traditionally associated with Flaubert, and Pascal has conducted some useful analysis of his practice: 'With Flaubert FIS [free indirect speech] is used fully consciously, asserts its natural rights, and needs no warning lights; it is a natural form of expression' (*2*, 101).[59] Pascal notes the systematic use of the imperfect tense for the free indirect form in contrast with the preterite characteristic of the pure narrative mode, and he indicates that 'one of [Flaubert's] signal achievements' is to have 'extended FIS to embrace . . . those mental responses that are beyond (or beneath) verbal formulation and definition, that remain at the level of sentient and nervous apprehension' (108). Finally, he draws atten-

[58] Genette distinguishes between the two techniques: 'in free indirect speech, the narrator takes on the speech of the character, or, if one prefers, the character speaks through the voice of the narrator, and the two instances are then *merged*; in immediate speech [i.e. stream of consciousness, *a term Genette dislikes*], the narrator is obliterated and the character *substitutes* for him' (*2*, 174).

[59] John Rutherford draws attention to the skilful use of the device, in conjunction with other direct and indirect discourses, in Alas's *La regenta* (*4a*, 59–64).

tion to the 'usurpation' by which 'Flaubert substitutes his language and his perspective, and that means his mode of feeling, for hers [Emma Bovary's]' (110). Jauss perceptively links the prosecution case against *Madame Bovary* with a failure to understand the new technique (*2*, 42–4).

The special legacy of Mikhail Bakhtin's largely posthumous work[60] is to have enhanced enormously our awareness of the linguistic and stylistic resources of the novel. 'The novel demands', he declares, 'a broadening and deepening of the language horizon, a sharpening in our perception of socio-linguistic differentiations' (*2*, 1981: 366). Unlike much criticism deriving from linguistic criteria, Bakhtin does not underestimate the social importance of fiction but rather incorporates spoken language (*skaz*) as an essential vibrant element of the conflictive life of society:

> The prose art presumes a deliberate feeling for the historical and social concreteness of living discourse, as well as its relativity, a feeling for its participation in historical becoming and in social struggle; it deals with discourse that is still warm from that struggle and hostility, as yet unresolved and still fraught with hostile intentions and accents; prose art finds discourse in this state and subjects it to the dynamic unity of its own style. (*2*, 1981: 331)

For Bakhtin, then, the diversity of stylistic discourse—what he calls *heteroglossia*—stands at the fountainhead of the structure of the novel. This diversity gives rise to 'dialogism' (in other contexts it is called 'polyphony'), that is to say, the 'constant interaction between meanings, all of which have the potential of conditioning others' (*2*, 1981: 426). As opposed to an inert 'authoritative discourse'—the old omniscient, reliable narrator—and a represented or objectivized discourse—the direct speech of characters subordinate to the author's voice—this third phenomenon, Bakhtin explains, distinguishes the novel from monoglossic genres like poetry:

> Heteroglossia . . . is *another's speech in another's language*, serving to express authorial intentions but in a refracted way. Such speech constitutes a special type of *double-voiced discourse*. It serves two speakers at the same time and expresses simultaneously two different intentions: the direct intention of the character who is speaking, and the refracted intention of the author. In such discourse, there are two voices, two meanings and two expressions. And all the while these two voices are dialogically interrelated, they—as it were—know about each other . . . it is as if they actually hold a conversation with each other. Double-voiced discourse is always internally dialogized . . . A potential dialogue is embedded in them, one as yet unfolded, a concentrated dialogue of two voices, two world views, two languages . . . this double-voicedness sinks its roots deep into fundamental, socio-linguistic speech diversity and multi-languageness. (*2*, 1981: 324–6)

[60] I refer particularly to *Problems of Dostoevsky's Poetics*, first published in Russian in 1929 and revised in 1963 (English trans.: *2*, 1973), and 'Discourse in the Novel (1934–5)', included in *The Dialogical Imagination* (*2*, 1981). For a general appreciation see Todorov, who calls him 'le plus grand théoricien de la littérature au XXe siècle' (*2*, 1981: 7).

Double-voiced discourse is an 'internally persuasive discourse', '*not finite*', but '*open*' (346), which takes various forms. One such form, 'hybridization', not only mixes linguistic signs—the markers of two languages and styles—but demonstrates 'the collision between differing points of view on the world that are embedded in these forms. Therefore an intentional artistic hybrid is a *semantic* hybrid . . . a *semantics that is concrete and social*' (360). 'Stylization' is the 'clearest and most characteristic form of an internally dialogized mutual illumination of languages . . . it permits the maximal aestheticism available to novelistic prose' (362–3) and was therefore prevalent among such stylistically conscious writers as Mérimée and Anatole France. If the signifier and the signified are at odds, the result is 'parodic stylization'. A particularly valuable concept is that of 'hidden polemic', which 'senses its own listener, reader, critic, and reflects in itself their anticipated objections, evaluations, points of view. In addition, it senses alongside itself another discourse, another style.'[61] An analogous, but not synonymous, form is 'hidden dialogicality', where the second element of a dialogue is omitted, but is invisibly present.[62] We shall see in due course how these general concepts have specific application to the realist novel and, in a very high degree, to Galdós's work.

Some modern linguists have drawn attention to the differentiation made in certain languages (Turkish and Aymara among them) between factual statements and 'evidential' statements, based on second-hand evidence. From this they have postulated an ontological difference between direct and indirect knowledge, in that direct sense-experiences reach our consciousness by non-linguistic means (the so-called Ogden–Richards triangle, extended to the Baldinger–Heger trapezium) while reported experiences are conveyed by a relatively lengthy process of definition depending on linguistic means alone: a difference similar to that between intuition and intellectualization or between *saber/savoir* and *conocer/connaître*. On this basis, Roger Wright concludes that 'The distinction between two types of subject matter, direct and indirect, seems to be a more important distinction for analysis of natural language—that is, it has a greater ontological basis in human nature—than that between the fictional and the historical, the false and the true' (*2*, 17). This would seem to me to help explain the effects of immediacy sought by some writers—Galdós clearly among them—through a re-created contemporary world characterized by unmediated involvement through direct speech, dialogue, and even, in narrative settings, by free indirect speech. On this purely linguistic plane, then, the distinction between fictional and historical has little or no meaning (for this reason historical revision or 'poetic licence' is acceptable); it

[61] I quote from Lodge (*2*, 1990: 86), who is particularly enthusiastic about this concept. The translation is from Caryl Emerson's edition (*2*, Bakhtin [1973] 1984: 196).

[62] In 'Discourse in the Novel' (*2*, 1981: 262) Bakhtin enumerates five types of 'stylistic unities', in which the categories listed above are dealt with somewhat differently; the 'hidden polemic' is not included.

is only when one leaves aside a purely intrinsic, that is, linguistic, criterion, as I think one must, as one aspect of our critical perspective, that the distinction between history and fiction resumes—for me—its validity once more.

Also positive is the emphasis placed on the continued momentum of the finished text[63] by reader-response criticism exemplified by the Constance school of Wolfgang Iser and Hans Robert Jauss. Fundamental for reader-response theory is the concept of 'indeterminacy' which Iser refines from Roman Ingarden. Thus,

> literary texts are full of unexpected traits and turns, and frustration of expectations. Even in the simplest story there is bound to be some kind of blockage, if only because no tale can be told in its entirety. Indeed, it is only through inevitable omissions that a story gains its dynamism. Thus whenever the flow is interrupted and we are led off in unexpected directions, the opportunity is given to us to bring into play our own faculty for establishing connections—for filling in the gaps left by the text itself. (*2*, 1974: 279–80)

'... without the elements of indeterminacy,' Iser goes on, 'we should not be able to use our imagination' (283). Later, he refers to the 'three important aspects that form the basis of the relationship between reader and text: the process of anticipation and retrospection, the consequent unfolding of the text as a living event, and the resultant expression of life-likedness' (290). Jane Tompkins, in the volume she edits on reader-response, notes that Iser 'does not grant the reader autonomy or even partial independence from textual constraints' (*2*, xv). Stanley Fish moves beyond this; he rejects not only seeking the meaning of a text in the author but the text itself as the source of meaning, which is to be found only in the experience of reading itself: 'It is the experience of an utterance—all of it and not anything that could be said about it, including anything I could say—that is its meaning' (*2*, Tompkins 78). My own position is to accept that the reader has a substantial role in interpreting the text without abandoning either the initial part of the author or the primacy of the text in determining meaning.

Linked with the indeterminacy of a text is the question of closure or lack of closure.[64] The traditional function of the ending of a literary work is well established: the culmination of 'what would happen next' (*2*, Forster 34–5), 'the arrival of a visibly-appointed stopping-place' (*2*, James 425). Yet conventions pall very quickly, so that the well-rounded, completed story, with its habitual happy ending, 'a distribution at the last of prizes, pensions, husbands, wives, babies, millions, appended paragraphs, and cheerful remarks', as Henry James put it (*2*, 190), comes increasingly to be seen as artificial. Moreover,

[63] As Frank Kermode puts it succinctly, 'Everybody knows that competent readers read the same text differently, which is proof that the text is not fully determined' (*2*, 1983: 128).

[64] See in particular Kermode's influential *Sense of an Ending* (*2*), despite its extreme concentration on eschatological questions.

post-structuralist critics have perceived an organic incompatibility between the essence of narrating and a final closure. So, according to D. A. Miller, 'the narratable inherently lacks finality' (*2*, xi), hence the title of his book, *Narrative and its Discontents*, and its analysis of the tensions of his characteristic nineteenth-century authors, Austen, Eliot, and Stendhal. Marianna Torgovnick takes a more traditional point of view, finding no inherent superiority either in the novels René Girard (*2*) classifies as 'romanesque' (novelistic) rather than 'romantique' (romantic) or in the 'open-ended' conclusions endorsed by Alan Friedman (*2*).[65] We should bear in mind that the terms 'closed' and 'open' are themselves ambiguous and question-begging and that no question of hierarchy should arise between them. The development of reader-response theory has given a new emphasis to 'open-ended' solutions, which make more direct demands for the readers to participate. As a consequence it has become increasingly apparent that many classic realist texts (*Middlemarch* is a particular favourite) are far more equivocal than has hitherto been recognized. For a realist novelist an ending should be consequential, but in accordance with the uncertainty of the lives he is attempting in some degree to replicate he may strive for it to be at the same time indeterminate.[66] All these factors have important repercussions for our purposes which will be discussed in the next chapter.

Some recent theoretical systems of narrative have taken a pleasingly broad view of the range of narrative possibilities. As an example which is useful for our purpose, let us take the graph that Robert Spires puts forward, as a revision of an earlier one by Robert Scholes (*2*, 1974: 117–41, especially 135), which was in its turn an elaborated version of the well-known scheme of five fictional modes constructed by Northrop Frye (*2*, 33–67). The latter's archetypal modes (myth, romance, high mimetic, low mimetic, irony)[67] became in Scholes a V-shape graph, now with history at its base, as in Diagram 1. Scholes's series of dots traces the development of the novel right down the middle of his pattern and the shaded area indicates the plenitude of the realist novel. 'Stendhal, Balzac, Flaubert, Tolstoy, Turgenev, and George Eliot', he notes, 'all work near the center of this area. Dickens, Thackeray, Meredith and Hardy tend more toward the edges and corners' (*2*, 1974: 137).

Finally, Spires finds Scholes's treatment too diachronic and consequently inadequate as far as metafiction is concerned. His modal construction is linguistically grounded and ahistorical, but he does not overlook historical considerations. He places 'novelistic theory' and 'history' at the two extremes,

[65] Torgovnick's lucid study (*2*) provides a useful detailed classification of types of closure.

[66] For a sound general consideration of the question of endings in the French and Spanish realist novel consult Bauer (*4a*).

[67] Note that history is conspicuously absent from this scheme. Eagleton sardonically comments, 'The advantage of Frye's theory, then, is that it keeps literature untainted by history in New Critical fashion' (*2*, 1983: 92).

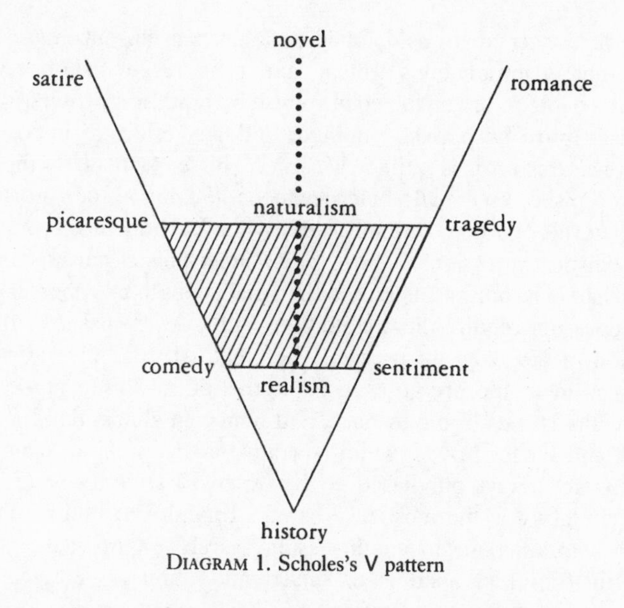

DIAGRAM 1. Scholes's V pattern

novelistic theory

----

metafiction

satire romance

picaresque tragedy

comedy sentiment

reportorial fiction

----

history

DIAGRAM 2. Spires's circular structure

outside the scheme itself, with 'metafiction' and 'reportorial fiction' as the two corresponding fictional forms within the circular framework (*4a*, 9).[68] (See Diagram 2.) By eliminating the generic criteria which characterized Frye and Scholes's paradigms, Spires creates a spiral-like structure which has more inner communication, but which blurs the place of historically defined genres such as the realist novel. While this genre cannot be entirely unconcerned with novelistic theory and metafiction, it is clearly also associated with history (though less closely than reportorial fiction is) at the other end of the scale; it is difficult to link the realist or historical novel in any fixed way with any of the collateral modes. Placing Balzac or Tolstoy or Galdós is therefore more difficult than in Scholes's diagram.

To conclude this section: I consider any attempt to erect absolute criteria to

[68] Both Scholes's and Spires's graphs owe a great deal to Roman Jakobson; see 'Linguistic and Poetics' (*2*, De George 85–122).

determine literary creativity as doomed to failure; any all-encompassing theory faces insoluble contradictions which cannot be reconciled, but which are perforce amenable to pragmatic compromise by practising writers.[69] No genre is chemically pure or perfectly unified, and any effort to impose absolute prescriptive norms on it is ultimately futile. This is particularly the case with the once despised, essentially heterogeneous and innovatory prose narrative called the novel. We may certainly agree with Bakhtin that some degree of unity of stylistic expression is aimed at by most poetic genres (I would say, however, that it is normative rather than indispensable), whereas the novel 'not only does not require these conditions but . . . even makes of the internal stratification of language, of its social heteroglossia and the variety of individual voices in it, the prerequisite for authentic novelistic prose' (*2*, 1981: 264). Diversity is thus of the essence, and hence an almost limitless freedom.

Scholes and Kellogg put it well: 'Narrative art is an art of compromise in which gains are always purchased at the expense of sacrifices' (*2*, 258), and 'narrative literature is the most restless of forms, driven by its imperfections and inner contradictions to an unceasing search for an unattainable ideal' (282). If the novel as such is of mixed and dubious pedigree (albeit of exhilarating potential), the historical novel is even more so. As Fleishman rightly declares: 'The genre is unashamedly a hybrid: it contemplates the universal but does not depart from the rich factuality of history in order to reach that elevation' (*2*, 8). Any study of the novel of historical content, therefore, must beware of the Scylla of linguistic self-sufficiency and the Charybdis of historicist takeover.

The question finally arises of how a historical novel is to be defined. Avrom Fleishman's excellent book, *The English Historical Novel*, already extensively quoted, opens broad parameters. We may readily agree that the form 'is pre-eminently suited to telling how individual lives were shaped at specific moments of history, and how this shaping reveals the character of those historical periods' (*2*, 10). Harry Levin amplifies: 'Preserving for us the quintessence of history, they [historical novelists] present facts as feelings. Instead of the taxes and treaties, they record the values and textures of a period, the dates that frame individual lives, the events that touch immediate sensibilities' (*4b*, 27). As for temporal scale, Fleishman argues that a historical novel should be set in the past, beyond an arbitrary number of years, say forty to sixty years (i.e. two generations), must cover a number of 'historical' events, particularly those in the public sphere (war, politics, economic change, etc.), and include at least one real personage: 'The historical novel is distinguished among novels by the presence of a specific link to history: not merely a real

[69] As Proust says, 'A work in which there are theories is like an object which still has the ticket that shows its price' (*4b*, ii. 1009; *2*, quoted in Genette, 259).

building or a real event, but a real person among the fictitious ones' (*2*, 4).[70] This represents a distinct advance in precision. And if the specific requirement of public history might seem unduly limiting (a historical dimension does not depend exclusively on public events or require the presence of a historical figure), due acknowledgement is made of the need for a real historical presence.

If we accept, as we evidently should, Walter Scott's crucial role in the development of the historical novel, it would seem obvious that all his novels of clear historical content should, despite the implications of Lukács's approach, fall within the norm, embracing as they do a lengthy period of time from Richard Lionheart (*Ivanhoe*) to Edinburgh in the mid-eighteenth century (*The Heart of Midlothian*) and varied scenarios in Scotland, England, and France. Auerbach has justly pointed out that any historical novel has a relationship with the present (*2*, 444).[71] Although this link obviously becomes more tenuous and somewhat different in nature when the subject of the novel is remote in time and place, a connection with present preoccupations is not necessarily absent. Thus, a text as chronologically distant as *Ivanhoe* deals with a theme of evident concern for the whole course of English history: the conflict between Anglo-Saxons and Normans. Even if this relationship is not so apparent in some of his successors, stories of distant countries and ages may provide general lessons of history, regarding such issues as power and authority, social or economic competition, civic responsibility, social integration, class and sexual relationships, etc. The classic example, as previously indicated, is *Salammbô*, but such once popular works as Bulwer-Lytton's *The Last Days of Pompeii* and *Rienzi* fall into this category. Like Fleishman (*2*, 34), I do not doubt that Lytton's intention was to provide historical examples or models of political conduct relevant for his own times. Even the many Spanish imitations of Walter Scott, set habitually in Spain but at a considerable distance in time, have potential application to modern times, however incapable, in most cases, their authors were of exploiting it: this is simply the measure of their failure. In short, I see no reason for denying the blanket title of 'historical novel' to all these works.

The popularity of Scott's novels and of the endless imitations they spawned all over Europe and America, together with the melodramatic qualities of Dumas's and Hugo's works, inevitably discredited the genre. By 1846 G. H. Lewes dismissed 'historical romance' as 'false History, and a bad story,

[70] Thomas Hardy quotes Leslie Stephen on historical novels: 'I can only tell you what is my own taste, but I rather think that my taste is in this case the common one. I think that a historical character in a novel is almost always a nuisance; but I like to have a bit of history in the background, so to speak; to feel that George III is just around the corner, though he does not present himself full front' (*4b*, F. E. Hardy 127).

[71] Benjamin makes a similar assertion, more aggressively, arguing that (old-fashioned) historicism always favours the victors: 'In every era the attempt must be made anew to wrest power away from a conformism that is about to overwhelm it' (*2*, 257).

palmed upon us for a novel' (*4b*, 35). Of Flaubert Anne Green says: 'On the whole, Flaubert's reaction to the historical novel of his century was one of dissatisfaction and impatience' (*4b*, 13). Moreover, historical novels were considered a cheap and easy way to acquire a smattering of history; hence the ironic entry in the *Dictionnaire des idées reçues*: 'Seuls les romans historiques peuvent être tolérés parce qu'ils enseignent l'histoire' (*4b*, *Bouvard et Pécuchet* 442).

What of the criterion of a contemporary or near contemporary setting? Kathleen Tillotson has discerned in the 1840s an important and flourishing subgroup of British novels about the recent past (she is thinking of a twenty- to sixty-year period: 'neither historical nor contemporary', *2*, 92), overlapping therefore with the limits proposed by Fleishman.[72] These works include, of course, *Waverley*, as well as *Middlemarch*, *A Tale of Two Cities*, and *Vanity Fair*. Are these novels, or any of them, historical novels? The line, as Tillotson says, 'may not always be easy to draw' (94). Some—*Waverley*, *A Tale of Two Cities*—would certainly qualify, and *Vanity Fair*, perhaps, with certain reservations; *Middlemarch* less so. Both Lerner (*2*, 156–7) and Fleishman (*2*, 136–46), in greater detail, make a good case for the slightly later *Henry Esmond* (1852) as a historical novel. It is prominent, moreover, for deflating the Carlylian Great Man concept of history: 'I would make History familiar rather than heroick' (*4b*, 2). Tillotson's subgroup thus lies on the borderline of what we may term historical novels: those that have their action in the immediate past (less than forty years) deserve a different designation. Here I would follow Fleishman in reserving the term 'historical novel' for works of a time-scale just outside, or beyond, normal living memory. Accordingly, *pace* Lukács, I should not classify most of the novels of Balzac's *Comédie humaine* as historical novels. Without minimizing their historical content, I am reluctant to assign that term, rather restrictive and often ill-defined, to novels which largely unfold their powerful studies of human interreaction on a purely fictional level and which have a contemporary or near contemporary viewpoint. Such a requirement would also exclude Flaubert's *L'Éducation sentimentale*, so replete with immediate historical references.

War and Peace poses a more difficult problem. Tolstoy's own reluctance to call it a novel, because of its lack of 'fixed limits to the characters I have invented' (*4b*, quoted in Christian 113), has to be taken into account, but no one today would hold that an open-ended form precludes its being considered a novel. Its so-called 'epic' qualities have been constantly stressed, but Tolstoy's attitude to heroism is too complex to allow such a term to stand unreservedly. The historical content, at a suitable chronological distance, is indeed strong, but the web of all-embracing personal and family relationships

[72] Compare Stern's establishment of a 'Middle Distance' as a criterion for the realist novel (*4b*, 113–28).

is no less evident. Given the magnificent fashion in which the fictional characters are interwoven into a consistent if loosely organized social whole, my inclination is to treat it as a novel—a truly great novel—*tout court*.[73]

At the same time there appear to me grounds for a further subdivision which would separate those historical novels with no direct link with the present—novels of the remote past like *Salammbô*, *The Last Days of Pompeii*, or of less distant times like *Ivanhoe*—from novels of the conscious past, according to Tillotson's criterion. Such novels may stretch back to what might be called the furthest reaches of the historical consciousness of the present, well beyond an individual's normal life-span, to say seventy-five years or even a full century: 'memory at one remove', in Tillotson's phrase (*2*, 99); the latter I shall call 'modern historical novels'. Thus, *Waverley*, the first of Scott's novels, of crucial importance within historical fiction,[74] would comfortably fit into this new category, and so, just, would the novel of the 1715 rising, *Rob Roy*—so important for the subsequent development of Scotland—as could Balzac's *Les Chouans*, set in the Revolutionary period, and—to anticipate for once—all the *Episodios nacionales*. Two important Spanish sequences, which we shall discuss incidentally later in this book, *El ruedo ibérico* and the *Memorias de un hombre de acción*, could be accommodated towards the end of the time-limit.

We have earlier established that historical figures or scenes which refer to a given outside the frame of the literary work cannot be ignored. The presence of significant historical events within a work of fiction gives it a special configuration. Outside references retain their reality beyond the frame of the work, and this reality cannot fail to influence the reader. If a novelist wishes to construct a historical work of general scope, he can turn to an imaginary Ruritania—or in a Hispanic context, to a Santa Fe de Tierra Firme or a Macondo—in which his talent can exercise itself in constructing the plot which suits him best, although even such an undetermined historical context requires the imposition of a series of norms—such as surroundings, culture, geopolitics—that have imaginative plausibility and solidity. If on the other hand it is a question of utilizing the real history of a country or specific historical figures from it (a Napoleon—or a Prim or a Cánovas), one has to take into account the external Napoleon, Prim, or Cánovas, bearing with them all their biography and bibliography, as well as the character, more or less close to him, but never identical, whom the author has created within his work. This external dimension influences even in the structural distribution of the work, for historical facts condition the author, even when he does not accept them. Of just as great moment is the effect produced on the reader, who is equally conscious of the external reality and the fictional one.

[73] '. . . to call it an historical novel or an epic novel or a psychological novel or any other sort of novel is to obscure rather than to illuminate' (*4b*, Christian 120).

[74] Its famous subtitle '*'Tis Sixty Years Since*' clearly marks both distance and relevance.

A reference to Napoleon—for example—conjures up a series of associations drawn from common experience outside the work. At the same time, the action of 'Napoleon' within the work of fiction corresponds to a structure assigned to the work as a whole with 'Napoleon' merely as a component part. In these circumstances he is not the first-hand, unmediated Napoleon of his own utterances (which in any case would be selective), nor the second-hand Napoleon of scrupulous historical reconstruction, but a third-hand Napoleon of expeditious literary interaction with fiction, or even a fourth-hand one corresponding to a passing biographical reference or fleeting association.[75] In other words, several levels of gradation are possible for historical characters or situations introduced into narrative fiction, all of them subject to the sort of structural analysis required of all fiction. Major historical personages, as we have established, are not suitable as protagonists for the historical novel; when they appear prominently in Scott's novels, it is 'with their personality complete' (*4b*, Green 8–10). As Lukács explains in praising Scott's '"middling", merely correct and never heroic "hero"' (*2*, 33), he 'lets his important figures grow out of the being of the age, he never explains the age from the position of its great representatives, as do the Romantic hero-worshippers. Hence they can never be central figures of the action' (40). On the other hand, it is important to throw indirect light on the great events of the time, and Lukács makes a perceptive criticism of Prosper Mérimée's *Chronique du règne de Charles IX*:

He deprives the leading historical figures of their heroism with rightful scepticism. But in so doing he makes the course of history private . . . there is no real organic link between the great historical event which Mérimée wishes to portray—the night of St Bartholomew—and the private destinies of the principal heroes. (91)

An important question which arises regarding both history and literature is whether or not events or personages are conditioned by a rigid chain of cause and effect, in other words, are determined; in quoting Collingwood earlier, I drew attention to the deterministic tone of his comments. To put it another way, what place, if any, does chance have in human conduct—what E. H. Carr[76] called 'two savoury red herrings . . . one labelled "Determinism in History, or the Wickedness of Hegel"; the other, "Chance in History, or Cleopatra's Nose"' (*1a*, 119)? Pure determinism is a result essentially of the application of nineteenth-century positivism to historical study and the consequent creation of a so-called scientific school of history. Sir Isaiah Berlin, in *Historical Inevitability* (*1a*), has argued forcefully that to accept historical determinism is to deny free will and condone the crimes committed by Napoleon, Hitler, or Stalin. As Pieter Geyl puts it, 'To represent the

[75] Further degrees of distance might be established. The pig Napoleon of Orwell's *Animal Farm* provides one such instance.

[76] In the G. M. Trevelyan Lectures of 1961, entitled 'What is History?'

historical process as a concatenation of events, one following upon the other inevitably, caused as they all are by a superhuman force or by impersonal forces working in society independently from the wishes or efforts of individuals—this is the fallacy' (*1a*, 1955*b*: 238). Yet, as Carr asserts, the question is far from being clearly focused: 'It is not that some human actions are free and others determined. The fact is that all human actions are both free and determined, accordingly to the point of view from which one considers them . . . cause and moral responsibility are different categories' (*1a*, 124). He points out that 'In practice, historians do not assume that events are inevitable because they have taken place. They frequently discuss alternative courses available to the actors in the story on the assumption that the option was open, though they go on quite correctly to explain why one course was eventually chosen rather than the other' (125). Moreover, the zeal to deny the quality of inevitability in history refers essentially to contemporary history 'since people remember the time when all the options were open, and find it difficult to adopt the attitude of the historians, for whom they have been closed by the *fait accompli*' (128). The novelist's task is precisely to create the potentiality of choice within a historically closed situation.

What of chance, as exemplified by Cleopatra's nose? First of all, it should be said that this distinctive physical charm was presumably itself a cause of Antony's infatuation: much under the heading of chance is the result of unrecognized cause. This said, chance still has a role. As Huizinga says, arguing against evolution and in favour of chance: 'knowing in the historical sense rarely if ever means indicating a strictly closed causality. It is always an understanding of contexts . . . an open one' (*1a*, 39).

The equivalent of the scientific mode of history in literature is naturalism. The unequivocal approach of a Ranke is matched by the conviction of a Taine[77] or a Zola that by determining the biological, environmental, and temporal circumstances of an individual you can fully account for his actions. This criterion effectively excludes moral consideration, as is shown in Taine's famous remark, 'le vice et la vertu sont des produits comme le vitriol et le sucre' (*4b*, vol. i, p. xv). By contrast, Carr and Hayden White[78] coincide in associating causation in history—as opposed to unique accidents—with moral values, that is, the norms under which a given society operates, thus bringing in an important new dimension to narrative history and fiction. Since the capacity for selecting events and adding fictional support is greater in the

[77] Harry Levin observes that 'Taine's determinism . . . is simply an intensive application of the intellectual curiosity of the age . . . It is simply a historian's awareness of what the past already has determined' (*4b*, 9).

[78] 'Could we ever narrativize without moralizing?' (*1a*, White 1987: 25; also in *2*, Mitchell 1981: 23). Louis O. Mink, arguing that narrativization, not a 'moral impulse', is primary and not derivative (in *2*, Mitchell 1981: 239), disputes the claim that such moralizing, however frequent, is an essential attribute of narrativization. Whatever the merits of this argument, the close association of narration and moralizing is not disputed.

historical novel, so the moral cohesion, and with it the dangers inherent in didactic decisions, increases in proportion.

Zola's professed aim, then, was to give the novel the objective quality of a scientific experiment. A quotation from *Le Roman expérimental* will remind us of this objective:

En un mot, nous devons opérer sur les caractères, sur les passions, sur les faits humains et sociaux, comme le chimiste ou le physicien opèrent sur les corps bruts, comme le physiologiste opère sur les corps vivants. Le déterminisme domine tout. C'est l'investigation scientifique, c'est le raisonnement expérimental qui combat une à une les hypothèses des idéalistes, et qui remplace les romans de pure imagination par les romans d'observation et d'expérimentation. (*4b*, x. 1183)

In fact, such determination operates essentially in three areas: biologically or genetically, environmentally or socially, and temporally, in period. These correspond to Taine's three factors, *race*, *milieu*, and *moment* (*4b*, vol. i, pp. xxii–xxxi). The last factor, time or *moment*, crucial for a novel like *Salammbô*, need not concern us much, since the novels which interest us deal with the near contemporary period or the recent past, but matters of heredity and environment, whether or not they have full determining force or merely predispose, are of crucial importance.

Tolstoy gives a special twist to determinism. His central thesis, as Berlin puts it, 'is that there is a natural law whereby the lives of human beings no less than those of nature are determined; but that men, unable to face this inexorable process, seek to represent it as a succession of free choices, to fix responsibility for what occurs upon persons endowed by them with heroic virtues or heroic vices, and called by them "great men"' (*2*, 27).[79] His urge to deflate great men, especially Napoleon, and with him momentous events such as the battles of Austerlitz and Borodino,[80] is intense, so that he systematically undercuts the heroism of his subject to a level of basic human survival.[81] He is also concerned to discount the role of chance as an explanation for historical occurrences: 'The terms "chance" and "genius" connote nothing that really exists and therefore lie beyond definition' (*4b*, *War and Peace* iii. 363). Not without contradiction (*2*, Berlin 29–30), Tolstoy reconciles the opposing claims of free will and necessity by seeing them already resolved once they have become history: 'History surveys a presentation of man's life in

[79] John Henry Raleigh notes that 'Tolstoy sees history as having several dimensions' ('Tolstoy and the Ways of History': in *2*, Spilka 211–24).

[80] Compare Fabrice del Dongo's quite marginal part in the battle of Waterloo in Stendhal's *La Chartreuse de Parme* and the indirect presentation of the same battle in Thackeray's *Vanity Fair*. Such avoidance of great events is analogous to the exclusion of eminent historical figures.

[81] I am thinking of Pierre's horrifying experiences as a witness at the battle of Borodino (XI, chs. 12–15). Christian remarks that 'To Tolstoy no war—not even a "just" defensive war—can be anything but a human tragedy' and cites 'Prince Andrei's thoughts on the eve of Borodino on the indescribable horrors of war' (*4b*, 111; XI, ch. 18).

which the union of these two contradictions has already taken place' (*Meaning of History* . . . , 181; iii. 455). He goes on to affirm that 'What we know of it [history] we call laws of necessity, what we do not know of it we call freedom' (iii. 466). Structurally, Tolstoy's arrangement of material in *War and Peace* seems unwieldy or even clumsy, since his substantial treatise on history is tagged on to the end of an already very lengthy fictional narrative instead of being integrated into it.[82] Indeed, his conception of history has attracted much adverse criticism from such seasoned writers as Turgenev, Flaubert, Henry James, and Lubbock, although in my view it is important to add that its central role in his literary creation cannot be ignored; it was impossible for him to exclude public issues as Turgenev and Flaubert advocated. Moreover, despite its philosophical contradictions and the ultimate predominance of fatality, it offers, for a novelist preoccupied with history, a viable practical formula. R. F. Christian puts it very well:

> In Tolstoy's philosophy the individual has freedom and control within a limited range of activity . . . But in any activity which affects people other than himself he is only one of a very large number of factors which make up the totality of causes . . . The fact that so many of Tolstoy's characters act with apparent freedom, exult in the spontaneous creative forces of life, choose their friends, their loves, their pleasures and their duties, act impulsively or accomplish deeds of heroism, does not in itself refute his theories. The life of these people can be rich, many-sided, apparently self-determined—without the conviction on their part that they are free agents life would be intolerable—and yet the consciousness of freedom which permeates it can still be an illusion. (*4b*, 110)

It is a commonplace of historical commentary to proclaim the importance of minor, underground currents at the expense of wars, reigns, etc.: i.e. private history versus public history. Lévi-Strauss makes a valuable point, particularly apposite for the historical novelist:

> Biographical and anecdotal history, right at the bottom of the scale, is low-powered history . . . [It] is the least explanatory, but it is the richest in point of information, for it considers individuals in their particularity and details for each of them the shades of character, the twists and turns of their motives, the phases of their deliberations.[83]

A well-known distinction made by Sir Isaiah Berlin in his stimulating essay on Tolstoy's historical perspective, *The Hedgehog and the Fox*, may help at this juncture. In his essay Berlin discusses the relationship between what he regards as two opposing and irreconcilable forces which mark 'one of the

[82] We may recall Henry James's famous remark about 'large, loose baggy monsters' with regard to *The Newcomes*, *Les Trois Mousquetaires*, and *Peace and War* (*sic*) (Preface to *The Tragic Muse*: *2*, 84). Consult also 'Ivan Turgenev', where Tolstoy is described as 'a monster harnessed to his great subject—all human life!—as an elephant might be harnessed, for the purposes of traction, not to a carriage, but to a coach-house. His own case is prodigious, but his example for others dire: disciples not elephantine he can only mislead and betray' (*2*, 317).

[83] 'History and Dialectic' (*2*, in De George 227). Compare also 'the historian loses in information what he gains in comprehension or vice versa' (228).

deepest differences which divide writers and thinkers, and, it may be, human beings in general'. The difference—'a great chasm'—lies between

> those, on one side, who relate everything to a single central vision, one system less or more coherent and articulate, in terms of which they understand, think and feel—a single, universal, organising principle in terms of which alone all that they are and say has significance—and, on the other side, those who pursue many ends, often unrelated and even contradictory, connected, if at all, only in some *de facto* way, for some psychological or physiological cause, related by no moral or aesthetic principle . . . The first kind of intellectual and artistic personality belongs to the hedgehogs, the second to the foxes. (*2*, 1–2)

The distinction seems to me a valuable one, but it has clear limitations. It is far too rigid and too absolute: unity and variety are not irreconcilable, and indeed it is difficult to envisage most great literature without a considerable measure of each quality. In fact, the important thing is the structural relation between the two factors: the degree to which one or other predominates and the interplay between them tell us a great deal about the purpose, internal organization, and impact of the work concerned.

In some degree we may identify Berlin's distinction with the contrast we have just made, between what has been called in French or Spanish *la grande histoire, la historia grande*—notable external events—and *la petite histoire*, or *la historia pequeña* or *chica*, that is to say, the trivial daily incidents which reveal the spirit of the time.[84] The latter, with its abundance of incident and anecdote, corresponds to the first aspect, that of the fox, while the former, although it does not necessarily provide the sweep of historical vision enshrined in the hedgehog metaphor, is set firmly in chronological development and accordingly offers a certain coherence and immutability. Just as Berlin's distinction should not be taken as exclusive, so the essential point about public and private history is not simply the existence or coexistence of each, but their constant and mutually illuminating interaction.

We shall discuss in Chapter 1 how Galdós reacts to the various issues described in this Introduction: the relation between history and narrative; the qualities and limitations of the historical novel and its relations with the realist novel; the use of historical fact in fiction; the discursive structure of fiction; the applicability of scientific laws; the opposition or reconciliation of public and private history in narrative.

[84] Compare Charles Reade: 'Not a day passes over the earth, but men and women of no note do great deeds, speak great words, and suffer noble sorrows. Of these obscure heroes, philosophers, and martyrs, the greater part will never be known . . . Here, then, the writer of fiction may be of use to the public—as an interpreter' (*The Cloister and the Hearth*; quoted in *2*, Fleishman 153).

1

Galdós's Contemporary Novels and Episodios nacionales

ONCE we turn our attention to the novelist who is the subject of our study, we are immediately struck by a strange fact, which through being so obvious is often overlooked: Pérez Galdós wrote not just one but two types of narration, clearly differentiated under the headings *novela contemporánea* and *episodio nacional*,[1] both of which seem to fall within the historicist definition of the nineteenth-century novel discussed in the Introduction.[2] There is little doubt that history for Galdós in general meant what it did for Ranke, but the freedom of commentary which even Ranke indulged in was much greater in Galdós, in part as a result of his fictional form, in part because of his conception of human fallibility, especially where language is concerned. This idea will be developed later.

Let us first attempt to trace succinctly the evolution of this curious bifurcation. It is no accident that Galdós, like Balzac, started his effective career with an explicitly entitled *Novela histórica original, La Fontana de Oro*, written for the most part before the September revolution of 1868 and completed immediately afterwards.[3] It deals with the 'trienio constitucional' between 1820 and 1823, when, following the revolt led by Colonel Rafael del Riego, the liberals were briefly able to impose a constitutional system upon the absolutist monarch Ferdinand VII. Significantly, in the *Preámbulo*, dated December 1870, Galdós sees a continual line of development between 'uno de los períodos de turbación política y social más graves e interesantes en la gran época de reorganización que principió en Cádiz en 1812 [i.e. *the liberal constitution of that year*] y no parece próxima a terminar todavía' and 'la crisis

[1] I shall retain the term *episodio* (*nacional*) in Spanish, since I am arguing that it is a distinctive form to which the English translation would do scant justice.

[2] Menéndez Pelayo's words, in his reply to Galdós's reception address to the Real Academia, have still not entirely lost their validity: 'Sin ser historiador de profesión, ha reunido el más copioso archivo de documentos sobre la vida moral de España en el siglo XIX' (*3a*, 93).

[3] It was financed by his family. In his review of the novel, José Alcalá Galiano describes him as follows: 'Su autor, cuya independiente posición [!] no le obliga a hacer negocios literarios, es un joven modesto, amante de las letras hasta el entusiasmo, adornado de una erudición vasta y sólida, de un talento superior, y poseedor de una pluma que, manejada ya en el periodismo político, ya en la crítica y ya en las creaciones de la pura fantasía' (*La Revista de España*, 20 [1871], 149). Galdós wrote a prologue to Alcalá Galiano's *Estereoscopio social* in April 1872 (*3a*, *Prólogos* 45–51). They later became bosom friends.

actual' of his own time (IV: 10–11): a similarity, that is to say, between the period 1820–3 and September 1868–December 1870, when Prim was scouring Europe in search of a suitable constitutional monarch. Prim was to die as a result of assassins' bullets at the end of the same month as Galdós dates this *Preámbulo*. The novel thus seeks to establish the direct link with the present advocated by Lukács and Auerbach. If the clear historical precedent were an infallible guide, the revolution would be condemned to succumb to a reign of terror like Ferdinand's 'década ominosa', but what Galdós is no doubt advocating to his Liberal contemporaries is caution, moderation, and unity, in order to avoid the mistakes of their predecessors.

The novel has two alternative endings,[4] one—that of the standard edition—in which the hero Lázaro withdraws from politics in quest of a peaceful life; and another version, dated 1871, in which the hero is ambushed and killed by his absolutist uncle *Coletilla*. The order in which the two versions were written or published is still in dispute. Pattison has proved that a 'first edition' of 1870, with a relatively 'happy ending', is a forgery based on the 1892 Guirnalda edition. He claims, as a consequence, 'that Galdós changed his ending once, and only once, between the edition of '71 and the second edition (Guirnalda, n.d., but probably 1874), which is described on its title page as "notablemente corregida"' (*3c*, 6). Gilman (*3c*, 47) continues to believe in a lost first edition with a 'happy ending' and supports Gimeno (*3c*, 1976: 62) in his persuasive argument that Galdós changed to the tragic ending after Prim's death,[5] only to revert to the peaceful conclusion about 1874.

The rejected version has another fascinating feature. The tragic outcome, as Linda Willem points out, is only one of the two alternative solutions it offers. The narrator breaks into the story to explain that he had already written the 'happy ending' (a possible argument in favour of a lost first edition) when 'la colaboración de un testigo presencial de los hechos que vamos refiriendo [Bozmediano], le obligó a desviarse de este buen propósito dando la historia el fin que realmente tuvo' (*3c*, Gimeno 1976: 57). This abrupt break with the conventions of the fictional world presents an Aristotelian contrast between a version which is twice described as 'más natural y lógico'/'más lógico y artístico' and what actually happened. The narrator does not, however, entirely abandon his individual stance, but insists on leaving the choice to the reader: 'Puede el lector aceptar el que mejor cuadre a su gusto y

[4] For details consult Smieja (*3c*), Gimeno (*3c*, 1955; 1976), and Pattison (*3c*).

[5] Gilman, who examined the manuscript (still in the possession of Galdós's grandson Don Benito Verde), indicates that it does not contain this 'pregnant interruption' or the tragic ending. The chronology poses difficulties, but Pattison (*3c*, 7) may be too categorical in asserting that Gimeno's thesis is untenable, especially if Galdós's recollections in *Por esos mundos* (1910) that the novel appeared with an anticipated date (1869 for 1870) were mistaken only in the year: 1869/1870 instead of 1870/1871. My own surmise is that José Alcalá Galiano's strong disapproval of the tragic ending in his review of the novel (*La Revista de España*, 20 [1871], 158, quoted in Pattison, 7) was decisive in the definitive change or reversion to the 'happy ending.'

sentimientos.' As Willem indicates, 'The alternative conclusion format asks the reader to consider the very conventions underlying narration and how the demands of narrative fiction differ from those of narrated history' (*3c*, 53). It is thus a remarkable anticipation, not only of the distinct claims of story and discourse, but of the auto-reflexive mode and reader-response theory as well.

Among other characteristics of the novel are a certain *costumbrista* delight (*3c*, Hafter) in the presentation of the Carrera de San Jerónimo, a similarity with political journalism and cartoons, perceptively noted by Gilman (*3c*, 37), juxtaposed rather than integrated characters (*3c*, López-Morillas 47) and an adverse judgement on *el pueblo* (*3c*, Dérozier). Certain of these characteristics subsist in some degree in later works. In addition, the novel contains a chapter (XLI) devoted to an explicit and very hostile description of the king:

Fernando VII fue el monstruo más execrable que ha abortado el derecho divino. Como hombre, reunía todo lo malo que cabe en nuestra naturaleza; como rey, resumió en sí cuanto de flaco y torpe puede caber en la potestad real . . . No fue nuestro tirano descarado y descubiertamente abominable: fue un histrión que hubiera sido ridículo a no tratarse del engaño del pueblo. (**IV**: 173)

Galdós does not use this sort of obvious frontal denunciation again. Henceforth, he is at pains to show Ferdinand and other historical figures in action or discussion and not in unmediated and explicit description. For example, in *Memorias de un cortesano de 1815* Ferdinand's despicable actions and qualities are presented sarcastically in a nominally favourable light, by means of the scheming and hypocritical narrator Juan Bragas. The novelist works in a detailed portrayal of the king in all his baseness during one of his clandestine nocturnal escapades; since the monarch's identity is not acknowledged, the portrait is not distorted by Bragas's sycophancy. The reader, on the other hand, has no difficulty in recognizing the description (compare the famous portrait by Goya):

Sus ojos eran negros, grandes y hermosos, llenos de fuego, de no sé qué intención terrible, flechadores y relampagueantes. Bajo sus cejas, semejantes a pequeñas alas de cuervo, centelleaba, deshecho en ascuas mil por las movibles pupilas, el fuego de todas las pasiones violentas. Su nariz era desaforadamente grande, corva y caída; una especie de voluptuosidad, una crápula de nariz. La carne, superabundante, había crecido, representando con fértil desarrollo su preponderancia en aquella naturaleza. El labio inferior, que avanzaba hacia afuera, parecía indicar no sé qué insaciabilidad mortificante. La personificación de la sed habría tenido una boca así. Una línea más de desarrollo, y aquel belfo hubiera tocado en la caricatura. Observándole bien, se veía en la tal fisonomía peregrina mezcla de majestad y de innobleza, de hermosura y ridiculez. (**I**: 1321)

The technique is not without its defects: there is almost grotesque exaggeration; moral qualities are too readily related to physical characteristics; and the effect of non-recognition is spoilt by the last sentence: 'En suma: el perfil de aquel hombre solía verse en las onzas de oro' (I: 1321). What is more, Galdós's animosity continues in rather too evident a manner in other *episodios*. Thus, at the time of the famous confrontation between the infanta Carlota and Calomarde, Ferdinand's prolonged suffering is treated as an act of justice and retribution delivered from outside: 'por fin le tocaba un poco de potro' (*Los apostólicos* II: 211).

After *La Fontana de Oro* he demonstrated his early interest in the fantastic tale in *La sombra* (1870). Then, in the summer of 1871, comes *El audaz*, with the subtitle *Historia de un radical de antaño*. This time the action goes back to 1804, shortly before the French invasion. Despite a melodramatic plot, weak narrative structure, and some incongruous *costumbrista* scenes, its theme is appropriate for its period: the struggle between the *ancien régime*, exemplified by the Inquisition, and the overwhelming impact of the French Revolution; its more menacing aspects are broached in the figure of the mad José de la Zarza, known as *el tío Robispier*,[6] through whom the Terror is effectively recalled. However, the alliance between the revolutionary hero Martín Martínez Muriel and the *fernandistas* in opposition to Godoy[7] is implausible, just as the love-relation between Muriel and Susana is far-fetched and extravagant. Muriel himself seems uneasily poised between being a prophet before his time[8] and a deranged fanatic who eventually identifies himself with de la Zarza.

La Fontana de Oro and *El audaz* have the form of the conventional historical novel and fall comfortably within the time-scale outlined in the Introduction for the 'modern' historical novel. Evidently and rightly, the formula did not satisfy the budding novelist. He was trying, not too effectively, to supply historical facts and give at the same time a socio-historical interpretation[9] within a fully developed fictional, indeed *costumbrista*, plot. As a result of his dissatisfaction he proceeded to divide his interests in two directions.

First, he allowed himself more scope for basic historical description and analysis by developing a new form, the *Episodio nacional*. I shall not attempt to

[6] The cat in *La Fontana de Oro* is called 'Robespierre'.

[7] Francisco Godoy, 1767–1851, created *Príncipe de la Paz* in 1796, was the lover of the queen María Luisa and all-powerful favourite until 'the mutiny of Aranjuez' in Mar. 1808. It is widely believed that he was the father at least of María Luisa's younger children Francisco de Paula and María Isabel. Baroja further claims that Godoy, not Charles IV, was the father of all María Luisa's progeny ('Una familia ejemplar': *4a*, v. 771–4). Seco Serrano (*1b*, 97–106) staunchly defends María Luisa's reputation.

[8] 'En él estaba como en depósito la idea que más tarde había de expresarse en hechos' (IV: 235; *3d*, quoted in Hinterhäuser 33). As Montesinos points out (*3c*, i. 138), Muriel anticipates Santorcaz in *La segunda casaca*. Clark Zlotchew (*3c*) disputes Montesinos's approximation of Muriel with the young Galdós (*3b*, i. 70) and indicates the Inquisitor Albaredo as a figure of moderate reform.

[9] Cf. Hegel's comment on the ambiguity of history, which 'unites the objective with the subjective side and denotes the *historia rerum gestarum* quite as much as the *res gestae* themselves' and 'comprehends what has *happened* no less than the *narration* of what has happened' (*1a*, quoted in White 1987: 29).

add anything substantial to the information supplied by Hinterhäuser (*3d*, 33–9), Cardona (*3d*, 1968: 120–1), and others on how this form emerged. Suffice it to say that various influential figures may have played a part in his crucial decisions. In his extensive review of *La Fontana de Oro* (*Revista de España* 152) already referred to, José Alcalá Galiano, as well as criticizing the tragic ending, recommends young novelists to create a *novela nacional*, on the model of Erckmann–Chatrian.[10] Eugenio Ochoa, commenting on the first two novels in 1871, mentions Trafalgar as the key event in the collapse of the *ancien régime*. The name for the series appears to have been suggested by Galdós's friend and mentor José Luís Albareda (*Memorias* VI: 1660). Despite the existence of earlier historical narratives of near contemporary subject-matter in Spain,[11] it should be borne in mind that Galdós was proud of what he saw as an invention of his own: in the second prologue to the Illustrated Edition (1885) he proclaimed that his success depended in part 'de la circunstancia, feliz para mí, de no existir en la literatura española contemporánea novelas de historia reciente' (*3a*, *Prólogos* 56). I would add that in their theme of war, bravery, and patriotism[12] and their evident appeal to the young[13] the works of the first series have something of Fenimore Cooper (an acknowledged influence on Balzac) about them. He had found his formula for success as a popular writer[14] who aspired, against the odds, to financial self-sufficiency.

Trafalgar, written early in 1873, still within the revolutionary period, marks the beginning of this effort to capture in various volumes—rapidly consolidated into ten[15]—the military struggles and political development of

[10] It is very possible that Erckmann and Chatrian's *Romans nationaux* may have triggered off the *episodios*, but, as Hinterhäuser (*3d*, 46) judiciously concludes, it is unlikely that the influence went further than that. Galdós's library contains a copy of two of their novels bound together: *Histoire d'un conscrit de 1813* and *Waterloo*. Lukács's comments on the Alsatian authors are pertinent: he finds them uninterested in causes, and by missing out the 'upper' world completely, they exclude the all-important 'world-historical individual' (*2*, 247–56).

[11] Hinterhäuser attaches special importance to *Fernando el deseado* by Diego López Montenegro and Víctor Balaguer (*3d*, 35–8).

[12] In spite of Hinterhäuser's scepticism (*3d*, 48–9), the military background of Benito's father, uncle, and brother Ignacio is relevant here. This does not mean, however, that Galdós's attitude to patriotism was not ambivalent (see *3d*, Bly 1986).

[13] See e.g. the letters in the Casa-Museo de Galdós (*3a*) that the young Jaime Quiroga Pardo Bazán (Doña Emilia's son) sent to Galdós. At the age of 9 (4 Dec. 1885) he is impressed by all the heroes of the Peninsular war and by the expletive '¡porra!', and writes patriotically: 'Cuando leo sus novelas de Vd siento mucho orgullo en ser español, pero así y todo aunque me gusta mucho, siento pena por no haber vivido en aquel tiempo y ser jefe de una partida. ¡Porra! solo para despachurrar a Napoleón.' Later letters reveal continued enthusiasm for both the *episodios* and the novels and protest against the election of Commelerán over Galdós to the Academy. For children and the early *episodios* see also n. 34 below.

[14] According to Beyrie (*3b*, i. 168–9), the success of *La corte de Carlos IV* enabled the printing of each *episodio* to be increased from 1,000 (*Trafalgar*) and 3,000 (*La corte*) to 4,000, which henceforth become the established printing-run for each new publication.

[15] The excerpt 'Antes de Trafalgar (Páginas de una novela)' (I: 207–17; with some variants), published prior to book publication in *La Revista de España* (30/120 [15 Feb. 1873]), declares that

Spain from the crumbling of the *ancien régime* until the end of the 'War of Independence'; a further ten volumes will take the story up to 1834, shortly after the death of Ferdinand VII.[16] *Trafalgar*, incidentally, just qualifies as being within reach of human memory, since Galdós was able to call on the memories of an ancient mariner, one Galán, who had fought in the battle as a child (*3b*, Berkowitz 1948: 98–9); the direct link with the present we spoke of in the Introduction is thus clearly if precariously achieved.

In the middle of this undertaking he also inaugurated the second new direction: adopting a more contemporary approach by writing four *novelas contemporáneas* beginning with *Doña Perfecta* (1876). These novels *de la primera época*, as they are now called, start by sharing certain fixed characteristics: a clear-cut confrontation between two political attitudes representing traditional Catholic values and modern liberalism respectively; an imaginary, provincial setting (Orbajosa, Ficóbriga, Socartes); and an unspecified time of action.[17] In dealing with a contemporary theme, Galdós has deliberately abandoned the specificity of time, place, and historical personalities found in *La Fontana de Oro* and the *episodios*. It was these novels, and especially *Gloria* (1876–7), which roused Pereda's antagonism[18] and produced the simplistic vision of the liberal Galdós, opposed to the traditional Pereda, which dominated criticism for so long.[19] Within these four novels there is moreover a clear gradation. If the description I have just given applies satisfactorily to the first two, *Doña Perfecta* and *Gloria* (1876–7), it is less applicable to the third, *Marianela* (1878), where the opposition is less stark, though social conflicts are far from absent.[20] The fourth work, *La familia de León Roch* (1878), does have intense confrontation, but the struggle is now specified; it takes part principally in the concrete contemporary world of Madrid. Significantly, many of its characters recur in later novels.

Shortly after he had published this last novel, Galdós also completed the

'Es el primero de una serie que se publicará periódicamente con el título de *Episodios nacionales*.' By the time *Trafalgar* appeared, as Cardona proves (*3d*, 1968: 120), he has decided on ten volumes and sketched in some tentative titles; by May he had the final titles (with one exception) fixed. In the second prologue to the Illustrated Edition of the *Episodios* Galdós speaks of his indecision until the success of the second volume *La corte de Carlos IV* (see *3a*, *Prólogos* 55; *3d*, Hinterhäuser 24).

[16] Dendle (*3d*, 1980: 137) has shown that the first edition of *Trafalgar* indicates that the first plan was to take the second series up to 1840.

[17] The time of the action can be deduced, but is not specifically given; see Aparici Llanas (*3c*, 51) for an account of these novels (except *Marianela*).

[18] Letters of 9 Feb., 17 Feb., and 14 Mar. 1877 (*3b*, *Cartas a Galdós* 47–58).

[19] Late examples are Hurtado and Palencia (*4a*, 1925; *3c*, quoted in Varey 2) and Allison Peers (*4a*, 1929; *3c*, quoted in Ribbans 1986: 5).

[20] For example, the brutalizing effect of work in the mines and the neglect which produced Marianela's stunted condition; as Geraldine Scanlon says, 'ultimately Nela is a victim of the class system' (*3c*, 1988: 48). See also Bly (*3c*, 1972: 49–66), who persuasively views the novel as a searing study of egotism. It is most emphatically not a romantic interlude, as Berkowitz (*3b*, 1948: 143) and Ricardo Gullón claim (*3c*, 1973: 66). See also Jones (*3c*).

second series of *Episodios nacionales*. On the last page of *Un faccioso más y algunos frailes menos* (1879) he claimed that the events from 1834 onwards—some forty-five years before—were too close to contemporary circumstances and too painful for him to continue:[21] 'Los años que siguen al 34 están demasiado cerca, nos tocan, nos codean, se familiarizan con nosotros. Los hombres de ellos casi se confunden con nuestros hombres. Son años a quienes no se puede disecar, porque algo vive en ellos que duele y salta al ser tocado con escalpelo' (II: 326).[22] Thus we have more or less the same strategic distance from events that we discussed in the Introduction. It is worth noting that when Galdós repeats the explanation in the epilogue to the Illustrated Edition (November 1885), he adds other factors: 'porque la excesiva extensión había mermado su escaso valor y porque pasado el año 34, los sucesos son demasiado recientes para tener el hechizo de la historia y no tan cercanos que puedan llevar en sí los elementos de verdad de lo contemporáneo.' It is clear therefore that in addition he is conscious of a risk of over-exposure and of the difficulty of dealing with the awkward limbo between the historical and the contemporary.

As important as the final statement of *Un faccioso más*, but less quoted, is the disillusioned prophecy about the future made by Salvador Monsalud on the previous page. In contrast with Benigno Cordero's optimism, Monsalud declares that 'no esperaba ver en toda su vida más que desconciertos, errores, luchas estériles, ensayos, tentativas, saltos atrás y adelante, corrupciones de los nuevos sistemas, que aumentarían los partidarios del antiguo; nobles ideas bastardeadas por la mala fe, y el progreso casi siempre vencido en su lucha con la ignorancia' (II: 325). In the long run, he believes reforms will come, but not in his lifetime; meanwhile, 'me parece que asisto a una mala comedia': this phrase, or such variants as 'sainete' or 'esperpento',[23] will recur again and again in the course of the *episodios*. In a similar prolepsis in the previous volume, *Los apostólicos* (1879), Monsalud had expressed like despair, expecting the task of liberalism to last a century: 'hemos de pasar por un siglo de tentativas, ensayos, dolores y convulsiones terribles' (II: 184).

In these passages Galdós is quite flagrantly using *a posteriori* judgement or hindsight. While he caters in a somewhat facile manner to the readers' awareness of these events, he puts at risk the plausibility of his character, who seems more clairvoyant than we might have expected. According to Monsalud's forecast, no hope of reform can be conceived of before the convulsions leading to the revolution of 1868, thirty-four years ahead; and its

[21] It should be noted that he explicitly claimed the proprietary right to go on using his fictitious characters from the *episodios* in the contemporary novels.

[22] Martha Krow-Lucal shrewdly comments: 'no hay que tomar demasiado en serio las famosas razones... Algo de esto podría haber, pero sabemos por su carta a Pereda que Galdós se encontraba ya harto del *género histórico*' (*3c*, 171); see letter of 20 May 1879 (*3a*, '28 cartas' 32), in which he expresses his weariness with 'este libro y esta colección' after finishing *Los apostólicos*.

[23] For the use of the word 'esperpento' see V. A. Smith and Varey (*3b*), and Amor y Vázquez (*3b*).

failure led to a restoration about which, in 1879, serious doubts could be entertained regarding both its benefits and its stability. And the time of writing is nearly half-way through the century Monsalud deemed necessary for effective reform; the full century would take us, incidentally, up to the Second Republic of 1931! At all events, his pessimism regarding the immediate future is clearly fully justified.

Monsalud's observations seem to me to add a further plausible explanation of why Galdós abandoned writing *Episodios nacionales*. He could see, I suggest, little point in going on recreating a period of frustrated ideals which was to last over thirty years before it attained near contemporary interest. And, in addition, we may note that Monsalud's decision to withdraw from politics coincides with one of the two solutions which determine Lázaro's fate in *La Fontana de Oro*; the other, of course, anticipates Pepe Rey's in *Doña Perfecta*.

Critics, led by Casalduero (*3b*, 1961: 43), have signalled the year 1879 as decisive in marking a new direction which is to produce momentous results. Certainly, Galdós then takes, for a writer dependent on his pen, a calculated risk. He abandons the sure market of the distinctive form he has created. At the same time he is evidently dissatisfied with the thesis novels; he revealed to Pereda his disappointment at the lack of success of the third volume of *León Roch* and his own dissatisfaction with it (*3a*, '28 cartas' 31). Now, by a conscious decision involving a most unusual period of deliberation lasting nearly two years, he sets out on a new venture. Its beginning is described in a letter to Pereda of 4 March 1879: 'Ahora tengo un gran proyecto. Hace tiempo que me está bullendo en la imaginación una novela que yo guardaba para más adelante, con el objeto de hacerlo detenido y juiciosamente.... Necesito un año o año y medio' (32). He needed all of that time, for the first part of *La desheredada* appeared twenty-two months later in January 1881. The 'gran proyecto' can be identified with what he calls, three years later (14 April 1882), in an important letter to Giner de los Ríos,[24] his '*segunda o tercera manera*'. And he assures Pereda, who had reacted so vehemently against *Gloria*, that 'Este asunto es bueno en parte político pero no tiene ningún roce con la religión' (32).

On abandoning the writing of new volumes of *Episodios nacionales* in 1879, Galdós did not lose interest in the genre or in his public, and he was very keen to maintain a steady and loyal readership.[25] Conscious like so many novelists within the historical genre[26] of the importance of illustration, he at

[24] 'Puse en ello especial empeño, y desde que concluí el tomo, lo tuve por superior a todo lo que he hecho anteriormente ... Yo he querido en esta obra entrar por nuevo camino e inaugurar mi *segunda o tercera* manera' (*3c*, quoted in Ruiz Salvador 53). Dendle (*3c*, 1980) considers it a 'logical development' from the twenty *episodios nacionales* which had preceded it.

[25] Several of his *episodios* and novels were published in 1879 and 1880 in the *folletín* of *El Océano*, which Galdós directed; the magazine seems to have been kept going in order to complete the instalments of *Un faccioso más*. See Letamendía (*3b*, 87).

[26] Compare Manzoni and the importance he attached to the illustrations of the 1840 edition of *I promessi sposi* (*4b*, ed. Titta Rosa 993).

once embarked on the ambitious project of an illustrated edition, which started to appear in 1882 (*4a*, Miller). He was no doubt aware, too, of leaving a gap: his assiduous readers had been left high and dry at 1834, and in so far as their historical interest was whetted, they would now have to move on to the 1860s or 1870s in the different context of the contemporary novels. Yet the change is not as abrupt as has often been claimed. Martha Krow-Lucal (*3c*) has convincingly argued that both *León Roch* and *Un faccioso más* are transitional works and that Galdós has learnt a new and valuable technique: the reuse of familiar characters. Recurring characters help to reinforce the continuity he desired, as does the reaching back, by means of historical references within the contemporary novels, through the reign of Isabella, to almost the beginning of the century. This procedure is, as we shall see, especially characteristic of the first of the new novels, *La desheredada*; it also applies to *Fortunata y Jacinta*. The new direction, then, is real enough, but does not entail a radical break with the past.

From now on Galdós will be free from any commitment to thesis novels or to straight historical narrations. The attachment to a definite, if imagined, situation and to the feigned actuality typical of the *episodios* will henceforth be carried over to the contemporary period. In this there is little doubt that he took a lead from Zola's naturalist novel (*4a*, Pattison 90–1) while adopting from Balzac[27] the idea of a multi-volume series of contemporary novels described in the Introduction; the predominance of the *Scènes de la vie privée* and the *Scènes de la vie parisienne* within *La Comédie humaine* corresponds to Galdós's own priorities.[28] A very similar relation between society and the individual—one which implies a historicist approach—thus exists in the two writers, despite their quite different political principles. Hemmings's description of Balzac as dealing with 'individual visions, ambitions and nostalgias acting on, colliding with, and modified by the solid-seeming yet disintegrating rock of society' (*4b*, 1967: xvi) could apply equally to Galdós.[29]

The series of twenty contemporary novels, undoubtedly his finest, is thus to command his undivided attention for some eleven years, until 1892, when as a result of a suggestion to put a version of the dialogue novel *Realidad* on the stage he started writing seriously for the theatre; most of his productions were at first adaptations of novels or *episodios* for the stage. He then alternated between the two genres until 1898, when he resumed the writing of *Episodios nacionales*.

[27] The impact Balzac made on Galdós is unquestionable and explicitly acknowledged in the *Memorias*, where he dates it from his first visit to Paris in 1867 (VI: 1431; *3a*, *Prólogos* 521). Cardona has pointed out that his acquaintance with Balzac goes back earlier than that, to 1865 (*3c*, 67).

[28] A character in *El doctor Centeno*, Arias Ortiz, is 'muy devoto de Balzac, lo tenía casi completo, y a los personajes de la *Comedia humana* conocía como si los hubiera tratado' (IV: 1382); many of these characters, including, significantly, Gobseck, are mentioned by name. Arias Ortiz also gave the nickname Gobseck to a moneylender, an antecedent of Torquemada (1372).

[29] I leave aside specific cases of influence by Balzac in Galdós's novels. One example described by Ricardo Gullón is *Les Employés* and its possible impact on *Miau* (*3c*, 1973: 304–10).

What induced Galdós to take the subgenre up again, nearly twenty years later, in 1898? Of the reasons given in 1879 and subsequent years for abandoning them, one—excessive exposure—had long since ceased to be a factor. The closeness to contemporary events has evidently diminished to near zero: the 'contemporáneo' has now become 'lo histórico', i.e. more than the forty years Fleishman (*2*) postulated for the historical novel. Only the chaotic and indeterminate nature of the intervening period still applies. He now has a longer and (at least, by the end) a more interesting period to tackle. It is worth noting that, without being distracted by other genres, he writes the third series—up to the beginning of the effective reign of Isabella II—'with clockwork regularity' (*3b*, Berkowitz 1948: 336),[30] completing its ten volumes in three years. What is more, he had an overall plan complete before he started: from the beginning he listed, accurately, the titles of the ten volumes which will make up the series.

In his short introduction to *Zumalacárregui* he claims that it is public demand which has caused him to break his vow.[31] This is not hard to reconcile with what appears to be the essential motive. Economically, he was going through a bad period and needed to extricate himself from the financial straits his lawsuit against his former publishing associate Miguel de la Cámara had got him into.[32] Galdós evidently thought of the *Episodios nacionales* as money-spinners;[33] the earlier series were still on sale, and the new ones would surely, to his mind, galvanize purchases. He further ventured on a new edition of his works rather garishly clothed in the national colours of yellow and red. In this way he was able to continue the appeal to a patriotic audience which the earlier works, full of conflict and adventure, had stimulated. We may even surmise that, nineteen years later, Galdós was hoping that his earlier teenage readers might be tempted to read the new series, especially as the subject was still in large part war and adventure.[34] Other reasons are evidently

[30] As Berkowitz shows (*3b*, 1948: 109), he wrote the first of the series, *Zumalacárregui*, at breakneck speed, in about a month.

[31] '. . . los amigos que me favorecen, público, lectores o como quiera llamárseles, me mandan quebrantar el voto, y lo quebranto'; dated 'abril de 1898' (II: 328). In the *Memorias* (VI: 1695) Galdós openly discloses the economic motive.

[32] See Berkowitz (*3b*, 1948: 325–6). His lawyer was Antonio Maura. For fuller particulars consult Guimerá Peraza (*3b*, 1967; 1977). Despite a very considerable income, Galdós never succeeded in regaining economic solvency; hence the failed National Subscription of 1914–16 (for details see *3b*, Botrel 1977). The letters to his administrator in Madrid, Gerardo Peñarrubia, described by Shoemaker (*3b*, 1984), give clear if incomplete indications of his penury.

[33] See Berkowitz (*3b*, 1948: 323). Botrel (*3b*, 1984–85) gives a fascinating account of the commercial fate of Galdós's work. Despite some early success, sales of the new series fell off rapidly from 1904 onwards. It is noteworthy that the early series of *episodios* and the first contemporary novels consistently outsold both the mature novels and the later series of *episodios* by a wide margin.

[34] Some years later, at the end of 1908 or beginning of 1909, he set out not to let a potential audience escape by preparing a special edition of the early *episodios* for children: *Episodios Nacionales, Guerra de Independencia, extractada para uso de los niños*. See Navarro González (*3d*).

not excluded; no doubt, as Berkowitz (*3b*, 1948: 334–5) and Dendle (*3d*, 1980: 33) claim, he was motivated by the Spanish crisis, especially once he had completed the third series, but this criterion can easily be exaggerated. As Ricard[35] has indicated, *Zumalacárregui* was completed before *el desastre*. Once his own memories come into play, well into the fourth series, there is no doubt about his personal involvement. Though the historical distance of about forty years remains, the period narrated is no longer past history to him. The reverse of his earlier criterion—to avoid the near contemporary as too painful—now prevails.

By this time, the flow of orthodox contemporary novels has come to an abrupt stop: *Misericordia*, the last of the great contemporary novels in most critical opinion, was already published in 1897. The approximation to the theatre, following the precedent of *Realidad* (1889) and *La loca de la casa* (1892), is continued in another dialogue novel, *El abuelo* (1897). His fictional output has now become very limited; there is a particularly startling eight-year gap between *El abuelo* and *Casandra* (1905), during which time he wrote and staged the sensationally controversial play *Electra* (1901). The two later narratives (*El caballero encantado*, 1909, and *La razón de la sinrazón*, 1915) explore the fantastic; *inverosímil* is the word Galdós applies to them both. In all four late novels the historical and political specificity of the earlier period has almost disappeared. At most there is an isolated historical allusion; the prior who personifies reconciliation in *El abuelo* is named Baldomero Maroto, an acknowledged reference to the participants (Baldomero Espartero and Rafael Maroto) in the *abrazo de Vergara* of 1839 (VI: 855); the wealthy Doña Juana Samaniego denounces the desintailment laws carried out by 'el maldito judío Mendizábal' (VI: 928). Otherwise, as for Don Pío Coronado's pupils Nell and Dolly in *El abuelo*, history has become confused with mythology (VI: 837; see *3b*, Beverley). The conflict between good and evil is once more clear-cut and Galdós offers, particularly in *Casandra*, a decisive denouement against reaction and in favour of progress; as Casalduero graphically put it, Casandra has killed Doña Perfecta (*3b*, 1961: 156–7).

It is intriguing to note that the two types of work do not run concurrently; only for a short time—at the period of the novels *de la primera época*—did Galdós write with an equal commitment to both forms. The reason for this division must be a matter of speculation. Is it largely coincidental? He had convincing reasons, as we have seen, both for abandoning the genre and for returning to it. The phasing out of the realist series corresponds to an increasing spiritualization and allegorization of his subjects, and at the same time he was obviously stimulated by his new interest in the theatre. Yet it does

[35] 'Tout ce qu'on peut dire . . . c'est qu'avant même la catastrophe Galdós se sentait inquiet de l'apathie dans laquelle sombrait son pays . . . Peut-être a-t-il pressenti le désastre. Mais le désastre n'a pas été la cause immédiate de son retour à l'orientation de 1872' (*BH* 59 [1957], 332, quoted in Hinterhäuser, *3d*, 52–3).

seem noteworthy that there should be no overlap at all. Perhaps he felt conscious of the differences of technique we shall discuss in a moment and was unwilling to undertake them both at the same time.

We are faced, then, with two distinct forms of narration of notable historical content which correspond to largely different periods of the author's life. Does their common historical concern mean that there is little difference between them? Some critics, indeed, such as Casalduero, have firmly denied that there is any distinction between the two forms: '*Episodios nacionales* es título con que [Galdós] agrupaba en colección una serie de obras que *fundamentalmente en nada se diferencian del resto de las novelas*' (*3b*, 1961: 43; my italics).[36] For all my respect for a pioneering study, I find this position untenable, and agree with Madeleine de Gogorza Fletcher's rather timid assertion, which I would take much further, that 'the *Episodios* do fulfill a different function from Galdós' other novels . . . It is the concrete practical aspect of the national problem which comes to the fore in Galdós' *Episodios* rather than philosophical-religious theorizing' (*4a*, 2).

My view is still more categorical: the whole range of the *Episodios nacionales* can be clearly differentiated from the contemporary novels in form and intention, and the relatively new-found awareness of the importance of history in the latter[37] should not lead us to blur the distinction. The fact that Galdós himself steadfastly maintains a separate denomination is in itself significant, and points to a degree of polarization within his work—that is to say, that a certain content and treatment of history were appropriate for one form and different ones for the other.

To some extent it is a matter of the perennial tension between content and form. As Cervantes has Cipión state in *El coloquio de los perros*:

> los cuentos unos encierran y tienen la gracia en ellos mismos; otros, en el modo de contarlos; quiero decir que algunos hay que aunque se cuenten sin preámbulos y ornamentos de palabras, dan contento; otros hay que es menester vestirlos de palabras, y con demostraciones del rostro y de las manos y con mudar la voz se hacen algo de nonada, y de flojos y desmayados se vuelven agudos y gustosos. (*4a*, 304)

Though the distinction is far from absolute (every content demands a form and the two—form and content—are basically inseparable), the *episodios*

[36] On the following page Casalduero claims: 'En algunos *episodios* y en algunas novelas el andamiaje histórico está presente; *en otros* episodios *se relega a un lugar completamente secundario, llegando casi a desaparecer*, y en las novelas, a partir de 1888, desaparece también casi en absoluto' (*3b*, 1961: 44). Between two arguably acceptable statements comes the one I have italicized, which is palpably false. Despite this, Casalduero treats the *Episodio nacional* as a completely distinctive form in his article 'Historia y novela' (*3d*). Amado Alonso, at pains to distinguish the *episodios* from the classical historical novel, also denies any essential distinction between the two forms Galdós used: 'en sus novelas sociales o novelas descriptivas, el asunto narrado es en parte verdadero y en parte inventado . . . Exactamente lo mismo que pasa en los *Episodios nacionales*: también aquí se mezclan de modo semejante lo fingido y lo verdadero' (*3b*, 212).

[37] See esp. Bly (*3c*, 1983), Gilman (*3c*), and Ribbans (*3c*, 1970).

incline towards Cervantes's first type of content-dominated narrative, since their material is preselected and judged of significance, while the novels tend towards the formalistic or stylistic orientation of the second. Cervantes's division has, incidentally, another relevant application, the difference between plain and elaborate style, something which greatly preoccupied Galdós[38] and which applies equally to the two forms. If the plain style roughly corresponds to 'summary', the literary elaboration and the importance of the spoken word (testimony, in Cervantes, of oral transmission) belong rather to 'scene'.

Where, then, do the differences lie? First, there is an essential divergence in historical density: the quantity of historical facts which have to be accommodated in an *episodio* is far more than in a conventional novel. This makes the nature of the former more inflexible and unchangeable.[39] Moreover, such an elementary fact as the titles consistently adopted makes their sustained purpose abundantly clear: not one of the forty-six titles of the *episodios* fails to refer explicitly to a historical event, date, or personage, whereas no single title of the contemporary novels after *La Fontana de Oro* includes any historical or political circumstance. In every *episodio* some major historical figure or development is shown in action, even though the historical figure is never the protagonist of the work: from the fourth series onwards, Sor Patrocinio, Isabella II, Narváez, O'Donnell, Prim, and Cánovas are conspicuous among such figures.

To take a concrete example: *Las tormentas del 48* is one of the *episodios* with the least direct historical content. The first work in the fourth series, it proceeds to outline the fictional structure which will hold for most of the series, offering a detailed autobiographical account of its leading invented character, José (Pepe) García Fajardo. Despite the predominance in this case of a fictitious plot, the historical background (Pius IX's election as Pope, the *coup d'état* against Louis-Philippe, the rise of Sartorius, the May rising against Narváez) is none the less always unobtrusively present. The next work, *Narváez*, runs straight on from *Las tormentas*, and places Fajardo, now reaping the rewards of his new-found social eminence, in a position directly to reflect leading political events. As a subversive element within the aristocracy, he calls into question the entire political and social fabric (chapters 2 and 3), and as an *aficionado* of history, he is preoccupied with the question of how history should be written (chapter 7).

In the contemporary novels, by contrast, individual historical events are

[38] See the discussion at the end of *Fortunata y Jacinta* (V: IV. 6. xvi, 544), and commentaries by Williams (*4b*, 264–5) and Ribbans (*3c*, 1986). Vigny's image (*4b*, 26) of polishing the rough edges of history to make an artistic creation is also pertinent.

[39] As early as 1875 Galdós speaks of 'el género novelesco-histórico, donde la acción y trama se constituyen *con multitud de sucesos que no debe alterar la fantasía*, y con personajes de existencia real' (*3d*, Smith 106; my italics). Compare the distinction established by Scholes quoted in the Introduction: 'The producer of a historical text affirms that the events entextualized did indeed occur prior to the entextualization' (*2*, 1981: 207).

used, frequently but not in any systematic fashion, to provide a kind of factual stiffening—not just background—for selected fictional occurrences, to indicate parallels between private and public behaviour, and to deepen characterization. Political figures and events from both the present and the immediate past are utilized in a shorthand fashion; moreover, the fictional characters are normally firmly locked into the overall political and social context. In this respect Galdós partakes of what has been defined as 'historism' and is comparable with Scott and Balzac, according to the discussion in the Introduction. Auerbach's remark on Stendhal's *Le Rouge et le noir*, that its characters are 'embedded in a total reality, political, social, economic, which is concrete and constantly evolving' (*2*, 463),[40] applies equally to Galdós.

Not all the major events of the time of action are included, nor are those which do find a place necessarily treated in more than a partial and limited way. What is conveyed, rather, is the immediate impact of certain crucial events and/or an acute awareness of the continuity of past and present: in Peter Bly's words, they are 'stepping-stones of historical data which lead to the novel's overall historical dimension.' (*3c*, 1983: 5) Historical recollections, evidently, do not reach back to years which are quite remote from the characters' present. Nor is history simply a question of major political happenings (*la historia grande*, again)—indeed it is by their very nature difficult to determine whether minor incidents (*la historia chica*) are not of greater fundamental importance. As the narrator in *Ángel Guerra* says:

> bien mirado el asunto, las ideas de Guerra sobre la supremacía de la Historia no excluían las de Dulce sobre la importancia de las menudencias domésticas, pues todo es necesario; de unas y otras cosas se forma una armonía total, y aún no sabemos si lo que parece pequeño tiene por finalidad lo que parece grande, o al revés. La Humanidad no sabe aún qué es lo que precede ni qué es lo que sigue, cuáles fuerzas engendran y cuáles conciben. Rompecabezas inmenso: ¿el pan se amasa para las revoluciones o por ellas? (V: I. 1. iv, 1209)

Both historical and domestic events have their function, which cannot immediately be determined accurately. The essential thing is that both must interact constantly and consistently in close metonymic relation.

Second, the *episodios* have greater structural coherence. From the point of view of the historical content, the subgenre demands an overall factual exactness and an easily intelligible chronology such as is not required in a non-historical novel. It unashamedly embraces the need for representation, a specific setting and, an extensive network of outside referentiality. Historical involvement on this scale evidently poses problems for modern theorists hostile to conventional history. As a consequence I have serious reservations

[40] Compare 'What realism appeals to and establishes is that social truth which insists on the interconnectedness of individual and society' (*4b*, Stern 77).

about Diane Urey's application of Derridean concepts to the subgenre.[41] What is present, but without breaking the immediacy of the presentation, is what Lukács, drawing on Goethe and Hegel, calls the 'necessary anachronism'[42] associated with any reconstruction of the past. The degree of selection the author can allow himself is sharply diminished; he cannot pass over major events or skip from one period to another without justification: restrictions which do not of course apply to the contemporary novels. Within this limited area they have an evident if secondary didactic aim of providing information and stimulating thought about the issues of the immediate past.[43]

Hans Hinterhäuser has rightly drawn attention to the priority assigned to political history over the fictional story, and offers some evidence that the story was normally added subsequently: 'Ahora estoy preparando el cañamazo, es decir, el tinglado histórico... Una vez abocetado el fondo histórico y político de la novela, inventaré la intriga,' declared Galdós in July 1910 (*Por esos mundos*; *3d*, quoted in Hinterhäuser 223). This does not, however, imply either a lack of concern for the fictional story (see second prologue to the Illustrated Edition: *3a, Prólogos* 54–61) or an absolute predominance of large political issues. As early as *El equipaje del rey José* (1875), the first novel of the second series, Galdós's narrator, in an 'importuna digresión', justifies the attention paid to obscure individuals, like his new protagonist Monsalud, as part of the historical process:

¿Por qué hemos de ver la Historia en los bárbaros fusilazos de algunos millares de hombres que se mueven como máquinas a impulsos de una ambición superior, y no hemos de verla en las ideas y sentimientos de ese joven obscuro? Si en la Historia no hubiera más que batallas; si sus únicos actores fueran las personas célebres, ¡cuán pequeña sería! Está en el vivir lento y casi siempre doloroso de la sociedad, en lo que hacen todos y en lo que hace cada uno. En ella nada es indigno de la narración, así como en la Naturaleza no es menos digno de estudio el olvidado insecto que la inconmensurable arquitectura de los mundos. (**I**: 1207)[44]

Since it is fiction which must supply these unsung heroes, we have here an implicit vindication of the historical novel, which is, as we shall see, at the same time a social novel.

[41] 'Derrida's critique of history is applicable to the *Episodios nacionales* overall because it questions any form of representation and the indissolubility of the linguistic sign' (*3d*, 1989: 176; see also 116).

[42] '... the writer would allow those tendencies which in historical reality have led up to the present... to emerge with that emphasis which they possess in objective, historical terms for the product of this past, namely the present' (*2*, 68).

[43] Note Galdós's own assessment of the first two series in the second prologue to the Illustrated Edition (Nov. 1885): '[su obra es] limpia de toda intención que no fuera la de presentar en forma agradable los principales hechos militares y políticos del período más dramático del siglo con objeto de recrear (*y enseñar también aunque no gran cosa*) a los aficionados a esta clase de lecturas' (*3a, Prólogos* 56; my italics). Hinterhäuser devotes an entire chapter of his book (*3d*) to 'Los *Episodios nacionales* como medio de educación política'.

[44] For a cogent discussion see Llorens (*3b*, 127–8). Compare the 'familiar rather than heroick' history of Henry Esmond discussed in the Introduction.

The beginning of *España sin rey* makes a further justification of unrecorded or overlooked personal events[45]—authentic or fictional—which form part of the totality of history:

Los íntimos enredos y lances entre personas que no aspiraron al juicio de la posteridad son ramos del mismo árbol que da la madera histórica con que armamos el aparato de la vida externa de los pueblos, de sus príncipes, alteraciones, estatutos, guerras y paces. Con una y otra madera, acopladas lo mejor que se pueda, levantamos el alto andamiaje desde donde vemos, en luminosa perspectiva, el alma, cuerpo y humores de una nación. (**III**: 785)[46]

Hinterhäuser (*3d*, 233–4) has noted the apparent casualness of 'acopladas lo mejor que pueda', but this merely indicates that Galdós's procedure is intuitive and impetuous rather than scientific.[47] Although, in Berlin's terminology, the hedgehog of historical perspective (*la historia grande*) gains in importance, the diversity of the fox, manifest in *la historia chica*, even if no longer fully autonomous, remains highly effective, co-ordinated and intertwined as it is with the major events. The *historia chica*, it should be emphasized, has two distinct aspects. First, the largely unrecorded undercurrent of steady but unobtrusive historical change: what Unamuno termed *intrahistoria*;[48] and, second, the representative fictional invention, which, as Vicente Llorens has acutely pointed out, is also historical (*3b*, 134–5). Pepe Fajardo, on returning from Italy to chronicle Isabelline Spain, characterizes the former aspect in a most effective metaphor:

El continuo engendrar de unos hechos en el vientre de otros es la Historia, hija del Ayer, hermana del Hoy y madre del Mañana. Todos los hombres hacen historia inédita; todo el que vive va creando ideales volúmenes que ni se estampan ni aun se escriben. Digno será del lauro de Clío quien deje marcado de alguna manera el rastro de su existencia al pasar por el mundo, como los caracoles, que van soltando sobre las piedras un hilo de baba, con que imprimir su lento andar. Eso haré yo, caracol que aún tengo largo camino por delante; y no me digan que la huella babosa que dejo no merece ser mirado por los venideros. (**II**: 1428)

The progress of *la intrahistoria* may not be much—the image of the snail's track is a very modest one—but it is continuous and accumulative. Later,

[45] In the discarded epilogue to *La batalla de los Arapiles* Galdós laments the lack of availability in Spain of 'documentos privados, memorias o historias individuales y anecdóticas' (*3d*, Smith 106).

[46] We may note that Galdós uses the image of the tree, with its associations of steady organic growth, to designate the fictional story of obscure individuals who represent *la historia pequeña* of everyday life.

[47] See Galdós's interesting self-criticism of irrepressible precipitation and impatience in the discarded epilogue to *La batalla de los Arapiles* (*3d*, Smith 106) and Hinterhäuser (*3d*, 245).

[48] See Unamuno, *En torno al casticismo* (*4a*, i. 792–8). It may perhaps be claimed that Foucault's episteme is a more all-embracing, more culturally orientated form of *intrahistoria*, but operating in a restricted period and cut off from other epistemes.

Tito, in undertaking to take over the historical narration, distinguishes between *la historia grande* and *la historia chica* in intimate domestic terms:

De asuntos privados, confundidos con los públicos, hablaré, para que resulte la verdadera Historia, la cual nos aburriría si a ratos no la descalzáramos del coturno para ponerle las zapatillas. ¡Cuántas veces nos ha dado la explicación de los sucesos más trascendentales, en paños menores y arrastrando las chancletas! (*Amadeo I*; **III**: 1023)

History, in the personified figure of (Mari)clio, will alternate between an exalted and a humble role, each role represented by the synecdoches *cotornos*: *zapatillas/chancletas*.

Likewise, the object of the *episodios* is clearly to offer a panoramic and comprehensive view which embraces society as well as politics,[49] and for this it depends on the fictional scaffolding. It is two-pronged: 'el género novelesco con base histórica', as Galdós calls it (*3a, Prólogos* 58). It has indeed been argued that each series consists really of only one novel. Given the variety of narrative patterns, I by no means accept this argument, but it is true that the successive volumes are accumulative and interdependent to a remarkable degree. As Galdós indicated in 1875, he intended the various books of the second series to be 'enlazados entre sí, pero sin violencia' (*3d*, Smith 107). Rodríguez discerns seven novels[50] in the fourth series, but in my view it is quite inappropriate to disregard the volume-division or to see each individual narration as having one distinctive protagonist. The truth is that various techniques are used to provide a thread of continuity between individual *episodios* and even across series.

One is the use of key figures. An outstanding early example is Jenara Baraona, so conspicuous in the second series through her unhappy marriage to Carlos Navarro and her amorous entanglement with Salvador Monsalud, among others (Ferdinand VII and Chateaubriand are included among her admirers), and as narrator of *Los cien mil hijos*.[51] She thus plays the important role of linking public and private events from Joseph Bonaparte's time right up to Espartero's regency and beyond, as is explicitly and nostalgically acknowledged at her death in *La revolución de julio*: 'Desde el azaroso tiempo de José hasta las dos Regencias inclusive, precursoras de Isabel II, y aún un poquito más acá, no había fragmento anecdótico relacionado con la vida pública que no existiera en sus archivos, y éstos solía mostrarlos, cuando

[49] 'Galdós quiere abarcarlo todo. El hecho histórico conocido y el incidente ignorado; las acciones militares y las intrigas políticas; hasta el cambio de modas y costumbres, de la vida literaria, de cuanto contribuye a dar el perfil y tono de una época' (*3b*, Llorens 130).

[50] According to Rodríguez, Fajardo, Santiuste, and Santiago Ibero occupy two *episodios* for their 'novels'; Lucila Ansúrez, Virginia Socobio, Teresa Villaescusa, and Diego Ansúrez have one each (*3d*, 140–50).

[51] Rubén Benítez (*3d*) amply demonstrates her importance as the first and only woman narrator of the *episodios*.

estaba de vena jovial, a sus buenos amigos' (III: 25).[52] In similar fashion, the birth of Beramendi's children, like Isabel Cordero's in *Fortunata y Jacinta*, is used to pinpoint historical events. Pepito is born in 1849, at the time of the scandal of *la monja de las llagas* and the *Ministerio Relámpago*; Feliciana in 1852, when Merino's assassination attempt against the queen takes place; a girl died in infancy at the period of Vicálvaro (1854); and finally, Agustín is born late in 1857, twenty days after the infante Alfonso (III: 691).

Another procedure which requires detailed attention is the building up of a firm and subtle network of interacting figures. This is partly achieved by the use of recurring characters. While in contemporary novels the device is optative and selective, in the *episodios* it is organically required. Once firm relationships have been established with real-life historical personages or situations, these have become part of the structure and cannot be realistically abandoned in subsequent volumes; they can only be slowly modified or replaced.

Thus a few families exercise a tentacular hold over the structure of several volumes. The most evident is the extended Ansúrez family, for which a symbolic or allegorical role has been claimed: it is to some extent a repository of primitive Celtiberian values, an aspect which is repeatedly asserted and at the same time ironically undercut by the extravagant archaeologist Buenaventura Miedes, with his Latinized forms of their names and pseudo-erudite commentaries.[53] Montesinos (*3c*, iii. 132–3) is right in my opinion to find this incipient and rather vague symbolism unconvincing, since it makes for woodenness and artificiality; and the development is not sustained enough to be called allegorical.[54] More than just symbolic, fortunately, such figures have an essentially representational function.

The old man himself, Jerónimo Ansúrez, stands for primitive rural values struggling against purely man-made obstacles—'Quien dice labranza dice palos, hambre, contribución, apremios, multas, papel sellado, embargo, pobreza y deshonra' (II: 1533)—together with an indomitable spirit of pride and independence; from his castle at the historical town of Atienza, he is intended to embrace all Spain, and for that reason his three wives come from three different regions (La Mancha, Aragon, Valencia). His offspring provide Galdós with a series of fiercely independent characters for several stories: 'todos los individuos de esa familia, de ese índice histórico, de ese resumen

[52] In *Un faccioso más*, no doubt because it was intended to be the last *episodio*, her history is truncated rather abruptly at this point: 'Esta mujer acabó para nosotros'. Her death is given as occurring 'hacia el último tercio del 68', with, as it happened, a very necessary escape clause: 'si no están equivocadas las crónicas' (II: 314–15). Doña Visitación, María Ignacia's mother, is related to her.

[53] Regalado (*3d*, 381–2) links him rather fancifully with Joaquín Costa, also a great champion of the study of the Celtiberians.

[54] Casalduero (*3b*, 1961: 155) and Fletcher (*4a*, 47) give these personages a completely allegorical interpretation, which I do not share.

étnico, son de una agudeza formidable' (III: 383). They are, in rising order of importance: Gil, adventurer and bandit; Rodrigo, musician and friend of Ibero; Diego, sailor and hero of *La Numancia*; Gonzalo, the renegade Moor El Nasiry of *Aita Tettauen*; Leoncio, the lover of Virginia Socobio, gunsmith, participant in the African war, and Ibero's friend and fellow conspirator; and, above all, Lucila,[55] the Celtiberian beauty who entranced Beramendi and Santiuste, became Bartolomé Gracián's lover, made two middle-class marriages (to the elderly Vicente, also called Juan, Halconero and then to Ángel Cordero) and was the mother of the second Vicente Halconero. Also symbolic in name and character are the two Santiago Iberos, father and son; from the 'Rioja alavesa'—a mixture therefore of Castilian and Basque—they represent for Galdós the honesty and tenacity of the primitive Iberian race.

Another family example, of a more adaptable sort, is provided by the Fajardos: as well as José (Pepe), who becomes the Marquis of Beramendi, his sister, the nun Catalina, his two brothers—the conservative Agustín and the moneylender Gregorio—the latter's wife Segismunda Rodríguez, and their son Segismundo have important roles. Interconnected traditionalist families are the Socobios and the Emperáns. Bruno Carrasco's progressive ideas give way to the opportunism of his daughter Eufrasia, beginning with her marriage to Saturnino Socobio; she participates in the whole gamut of political activities from the 1840s to the Restoration. Rafaela, the daughter of Carrasco's Progressive companion José del Milagro, has a different evolution, passing through promiscuity to clerical reaction. Teresa Villaescusa's family includes Centurión and Leovigildo Rodríguez, while her mother is from the Pez tribe, so numerous in the novels. All these groups are interlocked by ties of personal relationships or marriage or political affiliation.

In most cases true love, like sincere political principles, brings frustration: Lucila's passion for Gracián is thwarted by her lover's political and private irresponsibility; Teresa and Santiuste's attachment, by economic imperatives; the love of Vicente Halconero and Fernanda Ibero, by death. Only the most determined lovers, Virginia and Leoncio, who overcome poverty and ostracism, and especially Teresa and Santiago Ibero, win through in the end. The norm is conscious and cynical compromise: political wheeler-dealing on the one hand; prudent, loveless marriage on the other. Such alliances may in some cases be moderately successful: Pepe Fajardo and María Ignacia Emperán, Lucila's marriages, Vicente and Pilar Calpena; purely opportunist in others, often with more or less concealed adultery as part of the deal, as with Eufrasia and Valeria.

Links are provided with earlier series and between generations. Pilar is the

[55] See *Narváez* (II: 1532), where Fajardo declares that Lucila is 'la mujer más hermosa que yo había visto en mi vida', and *Aita Tettauen*, in which 'aquel primeroso renuevo del árbol celtíbero' has put on weight and become a prosperous countrywoman with seven children (III: 229).

daughter of Fernando Calpena, the protagonist of the third series; his wife, Demetria Castro-Amézaga, is Gracia's sister, married to the older Ibero. The younger Ibero's momentary love for Saloma provides a link with two earlier generations: her father, the conspirator Baldomero Galán, and her maternal grandfather, the unfortunate Adrián Uliburri, shot by Zumalacárregui (II: 329–33). In all the intricate web of fictional relationships historical underpinning is never lacking.

Within the novels, the author has a freer choice about whether or not to retain characters, but the technique remains a very useful one. It is clearly prompted by Balzac's sudden inspired decision, after *Le Père Goriot* (1834), to intertwine his characters in a complex web from one novel to another, in many cases retrospectively. Recurrence evidently enhances the plausibility and coherence of the mimetic world of fiction, and thus lends it a greater historical verisimilitude. It is not surprising, however, that it met with the disapproval, on the 'purist' grounds of going against the autonomous structure of each individual novel, of the Henry James school represented by Percy Lubbock (*2*, 207–10).

Galdós does not utilize the technique to the same degree as Balzac and does no rearranging of previously published work. As Shoemaker indicates in his thorough examination of the subject (*3c*, 1951),[56] only about 6 per cent of Galdós's approximately 3,800 imaginary characters (about 1,800 in the novels, short stories, and theatre, about 2,000 in the *episodios*) are genuinely recurrent, compared with nearly a quarter in Balzac.[57] Galdós does spread the device between novels and *episodios*, but only very sparingly, with a maximum of 42 cases. Typically, it serves to lend a historical dimension to characters in the novels, as the examples of Rufete and Cordero show. In the later *episodios*, type characters from the novels quite frequently appear.

Many (at least 25) of the really prominent characters in the novels are never mentioned in other works; they include, according to Shoemaker (*3c*, 1951: 206), all four leading characters of *Fortunata y Jacinta*; Ángel, Dulcenombre, and Leré in *Ángel Guerra*; Tristana; Benina, and Almudena in *Misericordia*, to give only a few examples. Sometimes, however, a minor character comes into his or her own as the protagonist of a later novel or novels: Rosalía de Bringas, Villaamil (*Miau*), Torquemada. The rapid intervention of a known character may produce a pleasingly appropriate effect: Doña Lupe introducing Torquemada on her deathbed to the world of Cruz del Águila, or Benina being mistaken for the already deceased Doña Guillermina. Most commonly,

[56] It is based, however, on Sainz de Robles's less than perfect 'Ensayo[s] de un censo de personajes' (*3a*, *OC* [*Episodios*] **III**: 1411–873; [*Novelas, cuentos, teatro*] **VII**: 1047–430).

[57] See Pugh's very thorough study, also relevant for its discussion of the advantages and disadvantages of recurring characters (*4b*, 82–7, 223–8, 461–75). Pugh claims that the statistics Shoemaker quotes (*3c*, 1951: 205) from Preston are 'of dubious validity' (xx), but this does not invalidate the contrast between Balzac and Galdós.

the recurring characters are minor figures with a distinct profession or social status: doctors like Augusto Miquis or Moreno Rubio, oculists like Teodoro Golfín, priests like Nones or León Pintado, lawyers like Muñoz y Nones, functionaries like Basilio Andrés de la Caña. They are often singled out by a literary catchword, or *muletilla*, a device which may have its origin in Dickens (*3c*, Chamberlin 1961: 297–8). Occasionally, there is a family cluster like the ubiquitous bureaucratic Pez tribe. Some serve as link-characters in a variety of situations, the most notable being José Ido del Sagrario, present in four novels and four *episodios*. He has a distinctive, almost grotesque, physical appearance, which remains basically unchanged: an excessively thin body, a turkey-like face, full of carbuncles, hair as stiff as a brush, eyes that water easily, a weak voice, a large Adam's apple, periodic twitching of limbs or eyelids. He has his appropriate *muletilla*: *francamente, naturalmente*; and he suffers from an alarming allergy to meat. Rather Dickensian in his caricaturesque appearance, he is Micawber-like in his constant bad luck. Whereas in the novels he has structural importance (as a butt for Polo and companion for Centeno; as a writer of a *folletín* version of the story of Sánchez Emperador's daughters; as the vehicle for the transmission of information about the false *Pitusín*; as the foil to both Juanito and Maxi), his activities in the later *episodios* seem much more gratuitous.

A third difference concerns the formal framework. The *episodios*, in contrast with the novels, have a distinctive regularity of form. As Alfred Rodríguez has indicated:

> It is their rigid formal outline, far more than any stylistic, thematic, or procedural criteria that might be applied, that distinguishes the *Episodios nacionales* from the rest of Galdós' work. Because they deal with a specific constant (historical time), and with a subject matter for which the novelist's perspective required a specific order of treatment, the *Episodios nacionales*, unlike any other segment of the novelist's work, are fixed into a set literary form: ten volume Series unit, progressive sequence of time, and prescribed quantitative limitations. (*3d*, 198)

We may add that there is a difference of extension: the *episodios* are relatively short (approximately 85,000 words in length, divided into about 30 chapters). These characteristics lend a certain homogeneity, within the serial structure, which the novels do not possess. The latter, by contrast, have a freer construction; they do not follow the standard format of the Balzac novel factory: regular length and a fixed type of development—long slow introductory description, quickening pace as the characters face their almost inevitable financial crises, to reach a conclusion which is not quite definitive (*4b*, Bellos 76); on the contrary, Galdós's novels may come in one-, two-, three- and four-volume narratives and the pace is generally leisurely; only occasionally, as in *Torquemada en la hoguera* or *Tristana*, are they as short or shorter than the *episodios*. The novels may be purely independent works or may be linked together in a loosely articulated group: *El doctor Centeno–Tormento–La de Bringas*, the Torquemada novels, or the spiritualized series *Nazarín–Halma–Misericordia*. In either case, they are distinguished by the variety of their construction and expression,[58] a sign of Galdós's quality as a restless experimenter.

Narrative structure constitutes a fourth divergence. Initially, in the first series, Galdós adopted a technique of autobiographical reminiscence, with Gabriel Araceli describing events from his old age. Despite the advantages of continuity, this procedure suffered from inevitable limitations. In the discarded epilogue to *La batalla de los Arapiles* (1875; *3d*, Smith), he chafed at the rigidity of narrating in the first person:

> establézcase la necesidad de que los acontecimientos históricos ocurridos en los palacios, en los campos de batalla, en las asambleas, en los clubs, en mil sitios diversos y de no libre elección para el autor, han de pasar ante los ojos de un *solo* personaje, narrador obligado e indispensable de tan diversos hechos en período de tiempo larguísimo y en diferentes ocasiones y lugares, y se comprenderá que la forma autobiográfica es un obstáculo constante a la libertad del novelista y a la puntualidad del historiador. (*3d*, quoted in Smith 106)[59]

Both the breadth and the vehemence of the text are striking.

First-person narration is therefore abandoned for the second series. 'Tales dificultades', he declares, 'obligáronme a preferir en casi todas las novelas de la segunda serie la narración libre.' He indicates also a second change, which emphasizes the novelistic qualities of the works: 'como en ellos la acción pasa de los campos de batalla y de las plazas sitiadas a los palenques políticos y al gran teatro de la vida común, resulta más movimiento, *más novela*, y por tanto, un interés mayor. *La novela histórica viene a confundirse así con la de costumbres*' (*3a*, *Prólogos* 58; my italics). The balance between the two constituent elements—history, fiction—of the *Episodio nacional* is thus clearly established.

On returning to the subgenre with the third series in 1898, Galdós continues to adopt a largely third-person narrator but, on occasion, seeking greater immediacy, he introduces a modified first-person form: the epistolary mode, already used in *La incógnita* (1888–9). Letters thus form the entire structure of *La estafeta romántica* and play a considerable part in *Luchana*, *Vergara*, and *Los ayacuchos*.

In the fourth series a mixed technique, providing maximum flexibility, of roughly alternating first- and third-person narrative is used. What is now

[58] See the fine study, emphasizing the all-pervading irony, by Urey (*3c*, 1982). Other noteworthy contributions are Engler (*3c*, 1977), Gilman (*3c*), and Rodgers (*3c*, 1987).

[59] Compare his comments in the Illustrated Edition: 'En la primera serie adopté la forma autobiográfica, que tiene por sí mucho atractivo y favorece la unidad; pero impone cierta rigidez de procedimiento y pone mil trabas a las narraciones largas. Difícil es sostenerla en el género novelesco con base histórica, porque la acción y trama se constituyen aquí con multitud de sucesos que no debe alterar la fantasía, unidos a otros de existencia ideal, y porque el autor no puede, las más de las veces, escoger a su albedrío ni el lugar de la escena ni los móviles de la acción' (*3a*, *Prólogos* 58).

sought is a method of conveying a contemporary situation: a historical reconstruction of the past which occurs before our eyes; diachronic time yields, at intervals, to synchronic time. As LaCapra says, 'Synchrony "stops" time at a given moment and displays its "sameness" or "nowness" (*Jetztzeit*, in Benjamin's term: *2*, 263), in a spatialized tableau' (*2*, 1987: 204). Historical and fictional figures are in the process of playing their role in the present—their present—and are therefore subject to the pressures and uncertainties of the moment. The immediate perceptual linkage Wright discerns in direct experience (see the Introduction, pp. 24–5) is artificially recreated. The narrator and the story, and so the narratee, are thus extremely close, but are remote in time from the implied author and implied reader.[60]

After the high-flown narrative tone in *Bodas reales*,[61] the last of the previous series, the fourth commences with Fajardo's first-person memoirs in the twin narratives *Las tormentas del 48* and *Narváez*. Then comes an intercalated third-person narration for *Los duendes de la camarilla* before we return to Fajardo's memoirs in *La revolución de julio*. By this alternation convincing immersion in an earlier contemporary reality is achieved. The Olympian narrator of *O'Donnell*, called in burlesque tone 'Doña Clío de Apolo', stresses ironically the apparently heroic proportions of the eponymous protagonist, as does the later narrative devoted to Prim. *Aita Tettauen* (= 'Tetuan War') has a complex structure; it starts with omniscient narration for the first two of its four parts; the third part is an epistolary narrative from the Moslem standpoint by one who turns out to be El Nasiry, Gonzalo, the renegade son of Jerónimo Ansúrez; the fourth brings Juan Santiuste to the fore and leads to Santiuste's own chronicle in *Carlos VII, en la Rápita. La vuelta al mundo en la 'Numancia'* reverts to an impersonal narrator, a technique continued in *Prim* and *La de los tristes destinos*. The third-person narration, by extensive use of 'summary', allows a faster narrative pace than the memoir and establishes a certain distance in time and space from the events while still mirroring the attitude of the leading characters, with whom the reader is already acquainted.

Similar techniques continue into the fifth series. The first volume, *España sin rey*, has an anonymous personal narrator. The second, *España trágica*, is centred on Halconero and includes lengthy extracts from his diary. *Amadeo I* marks the transition whereby the last four volumes return to the first person, narrated by Tito from a great chronological distance, deliberately acknowledged.

Whether he is narrator or simply protagonist, the prominent fictitious figure within the action, such as Beramendi, Santiuste, and Halconero, fulfils

[60] Ricardo Gullón has acutely pointed out that the reader of the *Episodios* is necessarily distinct from the reader of the novels (*3d*, 1974: 54–7).

[61] In the first sentence Galdós makes an interesting division between History, which retains all human occurrences, great and small, and Time, which is memoryless (II: 1307). Urey (*3d*, 1989: 75) interprets this as indicating that 'History, at least in this case, is not a source of instruction, but rather an assortment of insignificant details'. I prefer to remember the almost imperceptible traces of the snail (II: 1428).

essentially the role which Henry James defined as that of a 'reflector',[62] or what Genette calls 'inner focalization' (*2*, 189–94). Each of these characters is, as in James, an interested commentator, trustworthy and a participant in the events described, but he is more eccentric, more partisan, more involved. He is remote, in circumstance as in time, from the implied author, but is not unreliable.[63] A major exception from the second series, as we have seen, is Juan Bragas, who casts an ironic light on the reign of Ferdinand VII. Tito, for reasons we shall discuss in later chapters (5 and 6), has a quite ambivalent position.

In the novels the narrator, without being concealed, is invariably anonymous; typically the friend or companion of the main characters, he is not very prominent in the action himself, but very aware of what is happening and so fairly close to the story. Having lost his privileged status as omniscient narrator, Galdós's narrator is unmistakably an individual prone to human frailty, subject, like other human beings, to ignorance, prejudice, and foibles. He habitually speaks in a tone of familiar exaggeration or ironic hyperbole. Reliable within his personal limitations, he stands at some distance from the implied author. In variable degrees he may have an ironic role, through his unobtrusive but highly significant intervention in *La de Bringas* or through his initial categorical statement, in rhetorical oxymorons, at the beginning of *Torquemada en la hoguera*. The two first-person narrators—in *El amigo Manso* and especially *Lo prohibido*[64]—though very different, are unreliable, remote from the implied author, but very close to the story; the narratee is accordingly very distant from the implied reader and irony forms an integral part of the structure.[65]

Not infrequent are non-mediated techniques, when the novelist abandons diegesis in favour of mimesis. He makes moderate use of one of the standard forms of the traditional novel, the epistolary mode, as in the letters by Don Cayetano which conclude *Doña Perfecta*,[66] and experiments with one entirely epistolary novel, *La incógnita* (1888–9), which goes in tandem with a related experiment, the first dialogue novel *Realidad*.

[62] See Lubbock (*2*, 144–50), where he takes Strether, in *The Ambassadors*, as the quintessential reflector.

[63] Implied author, unreliable narrator: for these terms consult Wayne Booth (*2*, 1983: 158–9), Chatman (*2*, 147–9, 233–7), and the latter applied specifically to Galdós: Engler (*3c*, 1977: 137–83).

[64] For both novels consult Engler (*3c*, 1977) and Urey (*3c*, 1982: 70–9); for *El amigo Manso* see Kronik (*3c*, 1977) and, for its political implications, Blanco Aguinaga (*3c*, 17–48); for *Lo prohibido* consult Terry (*3c*).

[65] See Bieder on *El amigo Manso*: 'Es precisamente en los casos de narración irónica y de narradores no fidedignos cuando el papel del narratario se hace resaltar más—sobre todo en relación al lector implícito—y cuando más valioso y más necesario resulta ser el trabajo crítico' (*3c*, 253). For the use of irony consult Booth (*2*, 1974) and Muecke (*2*, 1969; 1982).

[66] It is worth noticing that in what appears to be an earlier version of *Doña Perfecta* straight narrative is used at the conclusion instead of Don Cayetano's letters: see Ribbans (*3c*, 1990).

A fifth element is the time-scale. The distance in time between the action and composition differs in a substantial way between the novels and the *episodios*. In the former—except in the case of first-person narratives—there exists an appreciable but not long chronological distance, of at least six years and usually longer, between the time of the action ('story time') and the time of the narration ('discourse time'). The implied author and the non-participating but discernible narrator are thus rather separate in time, though they are brought together by the close relation with the present, since the narrator habitually refers to the current reality of the characters.[67] The narrator is very close in time and space to the story; the narratee goes along with the narrator's opinions, understanding and apparently sharing them, in his judgement of situations which are no longer pressing, while the implied reader remains at a certain distance. In the *episodios*, on the other hand, the date of composition is invariably and inevitably much later than the action, with a time-scale of as much as sixty years. In most of them 'discourse time' is not allowed to interfere with 'story time', and readers are obliged to view events as contemporaneous; while the narratee is closely involved in the situation, the implied reader is very remote from the events taking place, though he may identify imaginatively with them. Once more, Tito, writing consciously thirty-seven years after the events, is the exception; in his case the gap between the narratee and the implied reader is enormous, and the latter has the utmost difficulty in identifying himself with the former and hence with Tito.

Sixthly, Galdós's extraordinary sensitivity to the subjective effects of language is common to both forms. The delight he takes in the creation of a distinctive idiolect for Almudena (*3c*, Lida 1961) and Torquemada (*3c*, Hall), to give only two of the most evident examples, is obvious and well known. A similar case from the *episodios* is the pseudo-Moslem chronicle of El Nasiry (III: 291–327). Yet human language is as volatile as all other human endeavours, so that all discourse is as ambiguous as its speaker. This may account for such a fascinating remark as 'El estilo es la mentira. La verdad mira y calla', made by Amparo during her bewildered meditation about the sum of money she has received from Caballero (IV: 1488). Language is suspect because it has an unrivalled capacity for creating misunderstandings, being manipulated, or becoming a vehicle for self-deception. I limit myself here to a single example which has something of all three.

When Juan Pablo Rubín marshals his ill-digested ideas for the benefit of his unsophisticated café admirers, his abstract, simplistically Hegelian

[67] Thus *El doctor Centeno* contains a sharp contrast between the time of action and the time of narration twenty years later. Similarly, in *Tormento* the narrator describes Bringas's deplorable state at the time of writing sixteen years later (IV: 1458), while *Fortunata y Jacinta*'s narrator declares that Don Baldomero and Doña Bárbara are still living as he writes well after the events of the novel (V: I. 2. iv, 26).

arguments are nothing but vacuous plays on words: 'cuando nos morimos pasamos a fundirnos en el grandioso conjunto universal', 'morirse es cumplir una ley de armonía', 'la Naturaleza que nos rodea, inmensa, eterna, animada por la fuerza'. They contrast with, but are no less specious than, the mixture of the concrete ('con nuestra defunción lo que hacemos es darle jugo a las plantas'), the superstitious ('Las ánimas benditas'), and the simply misunderstood ('esas *naturalezas*' = large breasts) characteristic of the simple women; they, at least, are seeking some sort of answer to the problem of existence or reality. When confronted with the blind pianist, whose last shred of comfort Rubín's philosophical notions destroy, he equivocates: '—Le diré a usted . . . hay opiniones . . . No haga usted caso. Si no fuera por estas bromas, ¿cómo se pasaba el rato?' (V: III. 1. vi, 308). All this occurs just before Juan Pablo is obliged, in order to benefit from the all-absorbing restoration, to make a complete reversal of all he had previously affirmed; his purely verbal resistance to Feijoo's proposal is gradually scaled down—'Yo no puedo permitir', '¿Y el decoro de los hombres?', 'no puede ser; me ofendería, sí, señor, me ofendería'—to an acceptance based fallaciously on the 'fuerza' of his fatalistic concept of Nature: 'Cómo ha der ser . . . paciencia. Tengo que ser alfonsino . . . a *la fuerza*. ¡Vaya un compromiso! . . . ¡Rediós, qué compromiso! . . .' (309; my italics). At every turn in this episode language turns out to be, in Bakhtinian fashion, full of hidden tension and deception.

Direct narration frequently veers towards the emotional position of the characters. Hence the very extensive use, in both forms, of free indirect style.[68] As Kay Engler has indicated, 'The primary function of indirect free style in Galdós' novels is the creation of an emotional overtone which draws the reader closer to a particular character and thus facilitates a shift in point of view' (*3c*, 1977: 75). Galdós's practice is in fact often much more subtle than that. At the beginning of the passage just cited, the narrator connives with Juan Pablo Rubín, ironically upgrading his revelations to a Messianic utterance, to take false comfort in the fact that Jesus—in whom he does not believe—preached the good news to simple and sinful folk: '¿Y qué? ¡Cuánto mejor no era sembrar la nueva doctrina en entendimientos sencillos y absolutamente incultivados! Pero el mismo Jesucristo, ¿no escogió por discípulos a unos infelices pescadores, hombres rudos que no conocían ninguna letra, y a mujeres de mala vida?' (V: III. 1. vi, 307).

The famous incident in which Luisito Cadalso apparently has a vision of God in *Miau* is a fine example of the hybrid dialogic effect produced by a narrative summary which is infiltrated by the character's ideas and vocabulary

[68] Alan Smith attributes Galdós's skilful use of free indirect style to the influence of its acknowledged master, Flaubert (*3b*, 33–6). Alas provides fine examples in *La regenta*, e.g. de Pas's voluptuous delight in his power as he looks down on Vetusta from the cathedral tower (*4a*, i. 107).

and combined with free indirect style.[69] Similar is the important passage describing the onset of the haemorrhage which kills Fortunata. In the interaction, well analysed by Salvador Bacarisse, between a string of preterites, which each refer to a fixed moment in time, and imperfects, which stress the duration of the experience, 'Galdós halts the linear progression of the narrative in order to present Time with as much immediacy as he had achieved by the avoidance of concepts' (*3c*, 245). The double-voicing is particularly subtle and the tension between what the narrator is describing and what the character is experiencing is a fine case of what Bakhtin calls 'hidden polemic'.

In the *episodios*, no doubt because of historical priorities, such features are not quite so common, but splendidly executed examples do occur. An outstanding one[70] is found in chapter XX of *O'Donnell* (III: 179–82), and I use this as my principal sample. The first paragraph is a long narrative passage which, though presented as direct narration and containing touches of narrative discourse, as in the formalized synecdoches 'el brazo férreo de Narváez y la despejada cabeza de Nocedal', is clearly focalized by a character in the action. It is full of exclamations, like the opening sentence: '¡No había caído mala nube sobre nuestra pobre España!'; rhetorical questions: '¿Era esto gobernar un país? Era esto más que una feroz política de venganza?'; and colloquialisms expressing subjective opinions: 'comiéndose los niños crudos', 'A los españoles que no eran del odioso *moderantismo*'. Only at the beginning of the second paragraph, when a piece of more or less straight summary begins, is the speaker identified as a standard type-character, Mariano Díaz de Centurión. His *tertulia* has been suspended, they have moved to a smaller house, abandoning Doña Celia's luxuriant plants (a synecdoche for prosperous times) in the process. After a short piece of direct speech, the narrative summary resumes; the fact that it is iterative is marked by the imperfect tense. A parenthesis deals with Manuela Pez's unprincipled ambitions for her daughter Teresa. In the partly focalized narrative that follows, Centurión sees no hope for the future (O'Donnell is as bad as Narváez) and condemns Segismunda's meanness to her brother Leovigildo. The focalization increases, with an augmented quota of dialogic devices like exclamations and interrogations, biblical references, and the use of Centurión's catch-phrase *cojondrios*. Finally, interior monologue takes over; rhetorical questions punctuated with negative replies[71] increase as he waxes more and more furious about the brutally repressed revolt at El Arahal

[69] The passage has a strong inner tension which recalls Bakhtin's 'hidden polemic', especially 'any speech replete with reservations, concessions, loopholes and so on' (*2*, in Matejka and Pomovska 188).

[70] Other notable examples are Fajardo's deliberations on his decision to marry María Ignacia (II: 1510–12) and Teresa's agonizing self-analysis on the death of Leal (III: 614–17).

[71] '¿Qué pedían los valientes revolucionarios del Arahal? ¿Pedían libertad? No. ¿Pedían la Constitución del 12 o del 37? No. ¿Pedían acaso la Desamortización? No. Pedían pan' (III: 182).

near Seville, a revolt which he claims was not intellectual or emotional but physiological: a demand for bread. Ruthlessness, the abandonment of *desamortización*, and savage repression on the political plane are linked with personal calamities and fictional events around him; and they come together at the level of basic human sustenance: food.

Galdós is never happier than when he is using the spoken word, especially in a context which allows everyday, colloquial language.[72] He uses direct soliloquy with some frequency and extremely good effect: Father Nones's philippic to Polo—a virtual monologue—in *Tormento* (**IV**: 1504–7) or the extraordinary stream-of-consciousness sequence by Moreno-Isla on his love for Jacinta in *Fortunata y Jacinta*[73] can serve as examples. A complex and impressive variant is the long discourse in which Guerra imagines his mother bitterly confronting him (**V**: I. 3. iv, 1235–7): it is a clear instance of what Bakhtin calls 'hidden dialogicality' (*2*, 1973: 163–4). Examples are found equally in the *episodios*, a representative case being Beramendi's long rebuke to his irresponsible friend Aransis (*O'Donnell*; **III**: 139–43).

Dialogue, the purest form of 'scene', is even more often employed, and with evident predilection, to lend greater spontaneity. What is particularly notable is the tendency to dispense with the conventional tags such as '*A* said', '*B* replied' (Chatman's free direct thought: *2*, 182–3), and to resort to theatre-like indications of speakers, supplemented as necessary with short stage directions; I shall call this device 'pseudo-dramatic dialogue'. As early as *La desheredada* (1881) it is used in an interchange between Isidora and Joaquín (**IV**: II. 12. ii–iv, 1120–7), but the most evident example is the *tour de force* between the two disguised figures which opens *Tormento* (**IV**: 1455–8). This episode provides a close link of continuity with the dialogue between the same speakers, Ido and Centeno, which closed the previous novel *El doctor Centeno* some four years ('story time') before. Furthermore, it introduces a second plane of reality concerning the Sánchez Emperador sisters, the subject of Ido's *folletín* novel as well as the straight narration. Such pseudo-dramatic dialogue continues as an important element throughout the novel (1565–6); Rosalía and Bringas's final indignant conversation about Amparo's 'immoral' elopement with Caballero is also in this form. It remains a constant technique of later novels (e.g. *Ángel Guerra*, **V**: III. 2. i–iv, 1454–66), but it reaches its culmination in the dialogue novel *Realidad* (1889), a form to which he returns in *La loca de la casa* (1892) and *El abuelo* (1897).

In the prologue to *El abuelo* Galdós offers an interesting defence of the dialogue technique. He sets out, he claims, to 'halagar mi gusto y el de ellos [mis lectores], dando el mayor desarrollo posible, por esta vez, al proce-

[72] 'The element of *skaz*, i.e. the orientation toward spoken language, is necessarily inherent in every narrated story' (*2*, Bakhtin 1973: 158). Bakhtin's (and Galdós's) enthusiasm for the spoken word makes Derrida's claim of the priority of writing seem extravagant and perverse.

[73] **V**: IV. 2. iii, 450–3; see also Sobejano (*3c*) on Moreno's death scene shortly after.

dimiento dialogal y contrayendo a proporciones mínimas las formas descriptiva y narrativa' (VI: 800).[74] Basing himself on an unanalysed concept of fiction as imitation of life, Galdós is defending, quite rightly, the validity of mimetic and dramatic values in the novel: the importance and effectiveness, that is, of 'scene'. In my opinion, however, he lets himself get carried away by his enthusiasm for the spoken word, which he mistakenly thinks fully capable of functioning adequately on its own. He underestimates the effectiveness of subjectivized narration as a technique and his own capacity for using it. He associates it, erroneously, with discursiveness, which is not inherent in narrative, even if it tends to be in Galdós's own practice.[75] His new criterion also points towards certain of the defects of his theatre, such as the overemphasis on speech to the detriment of other dramatic features such as action, decor, and staging, and the curious fact that the dialogue which makes up his plays is generally more stilted than that employed in the novels. These new emphases affect the few novels he writes after 1897 (and perhaps account for the absence of others), but do not have the same influence on the renewed *Episodios nacionales* from 1898 onwards, where length and discursiveness were less of a problem. Dialogue has a prominent part in them, and *Prim, La de los tristes destinos*, and *España sin rey* all end with a piece of pseudo-dramatic dialogue of immediate impact. There are no complete dialogue novels; in this respect these *episodios* are heirs to the realist tradition of the finest contemporary novels.

Finally, given the basic distinction between the two narrative forms, we should ask two questions, in the light of the discussion conducted in the Introduction: first, how is history treated in each of them, and second, in what sense can either or both of them be classified as historical novels? The *episodio* can hardly be denied the classification as far as historical content is concerned; the doubt concerns the extent to which these works are truly novels, a term Galdós on the whole avoids when he refers to them. For the novels, it is the degree of historicity which is in dispute. These are the issues—together with the way the historical content is handled—which I shall be concerned to elucidate in the rest of this chapter.

First, then, let us consider the value of the *episodios* as history.[76] Opinions differ widely on this point. Some critics have praised Galdós's research, his

[74] In the prologue to *Casandra* (1905) a few years later, he defends the form as '*novela intensa o drama extenso*', seeking a marriage, licit or illicit, of the two genres. Such a union will give, he claims, 'procedimiento analítico' to the theatre—the discussion of serious issues so lacking at this time—and greater concision to the novel: 'que sea menos perezosa en sus desarrollos y se deje llevar a la concisión activa' (VI: 906).

[75] It is the one constant criticism that Clarín makes against Galdós. See Beser: 'El pecado por exceso, que Clarín llama exuberancia y, más a menudo, prolijidad . . . lo señala como el defecto de mayor importancia en Galdós y los escritores realistas' (*4a*, 289).

[76] No attempt is made to define the sources of the various *episodios*, since this would go far beyond the scope of this study.

accuracy, or his balance: thus Rodríguez speaks of his 'impartial selection of textual sources, his undistorted and precise use of data they provided' (*3d*, 35). Hinterhäuser (*3d*, 55–71) systematically lists the written sources for the various series and individual volumes; these sources are considerably supplemented in Cardona (*3d*, 1968: 120–42). The latter scholar also rightly emphasizes the oral sources, especially the particular case of Ramón de Mesonero Romanos, whose correspondence with Galdós (*3a*, ed. Varela Hervías) gives ample evidence of the novelist's avid quest for vivid personal detail, particularly regarding Ferdinand VII. Galdós wrote a flattering personal portrait of Mesonero in his *Galerías de españoles célebres* (8 February 1866; **VI**: 1326–8) and included another some ten years later in *Los apostólicos* (**II**: 197). Less directly, he related him to Estupiñá, the purely oral chronicler of Madrid, by making the two figures share a birth-date and a physical resemblance.[77]

There is other evidence of Galdós's search for utilizable data: the detailed consultations he carried out concerning the African campaign for *Aita Tettauen* (*3b*, Ricard 1968); the replies he received to his enquiries regarding Montes de Oca and Oñate (*3d*, Cardona 1968: 136–7); and the manuscript with the title 'Anotaciones: datos históricos', which contains a series of questions about the personal life of Espartero.[78]

Not all critics, however, are in agreement. Brian Dendle, for instance, who has dealt so extensively with the *Episodios nacionales*,[79] takes a harsher attitude towards Galdós's historicity:

> Galdós's treatment of the past is selective and is strongly colored by hindsight... The Spain so vividly created in the *episodios* reflects the fertile imagination of a novelist, not the reasoned assembly of facts and arguments of an historian. Galdós's research for the *episodios* was hasty; he sought from sources and correspondence colorful details rather than an understanding of the past on its own terms. Historical recreation was often accessory to the exigencies of a fictional narrative rooted in a moral vision of Spain. (*3d*, 1980: 182–3)

In his more recent book on the early series he repeats the charge: 'Galdós's historical research was... both hasty and superficial... The result is fiction

[77] See Anderson (*3b*) and Palomo (*3b*). Both figures were born on 19 July 1803. Estupiñá is described as resembling Rossini, and Galdós makes the same parallel with Mesonero Romanos (**VI**: 1326–7). In several of his articles Mesonero has a character called Don Plácido Cascabelillo (*4a*, 'La comedia casera', '1808 y 1832', 'Las tres tertulias'), who has something in common with Estupiñá as well as his first name; he is the first creation of Mesonero's mentioned in *Los apostólicos* (**II**: 197). In the discarded Alpha version in the Harvard manuscript (*3a*), Estupiñá was born on the same day as Victor Hugo, but Galdós had the inspiration to change it to a much more suitable identification.

[78] These 'Anotaciones' in the Casa-Museo de Galdós in Las Palmas (*3a*) consist of a series of 'Preguntas' with their appropriate 'Contestaciones' from an unknown correspondent. The information was incorporated into *Vergara*. For further examples see Faus Sevilla (*3b*).

[79] In my view the fundamental defect of Dendle's well-documented books is their tendency to consider the *episodios* almost exclusively in the light of the politics of the time of writing.

masquerading as history, a narrative that is highly colored, dramatic, impressionistic, and politicized, and that focuses on a few scenes, or "episodes," from Spain's past' (*3d*, 1986: 155; also 161).[80] For Dendle, Galdós's qualities lie rather in the imagination: 'Rather than as novels of historical *vulgarización*, the *episodios* should be treated above all as highly creative works of the imagination' (1986: 161–2).

It seems to me that this assessment seriously underestimates the importance of the historical purpose of the *Episodio nacional*. Granted that, like all historical novels, the *episodio* is not history as such, it does nevertheless constitute a special attempt to convey historical notions.[81] Even those volumes with the smallest historical content—and in none is it negligible—are substantially conditioned by their starting-point, usually conveyed by the title; some attention is given, too, to events abroad, especially in France, such as the fall of Louis-Philippe in 1848, the Franco-Prussian war and the Paris Commune. In most cases, despite an impressive degree of individual characterization (Beramendi and Teresa Villaescusa are outstanding examples), the fictional characters actively convey political attitudes or are interwoven with political events: they are controlled by the history, not the history by them.[82] Galdós, as I see it, is deeply concerned with the actual historical events he is describing or demonstrating, for their own sake as well as, or more than, their implied relevance to the time of writing. The application to current events is there, without a doubt; the link with the present corresponds with the criteria outlined by Auerbach and, in a very different way, by Croce. It is very real, but is implicit rather than explicit. It is up to the reader to draw the parallel as he sees fit: one more aspect of the open-endedness or lack of closure of these narratives.

Where does this leave the *episodios* as novels? Diane Urey, in her excellent recent book, arguing forcefully for treating 'the historical novels as the literary texts they indeed are' (*3d*, 1989: 9), claims that Galdós displays in them 'an amazing artistic virtuosity' (10) and that his 'brilliant manipulation of language

[80] Even more extreme is Pedro Laín Entralgo's remark, written in 1945, out of excessive regard for the Generation of 98: 'Los *Episodios Nacionales* son una serie de cuadros de historia atravesados por el hilo unitivo de cierta acción novelesca elemental. La técnica de los *Episodios* puede ser reducida a sencillísima receta: Tómese la materia histórica contenida en un tomo de la *Historia* de Lafuente, redácteseia con mejor pluma, vístasela de ropaje novelesco—y si el ropaje es una simple hoja de parra, mejor: un muchacho de origen oscuro que va medrando de aventura a aventura, camino de su *happy end*—; hágase todo esto y se tendrá un tomo de Galdós' (*4a*, ii. 250).

[81] Carrasquer, discussing the historical novel in relation to Ramón Sender, argues categorically that in dealing with this form 'no puede hablarse de *historia* en primer término, ni siquiera de *historia novelada*'. The *episodios* are 'Un caso límite... pero aún aquí predomina neta y francamente el interés de la narración literaria sobre la narración histórica' (*4a*, 70). Though I have some sympathy with the argument, I do not feel that the two elements are so easily measurable or separable.

[82] 'La función histórico-ideológica de muchísimos de los personajes ficticios de los *Episodios* es palpable' (*4a*, Longhurst 253). See also Llorens (*3b*, 105–40).

could surely only be achieved by a keen perception of its intrinsic and inseparable role in every aspect of life, including what is called history' (11–12). With this I am in entire agreement. When, however, she goes on to assert that 'the novel and history are interchangeable and indistinguishable modes of discourse because they both necessarily rely on the same narrative strategies' (13), I harbour serious reservations. The *episodios* do share the same narrative strategies with both the contemporary novels and narrative history, even though the focus and the emphasis within these strategies, as I shall be arguing in the course of this book, inevitably vary considerably; the statement in favour of 'más novela' in the second prologue to the Illustrated Edition, quoted above, is particularly noteworthy. From this it does not follow, however, that there is no difference between fiction and history. As indicated in the Introduction (*2*, Scholes 1981: 207), it is essentially a question of outside referentiality and determinacy. Historical events and characters have a fixed, delimited, predetermined cast, which a historically orientated novelist has largely to respect if he is to use his material effectively. Whether or not his readers have any prior knowledge of the events, Galdós is concerned to maintain a minimum of historical completeness and accuracy. Fictional events and characters, wherever they occur, do not have this limitation. Galdós is free to choose, and in fact does choose, to portray the personality of Isidora Rufete in such a way that she ends up on the streets, to produce in Fortunata a reaction which will cause her death, to allow Torquemada to acquiesce in actions which result in his becoming a reluctant member of the new aristocracy. In the historical context of the *episodios*, however, once he is committed to the particular task of chronicling a given period, he has no choice but to take on board Merino's assassination attempt in 1852, Francisco Chico's murder in 1854, the 1866 sergeants' revolt, the expulsion of Isabella II in 1868, Prim's assassination in 1870, and so on. These factual limitations leave, to be sure, wide scope for interpretation, reassessment, fictional invention, individualized characterization, in all of which functions there is a very close similarity with the non-historical novel. To that extent Urey's point that 'However diligently the reader may pursue some *absolute* point of reference, some *sure* knowledge of the past in the novel or in history, in Galdós's novel histories this effort is constantly undermined' (13; my italics) has some justification. We as readers certainly cannot aspire to such absolute truth, such sure knowledge; no full mimesis of the past can be created; any re-creation is tentative, incomplete, uncertain, illusory; and such a re-creation is subject to the exigencies of narrative discourse. None of this means, however, that a dialogue on the subject of the significance of the past is not established. I do not agree therefore that 'An excessive pursuit of a historical truth or of a sure knowledge of the past is always destructive in Galdós's work' (235). The fact that ultimate truth is unknowable, or that such a 'would-be historian' as Fajardo, from within the fictional structure, is contradictory, does not prove that the historian's task is futile (122). It indicates, rather, that it is complex, incomplete, and inexhaustible: however inadequate, relative, or illusory his knowledge inevitably is, the broad parameters remain intact.

In fact, within this effort at historical reconstruction a different basic contradiction is concealed. The certain knowledge of what the future will bring poses a difficult problem for the novelist. However much he strives to dig deep into the conditions of the time and leave every possibility open, he cannot ignore the fact that he and the public know what will happen, which tendencies and individuals are destined to prevail. Thus he is trapped in a deterministic pattern far more rigid than that of Zola's experimental novel. As he selects his data to chronicle half a century of national life Galdós cannot overlook the *a posteriori* perception given him by the many years that have gone by. There is necessarily something *déjà vu* in the historical perspective of the *episodios*: it is related to the necessary anachronism syndrome we have spoken of earlier. Thus it is very noticeable, and at the same time inevitable, as we shall see in later chapters, that he should single out figures who in time are going to distinguish themselves or subjects which will grow in significance. What matters, therefore, is his ability to offer a distinctive combination with fictional material in order to bring a fresh interpretative stimulus to the story.

Galdós takes particular trouble, through fictitious characters carefully built up and integrated into real situations, to offer an extremely broad spectrum of opinions drawn from all parts of the contemporary scene. It is what Dendle calls his 'proteísmo' (*3d*, 1980: 200) and involves an essential 'indeterminacy': not just those criteria which are finally going to prevail, but all possible options open to the people of the time, who have to make a choice without knowing the outcome of their actions. Urey is one of the critics who has most lucidly established the emergence of various interpretations of historical events. Speaking of the *Ministerio Relámpago*, for example, she notes that 'Neither *Narváez* nor the fourth series ever provides an unambiguous interpretation or resolution to the problems' (*3d*, 1989: 120).[83] This is very much in accord with my perception that the *episodios* possess an open and unresolved character which is intrinsic to their structure. Such an open-ended quality makes the *episodio* very susceptible to such reader-response theory as Barthes's, in which the text is considered autonomous; unlike those classic books which Barthes (with no little arrogance, it seems to me) relegates as *lisible* (including the bulk of Balzac, *Sarrasine* excepted), the *episodio* at its best may be regarded as *scriptible*;[84] it makes the reader not a consumer but a producer of the text, without implying any authorial abandonment of control of the text. From a

[83] In the previous version (*3d*, 1983: 205) she puts it as follows: 'it is left to the readers to understand this history, to form an aesthetic response to this novel, and to define for themselves the truth of these events'. This is even closer to the point I am making.

[84] See *2*, 1970: 10. The concept is similar to Derrida's notion of *undecidability*, from which it probably derives. See Lentricchia (*2*, 142, 172).

very different point of view, Galdós's open-mindedness reflects E. H. Carr's concept of history. We shall encounter many examples of this special merit of the *episodios*.

We should now consider to what degree the *Episodio nacional* differs from the standard form of the historical novel. Much depends, of course, on how a 'historical novel' is defined. As has been indicated in the Introduction, clear definitions are hard to come by. To narrow the field, we may ask how the form fits in with Lukács's approach.[85] Rodolfo Cardona has argued, in a recent essay, that 'los *Episodios nacionales* constituyen, tal vez, una de las manifestaciones más claras de la novela histórica tal y como la concibe el crítico húngaro' (*3d*, 1988: 111). This I find difficult to accept. The factors we have seen—the quantity of historical data involved, the requirement of close factual coherence, the rigid format, and the continuity existing between one *episodio* and another—all militate against a simple identification with Lukács's equation of the historical and realist novel. If one takes a limited view of the term 'historical novel', especially if its Romantic version is allowed to dominate, then, as Alfred Rodríguez has argued (*3d*, 197), the *Episodio nacional* does not qualify. It is no less true that the weight of history is largely borne by fictitious characters, who in turn give these works their narrative plot. For this reason Francisco Ayala denies that they are historical novels:

> Los *Episodios* no pertenecen al subgénero de la novela histórica, que toma la historia como un marco externo donde encuadra el desenvolvimiento de unas vidas humanas; en ellos son las vidas humanas mismas las que, en su despliegue, tejen la historia; y ésta, sobre la única materia de las pobres existencias individuales, proyecta su sentido más allá de cada una, hacia un plano transcendente. (*3d*, quoted in Rodríguez 197)

We may accept Ayala's cogent observation wholeheartedly without adopting such a restrictive criterion. In line with the arguments developed from Fleishman and Tillotson, I see no need to exclude the *Episodios nacionales* from being classified as 'modern' historical novels according to the definition adopted in the Introduction.

There is no doubt, on the other hand, that, as a result of its dense historical content, the subgenre suffers in an acuter form from the same difficulty of how to merge historical facts and fictional story as is faced by the standard historical novel. The problems posed by Manzoni and backed up by Amado Alonso remain valid (even though Alonso excused the *episodios* on the grounds of their contemporaneity: *2*, 81). In particular, it is clear that Galdós, especially from the desolate post-1898 perspective we encounter from the fourth series onwards, harbours very serious doubts about how to cope with the two heterogeneous elements which make up the narrative: 'No puede usted figurarse lo difícil y desesperante que es para el escritor colocar

[85] Rafael Bosch applies Lukács's concept of the novel methodically to Galdós and finds it, with some exaggeration no doubt, a perfect fit (*3c*).

forzosamente dentro del asunto novelesco la ringla de fechas y los sucedidos históricos de un episodio,'[86] he complains in a letter to a friend in 1905. The consequences of this frustration will be examined in later chapters.

The *episodio* evidently succeeds in some areas better than others. It is specially concerned with political coverage, and indeed frequently indulges in inessential political detail. By contrast, Vicente Llorens has rightly deplored the relative absence of economic issues, and in particular, of awareness of the new Basque and Catalan industrial complex (*3b*, 118); little is said about the economic crisis of 1864–8 as a contributory factor in the revolution.[87] These factors compound the hybrid nature inherent in the historical novel as well as lessening its universality. Given the fact that all narrative work, as I have argued in the Introduction, is a compromise between incompatible demands, this is no great disadvantage: the *episodio nacional* provides as I see it an adequate but not perfect solution to the special problem it faces of the extension, locality, and inconclusiveness of its historical material. We have seen that in his change of technique from the autobiographical form of the first series to the *narración libre* of the second, Galdós was seeking more novelistic freedom. As a result of its political coverage, then, the *episodio* falls near the extreme boundaries of the novel form, a 'caso límite', as Carrasquer said (*4a*, 70), much as the 'reportorial novel' does in Spires's diagram (see above, p. 27). The *Episodio nacional* is not, however, a reportorial or documentary novel; it does not apply directly to the contemporary scene and does not implicate the author as a narrator-witness. Although it seeks overall historical accuracy, it makes great use of invented characters in a way that Truman Capote or Norman Mailer are at pains to avoid.[88] At the same time it does not involve the specific and urgent political commitment of the *novela testimonial*.[89] For these reasons, without denying their general status as modern historical novels, I prefer where possible to retain their distinctive title of *Episodios nacionales*. It is a subgenre that deserves to be considered in its own right and on its own terms.

I now turn to the novels and consider first their historical content. Tolstoy's conception of history, discussed in the Introduction, raises many matters relevant to Galdós. Despite its philosophical contradictions and the ultimate predominance of fatality, it offers, for a novelist preoccupied with history,

[86] Montesinos (*3c*, i. 81–2). Though it is true that there is a remarkable flowering of fictional plots in the series, I do not agree with Dendle, for reasons previously mentioned, that 'The treatment of history in the fourth series is . . . skimpy to the point of casualness' (*3d*, 1980: 79).

[87] See Carr (*1b*, 300) and Vicens Vives (*1b*, 1961: 314). Fontana (*1b*) attributes much of the economic crisis which leads to the revolution to the disastrous fall in railway shares.

[88] For a broad view of the documentary novel see Foley (*2*); for more restricted perspectives on the nonfiction novel see Hollowell (*2*) and Smart (*2*). It is irrelevant to my purpose to discuss whether or not such a work as Capote's *In Cold Blood* fits into this genre or subgenre.

[89] 'El testimonio es una forma de lucha', declares René Jara (*4a*, in Jara and Vidal 1). See also Barnet (*4a*, in Jara and Vidal 288).

a viable practical formula. In making his famous distinction between the hedgehog and the fox, Professor Berlin admits to a special problem in placing Tolstoy, offering, finally, the ingenious solution that Tolstoy 'was by nature a fox, but believed in being a hedgehog' (*2*, 4). In these terms, Galdós is without doubt predominantly a fox, that is to say, that he participates in similar qualities: unsurpassed 'powers of insight into the variety of life—the differences, the contrasts, the collisions of persons and things and situations, each apprehended in its absolute uniqueness and conveyed with a degree of directness and a precision of concrete imagery found in no other writer' (*2*, 40). Yet the consistent aims and organization of the *episodios* show a clear aspiration towards the qualities of the hedgehog, though, perhaps fortunately, Galdós is much less concerned with an overall philosophy of history than Tolstoy. Pragmatically, he seems, at least in part, to have embraced a substantially similar outlook. He is very conscious of the deterministic approach to history we have seen in Collingwood and others; he frequently flirts with its suppositions, without ever accepting them fully, 'skirting the borders of determinism', as Whiston felicitously puts it (*3c*, 1980: 126; *3c*, Goldman 1988: 145). Perhaps a striking sentence from *La desheredada*, in which both Providence and Chance play their part, most clearly expresses his viewpoint: 'La Providencia y el Acaso juegan al ajedrez sobre España, que siempre ha sido un tablero con cuarteles de sangre y plata' (IV: 1067).

Pattison makes the interesting suggestion that Galdós might just have read *War and Peace* in French translation before writing *Fortunata y Jacinta* (*3b*, 90–3). Accordingly, he postulates the 'discovery of a new kind of naturalism in Russian literature, particularly in Tolstoy's *War and Peace*, which caused him to pause and meditate' (90). Although the year's pause in production Pattison discerns after *Lo prohibido* is largely illusory,[90] Galdós certainly has his narrator expressing himself in a way which is somewhat reminiscent of Tolstoy:[91]

> Fue de esas cosas que pasan, sin que se pueda determinar cómo pasaron; hechos fatales en la historia de una familia, como lo son sus similares en la historia de los pueblos; hechos que los sabios presienten, que los expertos vaticinan sin poder decir en qué se fundan, y que llegan a ser efectivos sin que se sepa cómo, pues aunque se los sienta venir, no se ve el disimulado mecanismo que los trae. (V: III. v. 2, 362)

An escape is, however, provided from this apparently entirely deterministic approach, first, by the fact that it is an interested narrator who is speaking, and second, by the reiterated statements of ignorance about how the processes involved actually worked.

[90] Galdós seems to have started work on the first draft of *Fortunata y Jacinta* immediately after returning from his Portuguese vacation with Pereda in May 1885. For further discussion see my forthcoming book on the evolution of *Fortunata y Jacinta*.

[91] For a detailed study of the two writers see Turkovich, who finds the similarity 'astounding' (*3c*, 710).

As far as treatment is concerned, the first fundamental distinction which needs to be drawn is between those historical events which are actually occurring during the action of the novel and those figures or incidents which are recalled from the past.[92] Within the latter group, set in the past, there is a further subdivision, constituting three categories in all.

The first type consists of historical incidents of some weight which while forming part, at least to a limited degree, of the bustle of the action and the flow of historical time, also impinge on the fictional characters. It is a device typical of the *Episodio nacional*, which also occurs on occasion in the contemporary novel. The category has different facets: one of these embraces those cases of broad parallels established between characters in the novels and major historical developments. The two most notable examples are the parallel between the public life of Spain veering between law and disorder and the private life of Juanito Santa Cruz; and the dishonourable course, parallel to that of Spain, followed by Isidora in *La desheredada*, as she accepts a degradation similar to that suffered by the country. Also within this category come fictitious characters who play a full part, frequently or occasionally, in historical situations: the Bringases, involved in the fall of Isabella II within the royal palace, Villalonga as deputy during the crisis of the Republic and the new stability of the Restoration, Torquemada as the barometer of economic developments throughout the period, Villaamil in the vicissitudes of a *cesante*'s existence, Manuel Infante (*La incógnita*) representing the somnolent parliamentarism and *caciquismo* of the full Restoration period. The historical associations may be fraudulent ones, like Simón Babel's claims (V: I. 2. v, 1225), in *Ángel Guerra*, to be in correspondence with Don Manuel (Ruiz Zorrilla), preparing for a forthcoming revolution (the revolt in which Guerra participated has just failed) which would rule by decree, promulgated in true bureaucratic style in the *Gaceta*. One variant bridges the gap towards the second type. It is José de Relimpio's mental diary or *Efemérides*, which provides a comprehensive but schematic list of the key events, still very relevant to the reader, of the immediate past.

The second type no longer has the same active role; it comprises the historical recapitulations made by certain characters to provide a rapid synopsis of the salient events of recent history, with the function of inducing a general historical awareness or jogging the memory; they are in no way integrated into the plot. The best examples are provided by two characters from *Fortunata y Jacinta*, Plácido Estupiñá and Doña Isabel Cordero; they fulfil a function which may be compared to that of Jenara Baraona or Beramendi's children in the *episodios*. A lesser case from *La desheredada* is *la Sanguijuelera*, who remembers nostalgically all the finery of the public events

[92] It is a difference similar to that between the 'third-hand' and the 'fourth-hand' Napoleon I referred to in the Introduction.

throughout Isabella II's reign. As a soldier Father Nones, in *Tormento* (IV: 1503–4), witnessed Riego's rising and formed part of the firing-squad which shot Torrijos; then, as a priest, he confessed Merino before his execution. *Ángel Guerra* contains one Conejo, 'cigarralero, hombre tan sencillo como bruto', who recalls the past better than more recent events:

> Conocía personalmente a Espartero, a Serrano y a los Conchas. . . . hablaba de las cosas del año 38 como si hubiera sucedido la semana pasada, y apenas tenía vagas nociones del reinado de Isabel II y del de don Alfonso. Mejor sabía el paso de Luchana y la acción de Guadamino que la revolución de 68 y otros acontecimientos que ningún eco tuvieron en su espíritu. (V: II. 4. i, 1363)

Similarly, Frasquito Ponte Delgado, in *Misericordia*, serves as a marker of the past; sardonic references are made about him and his clothes as dating as far back as the time of Ferdinand VII and Riego; the years of the African war are alluded to in more detail: a companion of his speaks of 'cuando la Unión Liberal . . . Era ministro de la Gobernación don José Posada Herrera. Yo estaba en *La Iberia* con Calvo Asensio, Carlos Rubio y don Práxedes . . . Pues apenas ha llovido desde entonces' (V: 1975); the shoes he bought in Paris lasted until Prim's assassination (1922). Ponte's disastrous decline into carefully concealed indigence became evident in 1859 and reached its point of no return when, with the 1868 revolution, he lost the job he had obtained from González Bravo (1919). He spends his time entrancing the impressionable Obdulia with pseudo-historical anecdotes, especially about Paris; he speaks of various popular writers like Paul de Kock, Octave Feuillet, and Eugène Sue; he even finds a resemblance between Obdulia and the Empress Eugénie[93] and spends the peseta Benina gives him on a photograph of her.[94]

The third category is completely static: lumps of history unassimilated into the action and representing fragments of the past no longer functionally alive, though they may be symbolically. It includes many swift *off-camera* allusions, to use the term employed by Farris Anderson (*3c*, 12–17), in which passing mention is made of some eminent historical figure to fix a moment in time, to define a political attitude, or to link a political occurrence with a private happening. These no longer form part of any recognizable, ongoing historical process, but are petrified, isolated episodes from the irrevocable past, which

[93] Of Spanish origin, from a noble family from Granada, Eugenia, daughter of the Countess of Montijo, one of Madrid's leading hostesses, frequently appears in the *episodios*, both before her marriage, as when she reads Fajardo's confessions and practices Italian with him (*Las tormentas del 48*, II: 1456, 1470–1) and after she becomes empress (III: 174). The Carlist priest-cacique Don Juan Hondón always refers to Napoleon III as the 'marido de Eugenia' (III: 394). In his *Memorias* (VI: 1431–2) Galdós recalls seeing both Napoleon III and 'la bellísima Emperatriz' at the Universal Exhibition in Paris in 1867.

[94] Bly (*3c*, 1983: 174) considers these references to be 'The sterile pull of the historical past' and declares that the allusion to Prim 'has become a faint, off-stage echo' no longer of relevance to individual lives. I see in them a continuing desire to provide a bridge to the past, even though many more years have transpired.

may serve as models, warnings, or justifications of some current action. Prim is a favourite for this sort of reference, but there are others. A variant on this type indicates a physical likeness between a character and an eminent historical figure, normally foreign. Since the resemblance is purely physical, there is much scope for irony. The two most notable examples are Bringas, who is consistently and ironically identified with the French economist and statesman Thiers (*3c*, Varey), and the constant iconographical comparisons made between Mauricia *la dura* and Napoleon (*3c*, Braun 1977; *3c*, Ribbans 1977*c*).

Just as I was reluctant to classify most of Balzac's *Comédie humaine* as historical novels, I have similar hesitations, on three counts, regarding Galdós's fittingly named contemporary novels: first, the predominance of social interaction over direct assessment of historical events; second, their contemporaneity and, more especially, their tendency, via the narrator's voice, to merge into the present, which limits their interpretation of the past; and third, the very varied historical content, which even in the most intense cases is relatively small and incomplete, and is treated obliquely. On the other hand, there is to my mind not a shadow of a doubt that they are historically conscious novels.[95] The term used by Peter Bly as the title of his thorough and perceptive study of 1983, the 'novel of historical imagination' (*3c*), is perhaps the most suitable.

In the next five chapters I shall apply the criteria and the comparisons made in this chapter to Galdós's treatment of history from the reign of Isabella II onwards. In particular, those incidents which occur in both *episodios* and novels will be closely examined and a critical comparison made regarding the historical features of each, the integration of historical description, and the implications for the narrative structure in each case. These comparisons will, I trust, yield revealing results regarding both the historical occurrences and the differing literary techniques involved.

[95] As Butterfield pointed out long ago, a little history goes a long way towards adding a historical perspective to a narrative: 'Some well-known, historical character looms in the background, larger historical issues cast their shadows at times and perhaps at some point the narrow concerns of the individuals whose fate makes up the story, cross the path of these, and become interlocked with some piece of history' (*1a*, 1924: 79).

2

The Reign of Isabella II
Hopes and Disappointments, 1843–1858

DAUGHTER of Ferdinand VII and his fourth wife María Cristina,[1] Isabella was born on 10 October 1830. On her father's death on 29 September 1833, she succeeded as an infant to a disputed throne; the supporters of absolutism, called 'apostólicos' or Carlists, defending, according to Salic law, the rights of Ferdinand's brother Carlos,[2] rose up in arms in the first Carlist war. María Cristina's regency was marked by financial intrigues and scandal; the queen mother very soon contracted a secret morganatic marriage with a corporal in the Guards, one Fernando Muñoz.[3] Faced with the opposition of the *progresista*[4] general, Baldomero Espartero,[5] she abdicated in 1840. In all this turmoil, the young queen's education was grievously neglected in every respect, but especially in political affairs.[6] Not long after Espartero had fallen

[1] 1806–1878, daughter of Francis I, King of the Two Sicilies, and Maria Isabella, Ferdinand VII's sister. She was consequently her husband's niece.

[2] Carlos María Isidro ('Carlos V'), 1788–1855; married first to Francisca of Braganza and, after her death in 1834, to her domineering sister María Teresa, invariably known as the Princess of Beira.

[3] The relationship started shortly after Ferdinand's death and the secret marriage took place on 28 Dec. 1833, but was not legally sanctioned by the Pope until Oct. 1844, when Muñoz assumed the title of Duke of Riánsares. The couple, who had nine children, were notorious for their avaricious and unscrupulous business dealings. Marichal (*1b*, 150) comments that 'She left Spain in 1840 as perhaps the wealthiest woman in Europe'.

[4] The 'liberals', i.e. the supporters of Isabella's succession and some sort of constitutional rule, quickly divided into two main branches, the *moderados* or Conservatives, and the *progresistas*, or Progressives. I shall henceforth employ the English terms, with capital letters to distinguish the parties from normal usage.

[5] Baldomero Espartero, 1793–1879, Duke of la Victoria and finally, under Amadeo, Prince of Vergara. He made his name in the unsuccessful struggle to maintain the South American colonies at Ayacucho (1824). Hero of the first Carlist war, which concluded with the reconciliatory *abrazo de Vergara* (1839) between Espartero and Maroto; regent in 1841, he was the recognized if ineffectual leader of the Progressives, so esteemed that he was offered the vacant throne in 1868.

[6] The first chapters of *Los ayacuchos* give a lively impression of the extremely haphazard manner in which Isabella and her sister picked up casual information about the political process. Amid partisan intrigues her tutors Argüelles and Quintana, and her governess the Countess of Espoz y Mina, appointed in 1841 and expelled two years later on Espartero's fall, were unable to accomplish much in such a short time. Galdós places a fictitious character, Mariano Díaz de Centurión, whom we shall see again, in a strategic situation in the palace.

in turn in July 1843, Isabella was prematurely declared of age on 8 November, shortly after her thirteenth birthday.

Queen Isabella and her reign fascinated Galdós. In the novels the literary coverage given to the early period is inevitably uneven and spotty, for no contemporary novel is set before 1863. There are, however, interesting throw-backs to earlier historical occurrences; typically, these are references of the second type enumerated in Chapter 1.

The fourth series of the *episodios*, on the other hand, is entirely concerned with Isabella II's reign.[7] Written between March 1902 and May 1907, it deals with the principal events which occurred between late 1847 (when the queen was 17 years old) and her expulsion in September 1868. In this chapter we shall be concerned with the first half, approximately, of her reign and the first four *episodios*, *Las tormentas del 48*, *Narváez*, *Los duendes de la camarilla*, *La revolución de julio*, and part of *O'Donnell*. In addition, her effective reign is anticipated to some extent in the last *episodio* of the third series, *Bodas reales*, in which the events up to and including her marriage in 1846 are recounted. A long chronological distance (averaging over forty years) separates the date of composition from that of the events described.

A further important dimension is added by the account of the two visits Galdós paid, with his old friend and fellow Canarian Fernando León y Castillo, then Spanish ambassador to France, to the elderly exiled ex-queen in Paris in 1902 in quest of information while he was writing the series.[8] These impressions were published after Isabella's death in April 1904. In addition to these sources of information, revealing comparisons will at times be drawn with certain historical novels by Pereda, Baroja, and Valle-Inclán.

Let us take as our starting-point the best-known examples of historical recapitulation from *Fortunata y Jacinta*. Jacinta's mother, Isabel Cordero, manages to give birth to her numerous offspring punctually in unison with some of the main occurrences of the reign of her namesake. Significantly, her father is a character from the *episodios*, Don Benigno Cordero, the paradigm of those 'Hombres de costumbres pacíficas y sin ideal guerrero de ninguna clase [que] iban a familiarizarse con el heroísmo' (I: 1611). Cordero momentarily belies his pacific name and nature and becomes the hero of Boteros[9] at the head of the Milicia Nacional in the liberal struggle against Ferdinand VII's despotism. Subsequently he returns to a life of exemplary moral conduct and

[7] For Galdós's preliminary notes on the series see Cardona (*3d*, 1968: 121–2). It is worth noting that the titles changed considerably. *Las tormentas del 48* started as *El año loco*, *Los duendes de la camarilla* as *Bravo Murillo*, *Aita Tettauen* as *África El Mogreb*, *La vuelta al mundo* as *Méndez Núñez* and *La de los tristes destinos* as *Fin de un reinado*.

[8] From León y Castillo's letters to Galdós in the Casa-Museo (*3a*) we learn that the interviews were prepared from mid-1901, but Galdós appears to have put them off until the end of 1902.

[9] See *El siete de julio* (I: 1612–13). The street battle took place on that same day in 1822. Dendle (*3d*, 1972) has discerned the influence of Fernán Caballero's *Elia*, which has a character of the same name. For Benigno Cordero see Llorens (*3b*, 106–11).

material prosperity, though he does nurse the *Sí de las niñas*-like illusion of marrying a much younger woman, Sola.[10] A model none the less of the new emergent bourgeoisie, he is described in admiring tones in a passage of great historical density:

Hombre laborioso, de sentimientos dulces y prácticas sencillas, aborrecedor de las impresiones fuertes y de las mudanzas bruscas, don Benigno amaba la vida monótona y regular, que es la verdaderamente fecunda. Compartiendo su espíritu entre los grandes afanes de su comercio y los puros goces de su familia; libre de ansiedad política; amante de la paz en la casa, en la ciudad y en el Estado; respetuoso con las instituciones que protegían aquella paz; amigo de los amigos; amparador de los menesterosos; implacable con los pillos, fuesen grandes o pequeños; sabiendo conciliar el decoro con la modestia, y conociendo el justo medio entre lo distinguido y lo popular, era acabado tipo del *burgués* español, que se formaba del antiguo pechero fundido con el hijodalgo, y que más tarde había de tomar gran vuelo con las compras de bienes nacionales y la creación de las carreras facultativas, hasta llegar al punto culminante en que ahora se encuentra. (*Los apostólicos* **II**: 115)

The world of the mercantile class of the Arnáiz and Santa Cruz families in *Fortunata y Jacinta* is easily recognizable in this passage. It is evident that Galdós sought to forge a direct link between the liberal patriots of the 1820s and the new shopkeeper class; it is 'El tercer estado' which was born in Cadiz with the 1812 Constitution and which 'echando a un lado con desprecio estas dos fuerzas atrofiadas y sin savia [*the aristocracy and the clergy*], llegó a imperar en absoluto, formando con sus grandezas y defectos una España nueva'. (See *3c*, Rodríguez-Puértolas 1975: 22.)

In changed circumstances, Cordero's daughter displays heroism of a different order, at once commercial and domestic. To rub home this connection, the novelist introduces Cordero fleetingly ('un señor pequeñito con anteojos'), together with his friend and antagonist Father Alelí, prominent in *Los apostólicos*, into Arnáiz's *tertulia* in the 1820s (**V**: I. 3. i, 34).[11] Although in the *episodios* Don Benigno had no daughter named Isabel, Galdós seizes the opportunity to endow his heroine's mother with the queen's name. The grandfather's name is perpetuated in one of her many female offspring, Benigna.

The events signalled by Doña Isabel's fecundity are:

—Mi primer hijo —decía— nació cuando vino la tropa carlista hasta las tapias de Madrid. Mi Jacinta nació cuando se casó la reina, con pocos días de diferencia. Mi Isabelita vino al mundo el día mismo en que el cura Merino le pegó la puñalada a su majestad, y tuve a Rupertito el día de San Juan del cincuenta y ocho, el mismo día que se inauguró la traída de aguas. (**V**: I. 2. vi, 31)

[10] Galdós describes the first night of *El sí de las niñas* in *La corte de Carlos IV* (**I**: 260–6). On his admiration for Moratín see Cabañas (*3b*).

[11] There is no real anachronism as Ortiz Armengol (*3c*, 1987: 167) believes, since the *tertulia* takes place at the time of Bárbara's early recollections (i.e. the mid-1820s), not in the 1860s.

The associations reach back to September 1837, when the royal expedition of Don Carlos, led by Ramón Cabrera, reached the outskirts of Madrid, only to withdraw precipitately on the orders of the pretender (*La estafeta romántica*). They include the queen's marriage and Merino's assault—which we shall discuss—and an important economic development which Galdós frequently mentioned: the construction authorized by Bravo Murillo of the canal which brought water from Lozoya to Madrid in 1858.[12] To press home the historical identification, Doña Isabel dies at a moment of extreme crisis, on the same day in December 1870 as Prim; we shall consider the significance of this event in Chapter 4.[13]

In the same novel old Estupiñá, born in 1803 and identified to some extent with Mesonero Romanos, claims to have witnessed the main historical events from the early years of the century onwards.[14] In Galdós's words:

Una sola frase suya probará su inmenso saber en esa historia viva que se aprende con los ojos.

—Vi a José Primero como le estoy viendo a usted ahora.

Y parecía que se relamía de gusto cuando le preguntaban

—¿Vio usted al duque de Angulema, a lord Wellington?

—Pues ya lo creo —su contestación era siempre la misma—: como le estoy viendo a usted.

Hasta llegó a incomodarse cuando se le interrogaba en tono dubitativo.

—¡Que si vi entrar a María Cristina! . . . Hombre, si eso es de ayer . . . (V: I. 3. i, 34–5)

Worthy of notice are two features which were always important to Galdós: the emphasis placed on physical presence ('la historia viva') and the urge to endow past happenings 'evidentially' with a directly experienced immediacy, together with the telescoping effect of *eso es de ayer*.

A final paragraph accumulates an amazing quantity of live visual information:

Para completar su erudición ocular, hablaba del *aspecto que presentaba Madrid* el 1 de septiembre de 1840 como si fuera cosa de la semana pasada. Había visto morir a Canterac; ajusticiar a Merino 'nada menos que sobre el propio patíbulo', por ser él hermano de la Paz y Caridad; había visto matar a Chico . . . precisamente ver no, pero

[12] A rapid sketch of Bravo Murillo and his hydraulic plans is given in *Narváez* (II: 1592–4). On its importance see Galdós's chronicle of 2 Oct. 1892 (*3a*, *Prólogos* 468). It is worth noting that Doña Isabel Cordero's business sense enabled her to take advantage of the abundant new water supply to expand her trade in white linen.

[13] A similar case of historical coincidence was the death of Doña Asunción Trujillo, Doña Bárbara's mother, on the same day as General Diego de León was shot in 1841, against normal precedent, for rising up against Espartero and attempting to abduct the queen; see *Los ayacuchos* (II: 1207–20) and Carr (*1b*, 222) on León's legendary noble comportment. Manuel G. de Concha was also involved in the revolt.

[14] There is even a throw-back to the haberdasher's shop of Mauro Requejo at the turn of the century described in the first series, which took over the premises occupied by Baldomero I (V: I. 2. i, 18).

oyó los tiritos hallándose en la calle de las Velas; había visto a Fernando VII el 7 de julio cuando salió al balcón a decir a los milicianos que *sacudieran* a los de la Guardia; había visto a Rodil y al sargento García arengando desde otro balcón, el año 36; había visto a O'Donnell y Espartero abrazándose; a Espartero, solo, saludando al pueblo; a O'Donnell, solo, todo esto en un balcón; y por fin, en un balcón, había visto también, en fecha cercana, a otro personaje diciendo a gritos que se habían acabado los reyes. La historia que Estupiñá sabía estaba escrita en los balcones. (V: I. 3. i, 35)

Estupiñá's 'historia . . . escrita en los balcones' shows a remarkable concern on Galdós's part to recall the salient events of the century—to cover schematically the territory of the *episodios*—but the time-scale is jumbled so as not to produce a coherent diachronic structure. On the contrary, it is clearly and deliberately superficial, unchronological, and in some cases factually unlikely. No explanation of underlying causes is given. What is aimed at is to offer the same sort of wide but disconnected familiarity with past official history as might be possessed by the participants in the novel.

Such eminent figures as the Duke of Wellington and Joseph Bonaparte bridge the first two series (especially *La batalla de los Arapiles* and *Memorias de un cortesano de 1815* for Wellington, and *Napoleón en Chamartín* and *El equipaje del rey José* for Napoleon's brother), while the incidents of 7 July 1822[15] in the narrative of that name, the Duke of Angoulême's invading force in 1823 (*Los cien mil hijos de San Luis*), and Queen María Cristina's entry into Madrid in December 1829 (*Los apostólicos*) are related in the second series.

Other events coincide with the as yet unwritten third series: the Captain General of Castile, José Canterac, killed in the Puerta del Sol during the so-called *Motín de Correos* in January 1835 (*Mendizábal*); Rodil and the sergeants' revolt headed by one García at La Granja in 1836 (though they form part of the subject-matter of *De Oñate a la Granja* and *Luchana*, they are not mentioned in the narratives); Espartero's triumph following the rising of 1 September 1840 (*Montes de Oca*). All of these occur before we reach the period which concerns us, from the 1840s onwards. Leaving aside for the moment the last anti-dynastic phase which forms part of the aftermath of the September revolution, we shall in due course look one by one at the pertinent items listed by Estupiñá, as they are developed in substantially greater detail in *La revolución de julio* and *O'Donnell*.

From the very start of her effective reign, following the fall of Espartero,[16] the queen was the object of a tug of war between Liberals and Conservatives. In a

[15] The balcony scene mentioned by Estupiñá is, according to Ortiz Armengol (*3c*, 1987: 183), 'una leyenda liberal'.

[16] Despite the general's inadequacies, Galdós deplores Espartero's ouster (*Bodas reales*: **II**: 1319) and saw it as an early example of the regrettable lack of solidarity among the Progressive forces; in this case even Prim was at fault.

mysterious incident the Progressive Prime Minister Salustiano Olózaga,[17] barely tolerated by the Conservative forces, was accused of browbeating her into signing a dissolution of parliament behind locked doors.[18] Galdós devotes great attention in *Bodas reales* to this important occurrence, which destroyed the Progressives' chance of governing for a decade. In typical fashion, he begins by reporting contradictory popular versions lacking in reliability or first-hand information. The first is the racy picturesque account given by the bird-seller Juan López, known as *Sacris*, in which Olózaga is described as wanting to bring back Espartero and of drawing a knife on the queen. A second more sober version comes from a couple of neighbours, who claim that it was a plot of the traditionalists. For Bruno Carrasco, the devoted Progressive who had reluctantly accepted a post in his administration, 'lo de Olózaga era castigo de Dios', for his having conspired against Espartero. Then, the narrator comments with authorial voice on the exploitation of the situation by the Conservatives, supported cynically by the young González Bravo,[19] and the harmful effects it had on the adolescent queen:

El llevar al Congreso la acusación y darle forma parlamentaria fue la más escandalosa pifia de los señores moderados o palatinos; en vez de ahogar el escándalo en su origen, echando tierra sobre el error cometido, fuera obra de quien fuese, empeñáronse en desplegar ante el país toda la malicia y desparpajo de nuestros políticos, entregando la persona de la Reina a la voracidad de las disputas y al manoseo de las opiniones. ¡Bonito principio de reinado, bonito estreno de la majestad que, representada por una candorosa niña, debió ser resguardada de toda impureza y puesta en un fanal, a donde no llegara el hálito de las ambiciones! Por esto ha podido decir Isabel II que desde su tierna edad le enseñaron el código de las *equivocaciones*. (II: 1329)

Finally, Galdós gives an all but authoritative explanation by the esteemed politician Fermín Caballero, as recalled by Centurión; it reiterates Isabella's

[17] 1805–73. Olózaga had dramatically escaped from jail and certain execution in 1830 (*Los apostólicos*: II: 173). For a description seen through the eyes of Galdós and Baroja see Ortiz Armengol (*4a*, 179–93). Olózaga was the leading light of the 'tertulia progresista', famous for its conspiracies. Kiernan gives a short sketch of him (*1b*, 15–16) as a born parliamentarian, genuinely radical but with a susceptibility for rank and decorations.

[18] Raymond Carr comments that 'This incident is the delight of Spanish historians' (*1b*, 227). How far Isabella was manipulated by the opposing Conservative forces and how much manipulating she herself did, despite her tender years, remains in doubt. Ortiz Armengol remarks that 'en este confuso episodio hay opiniones para todos los gustos' (*4a*, 192). The American diplomat John Hay writes forthrightly about a 'clumsy lie, taught to this youthful queen in her cabinet' and comments: 'It would be hard to conceive of a coarser and more stupid plot than that to which Isabella II lent herself, in her vicious childhood, which drove Olózaga into disgrace and exile' (*1b*, 356).

[19] Luis González Bravo (or Brabo), 1811–71, a politician, notorious for his opportunism, who evolved from a demagogic radicalism to extreme conservatism. Galdós remarks on his 'audacia, rayana en el heroísmo', and goes on to declare: 'La Historia vacila entre admirar a este hombre o inscribirle con asco en sus anales . . . aquel cínico era simpático y airoso por extremo, que fuera de la política era un hombre encantador . . . ¡Oh! España, en todo fecunda, es la primera especialidad del globo para la cría de esta clase de monstruos' (II: 1329).

innocence, describes the Conservative conspiracy, together with the fallacies in their story—the quite normal signature, the lack of bolts on the door—and denounces the scandalous culpability, admitted by such eminent figures as the Duke of Rivas and Pastor Díaz, of exposing the young queen to such a bad example of double-dealing.

Two years after writing *Bodas reales*, Galdós was still curious about the incident in his interview with the ex-queen. He had little success in discovering anything new, but he suspects that the truth lies somewhere in between total innocence and full responsibility:

Resultaba la historia un tanto caprichosa, clara en los pormenores y precedentes, obscura en el caso esencial y concreto, dejando entrever una versión distinta de las dos que corrieron, favorable la una, adversa la otra a la pobrecita Reina, que en la edad de las muñecas se veía en trances tan duros del juego político y constitucional, regidora de todo un pueblo, entre partidos fieros, implacables, y pasiones desbordadas. (**VI**: 1191)

The political situation is consistently contrasted with the private affairs of Bruno Carrasco. Bruno sacrifices his family's well-being for illusions about being 'un hombre público'; the health and sanity of his wife, Doña Leandra, and his daughters' matrimonial prospects are seriously prejudiced by his obsession. The new government of González Bravo succeeds in suborning him with a tempting government post and other favours, but this does not last long. Though an intrepid politician, 'a quien sobraba de talento todo lo que le faltaba de escrúpulo' (**II**: 1341), González Bravo survives only long enough to carry through one enduring act: to found the Civil Guard.[20] With the return of María Cristina he falls, and Carrasco is once more out of a job.

The pattern of government by a series of generals with political ambitions to succeed Espartero now takes shape.[21] Galdós conveniently provides a list and brief description of the generals—Narváez, Serrano, Prim, Concha, O'Donnell—who are to dominate the next two decades, though as usual he is careful not to anticipate events.

Narváez supo ser el primer mandón de su época, porque tuvo prendas de carácter de que los otros carecían ... El *general bonito*, como llamaban a Serrano entonces, hombre afectuoso, presumido, de arranques gallardísimos en los campos de batalla, blando en las resoluciones, cuidándose principalmente de ser grato a todo el mundo, mujeres inclusive ... Prim, nacido del pueblo, tenía gustos y costumbres de aristócrata; aunque adelantado en su carrera militar, no había subido a las más altas jerarquías ... Concha, con extraordinario talento militar y más sagaces ideas que sus colegas, se reserva, sin duda, para mejores días ... y en la misma situación expectante se hallaba O'Donnell, cuya mente sajona entreveía, sin duda, empresas grandes que acometer en días normales. (**II**: 1315–16)

[20] It was planned with an essentially civilian basis, but Narváez gave it a fully military structure the next year (*1b*, Ballbé 141–54).

[21] For the early development of army intervention consult Christiansen (*1b*).

The most conspicuous is Narváez,[22] the only one at this early stage capable of dominating the situation: 'Podían ser éstos los hombres del mañana; pero el hombre de aquellos días era Narváez, no embrión, sino personalidad formada' (II: 1316).[23] He now led a particularly oppressive government, dedicated to shooting its opponents; it scored up more than 400 victims, the most notable being Martín Zurbano (21 January 1845). The Conservative tenor of his administration was consolidated by the reactionary constitution of 1845.

The key issue of the *episodio* is of course the queen's marriage—the date, as we have noted, of Jacinta's birth. The 'Spanish marriages'[24] had been a notorious subject of national and international intrigue, in which little or no account was taken of the wishes or interests of the teenage queen.[25] The French and British ministers, Bresson and Bulwer (Bullwer in Galdós), wrestled for influence in Madrid. The fortunes of the five leading contenders are amply aired in the actions and discussions within Bruno Carrasco's circle and juxtaposed with their private affairs.

The first claim to arise is that of the pretender Don Carlos's son Carlos Luis,[26] which runs parallel to the engagement between Tomás O'Lean and Lea Carrasco. What would be a fine match for the Manchegan girl with a cultured lieutenant-colonel who has an apparently brilliant future ahead of him comes to grief through political differences, Tomás insisting (his mother is from the traditionalist Emperán family) on the Carlist pretension, supported by Austria and the Pope, against Lea's arguments reflecting Carrasco's liberalism. It is a clear example of the parallel development of *la historia grande and la historia chica*. The fictional rupture comes as Francisco Javier Istúriz, an unconditional supporter of María Cristina, comes to power. Carrasco has to

[22] Ramón María Narváez, 1800–68, Duke of Valencia, was the most conservative and ruthless of the Isabeline political generals. Known as 'el guapo de Loja' (from his land-owning base in Andalusia), he led the Conservative hegemony during most of Isabella's reign. Galdós sums him up as possessing toughness and decision but entirely lacking in ideas or psychology: 'le corresponde un primer puesto en el panteón de ilustraciones chicas, o de eminencias enanas' (II: 1316). See Pabón (*3d*).

[23] In *O'Donnell*, speaking about 1854, Galdós returns to a brief, essentially negative, characterization of some of the key generals: Serrano is as weak as ever in political judgement, Narváez 'nada estable había producido'; nor has Espartero, or, as yet, O'Donnell: 'hasta ahora no era más que un enigma' (III: 149).

[24] For a comprehensive account of the discussions and intrigues in the courts of Europe concerning a suitable match for Isabella see Parry (*1b*) and Woodward (*1a*).

[25] Queen Victoria, in her correspondence with Lord Aberdeen, speaks of 'Poor little Isabel' and her wicked mother, who has 'wasted her time in frivolous amusements, and neglected her children sadly' (*1b*, Parry 14). The adolescent queen emerges in real life as undisciplined, greedy, wilful, and subject to a disfiguring skin disease (ichthyosis); once she reaches puberty, she becomes highly sexed (*1b*, Parry 160, 267, 285, 303); Galdós plays down these features for the moment.

[26] 1818–61; Count of Montemolín, 'Carlos VI' to his followers. He is very favourably described by his supporter Tomás O'Lean (II: 1354). Such a 'fusionist' solution, strenuously advocated by as well known a Catholic publicist as Balmes, would have settled the dynastic dispute, but the Carlist branch of the family was obdurate about its absolute rights to the succession.

endure the offence to his private reputation; to attempt to put it right would be, as he well understands, tantamount to embarking on another civil war parallel to the smouldering Carlist conflicts: 'Comprendía que España entera se lanzase a una nueva guerra civil para castigar tan desafuero' (**II**: 1369).

Carrasco initially supports the Prussian contender Prince Leopold of Saxe-Coburg, favoured by Great Britain. In lively discussion with her mother, Bruno's other daughter Eufrasia quotes the arguments she imbibes from her lover Emilio Terry, who is in close contact with the British embassy; but this potential alliance also turns out badly, as the couple eventually elope without marriage; at the end of the *episodio* her story, like that of the royal couple, is left inconclusive.

The lively, inconsequential chats of Doña Cristeta Socobio, the infanta Carlota's ex-lady-in-waiting, with her unsophisticated friend Doña Leandra are a mine of information, gossip, and prejudice pertaining to the remaining contenders, all Bourbons from the Neopolitan side: the Count of Trapani, María Cristina's younger brother, who initially enjoyed her support; and Carlota's two sons, Francisco de Asís, Duke of Cadiz,[27] and Enrique, Duke of Seville, the latter favoured by the Progressives, represented by Bruno Carrasco. The situation is complicated by the open intervention of Louis-Philippe of France and by the antagonism between Carlota and her sister María Cristina, caused by the latter's morganatic marriage to Muñoz and reflected idiosyncratically by Doña Cristeta.[28] Gradually Francisco de Asís's candidacy gains the upper hand, a result of his pious and passive nature ('por no meterse en dibujos', as Doña Cristeta says: **II**: 1364) compared with his brother Enrique, seen as dangerously activist. Her absurdly optimistic forecast, just as exaggerated as *Confusio*'s *Historia lógico-natural* is to be (see Chapter 7), has a strongly ironic tone: 'el casamiento de Isabel con un Príncipe español que ha de colmarla de ventura, de lo que resultará nueva hornada de reyes católicos, y una era como dicen los periódicos, una era de prosperidades y grandezas que devolverán a este reino su preponderancia entre los reinos de la Europa' (**II**: 1406). This first instance of what Genette calls a prolepsis (*2*, 67), or flash forward, turns out to be entirely false.

The final settlement is one which favoured the interests of Louis-Philippe; his minister Guizot has outmanœuvred the successive British governments of

[27] 1822–1902; Duke of Cadiz, son of Ferdinand VII's brother Francisco de Paula, and María Cristina's sister Carlota; the couple were thus first cousins on both sides. Francisco de Asís was derisively known as 'Paquita' on account of his effeminacy. Trend (*1b*) gives a lively account, in which Galdós is much quoted, of both Isabella and her consort.

[28] Cristeta claims to have been instrumental in bringing María Cristina into contact with Muñoz. She justifies the alliance on human grounds while condemning it on political ones (**II**: 1360). Baroja gives a scathing account of the two sisters in *Crónica escandalosa*, the penultimate of the *Memorias de un hombre de acción* (*4a*, iv. 997–1001, 1014–20, 1032–6). Parry (*1b*, 328) refers to an amazing report that Carlota had declared on her deathbed that Cadiz's father was Ferdinand VII, which would make Isabella's future husband her half-brother.

Aberdeen and Palmerston. The young queen is forced, very much against her will (II: 1373), into the worst of all possible choices: marriage to a man whom she thoroughly dislikes and who is widely regarded as impotent,[29] while on the same day her sister Luisa Fernanda marries Louis-Philippe's son Antoine of Orleans, Duke of Montpensier.[30] The potential consequences for the succession were all too evident. For the moment, Galdós puts a good face on it, despite the reported evidence of Francisco's clerical affiliations, his inclination to recognize the legitimacy of the Carlists, his lack of martial virtues (1372), and the hint of homosexuality shown in his fondness for dressing and undressing Christ figures (1366). After the failure of the reckless Galician rising led by Solís, Bruno comes to accept that a Spanish consort may be for the best, hoping, like his fellow Progressive Milagro, for his conversion to liberalism. He is not happy, however, about the alliance with Montpensier, and draws attention to the original Anglo-French understanding at the Château d'Eu, which would have delayed the second marriage until the queen had an heir, preferably a male one (1308).

The clearest indication that all is not well is found in the story of Doña Leandra. Frustrated in her desire to return to her beloved La Mancha, she dreams that Bruno has brought news that Isabella has married her cousin Francisco, despite an invasion of the palace by British and French troops on behalf of their candidates, and, with a touch of wishful thinking, that Eufrasia has married Terry at the same time. Later in a farrago of ever more raucous sentiments she denounces her daughters as sluts.[31] Finally, her resentment turns towards the royal marriage in colloquial direct speech which has more than a hint of prophecy about it: 'Que todos tenemos que gritar: "¡Vivan Isabel y Francisco!" ¡A mí con ésas! . . . Como he de gritar yo tal cosa si no me sale de dentro . . . y lo que me manda el corazón es lo otro, que no vivan, sino que mueran . . . pues ellos y su casamiento son la causa de que yo esté como me veo' (II: 1402). As Urey says, 'Leandra's vision and her death, Eufrasia's dishonor, and the marriage of Isabel and Francisco converge and become interchangeable in their historical, social, and moral codes' (*3d*, 1989: 93).[32]

[29] Parry (*1b*, 306, 324, 328) goes to some trouble to discount the fact, but the evidence is plain enough. Woodward (*1a*, 197) quotes Palmerston as saying that Don Francisco was impossible 'on account of his insignificance and the want of those qualities which the husband of the queen ought to possess.'

[30] 1824–90. Conspicuous for his skills at economic management, he coveted his sister-in-law's throne once her rule became problematical. He was a living epitome of the principle voiced by his father's minister Guizot: 'Enrichissez-vous!'

[31] '. . . todos arruinados, todos perdidos, y las hijas hechas unas . . . Soltó la palabra picante y soez, y repitiólo hasta tres veces: "Las hijas . . . *tales*"' (II: 1394). The missing word can only be *putas*.

[32] What for me is the legitimate uncertainty of the time about what would be the outcome, together with ironic hindsight about false predictions, Urey interprets as a flat rejection of the sense and form of history (*3d*, 1988: 118; 1989: 80).

Doña Leandra dies as the dual marriage takes place, and her funeral cortège has to be diverted because of the royal procession the next day. The ornate, archaic, and inflexible procession, 'llevando a las reales personas en urnas, como si fueran reliquias' (II: 1407), also resembles a funeral, as it makes its way to the ugly, ramshackle church of Atocha, where Prim's body is to lie in state many years later. Juxtaposed in oxymoronic contrast to these gloomy omens is the expression of optimistic hope for the future, with the potentialities for a new start in the person of the young queen:[33]

> Todo el regocijo de los corazones, toda la efusión de las almas era para la reina Isabel, para su juventud risueña y llena de esperanzas ... Entre el pueblo y ella había algo más que respeto de abajo y amor de arriba; había algo de fraternidad, un sentimiento ecualitario de que emanaba la recíproca confianza. Nunca hubo Reina más amada, ni tampoco pueblo a quien su Soberano llevase más estampado en las telas del corazón. (II: 1408)

At this stage the young Isabella can count on the love, trust, and expectations of the people, if only she is capable of making use of them with discernment. Yet the narrator is in no position to abandon the 'necessary anachronism' of knowing what is to come, and adds: 'En verdad que el pueblo ha querido de veras a la reina Isabel, así en sus tiempos felices *como en los desgraciados*' (my italics). And the ominous forebodings return in the last words of the *episodio*: 'las paredes de todos los edificios nacionales señaladas por feísimos y repugnantes manchurrones de aceite. Parecían manchas que no habían de quitarse nunca' (II: 1409). It is another case of prolepsis: advanced notice that the reign is indeed to be characterized by ancient structures marked with indelible stains (*3d*, Dendle, 1980: 78; *3d*, Urey 1989: 97).

The result of the marriage is a court full of discord and intrigue. After only four months the spouses are separated, though in the interests of propriety an uneasy accommodation is reached. The king-consort becomes the centre of clerical reaction, the famous *camarilla*. And Isabella's career of promiscuous relations quickly commences with the dashing young general Serrano, to be constantly renewed with other lovers throughout her reign.[34] Her liaisons are to produce nine pregnancies, the first (a son who died) in July 1850. Galdós

[33] Compare Centurión's reactions in *Los ayacuchos* in 1841—'Mis impresiones acerca del carácter y cualidades de la Reina no pueden ser más excelentes ... he formado el juicio que tendremos en ella una gran soberana' (II: 1217)—even if he does find her too generous and informal. He also discerns a link of sympathy with the common people. Doña Cristeta Socobio likewise sees her as the 'remedio de todos los males' (1357).

[34] For details of Isabella II's life from very different political standpoints see Pedro de Répide (*1b*) and Carmen Llorca (*1b*). They generally coincide about her real or possible lovers. These include Serrano; the Marquis of Bedmar (not in Répide); the musicians Mirall and Valldemosa (denied by Llorca) and Obregón (also not in Llorca); José María Ruiz de Arana; Enrique Puig Moltó (reputed to be the father of Alfonso XII); Miguel Tenorio (not in Répide); and Carlos Marfori. Baroja puts in another on his own account: the Conservative politician Sartorius (*4a*, iv. 1148). For an anecdotic account of Isabella II and other Bourbon monarchs see Sencourt (*1b*).

hardly alludes directly to her love affairs, no doubt largely through discretion; in this he differs from a later generation of writers like Valle-Inclán and Baroja. It is clear, though, that, with an unusual tolerance of what was considered sexual misconduct in women, he was much more concerned with other types of irresponsibility.

The most important means of access to the court and the government is initially provided by Pepe Fajardo, who has, as has been indicated, the important function of being the narrator and 'reflector'/focalizer of the first two *episodios* of the fourth series. If at first glance his lengthy account in *Las tormentas del 48* of his two-year residence in Rome, from September 1845 until October 1847, where he rebels against his intended calling as a priest, seems to have little to do with important historical occurrences, at least with Spanish ones, there is some, though perhaps insufficient, justification for this apparent diversion, full of mainly Italian historical and literary allusions.[35] His Italian period revolves round a major event, the election of Giovanni Mastei-Ferretti as Pius IX in 1846. Young Fajardo had favoured his election against the odds and reflects the intense expectation of a new patriotic and liberalizing policy, following Gioberti, in the Vatican. He is protected by none other than the famous Cardinal Giacomo Antonelli, but discovers, not surprisingly, that his dissertation entitled *Risorgimento dell'Italia una e libera*[36] is not to Antonelli's taste; a passing mention by a porter of the tireless democratic proponent of Italian unity Mazzini offers a glimpse of another dimension of the Italian struggle.[37] The background is thus set for the decisive change on the Pope's part, under Antonelli's guidance, to a rigid and absolute intransigence.[38] For the clerically dominated Spanish court support for the Vatican's opposition to the incorporation of its territorial possessions into a unified Italian state and hence to recognizing Victor Emmanuel's kingdom of Italy is to become a burning issue of principle.

[35] Lieve Behiels analyses and justifies these literary references very competently, concluding that by his marriage Fajardo '[se] despide definitivamente de sus ilusiones políticas y sentimentales, a las que la cultura italiana servía de fuente, de eco y de contrapunto irónico' (*3d*).

[36] The title may reflect Cavour's famous policy of *libera chiesa in libero stato*. Cavour is not mentioned in the *episodios*, but Don Basilio de la Caña, incongruously, bears a physical likeness to him (*Fortunata y Jacinta*: V: III. 1. ii, 296).

[37] A further ironic echo of the Italian problem occurs in Alejandro Miquis's drama *El grande Osuna*, as Federico Ruiz criticizes it: '¿ . . . de dónde saca este niño que Osuna quisiera unificar la Italia y hacer un gran reino, como el que mucho después soñó Cavour, contra los fueros de las dinastías reinantes y de la Iglesia? . . . El duque para este niño es el precursor de Víctor Manuel y un émulo de Garibaldi' (*El doctor Centeno* IV: 1440).

[38] His encyclicals proclaiming the dogma of the Immaculate Conception (1854) and the sinfulness of liberalism (*Quanta cura*, with its accompanying list of erroneous doctrines, the *Syllabus errorum*, 1864), and the First Vatican Council (1869–70), which asserted papal infallibility, reinforced the intransigent image of the Catholic Church. For an overview see Woodward (*1a*, 'The Catholic Church in the Nineteenth Century', 276–339). It is appropriate, and hardly surprising, that Fajardo's future father-in-law, Don Feliciano de Emperán, should resemble Pius IX (II: 1478). Rodríguez's comment (*3d*, 175) on the 'wise Antonelli' and the Renaissance atmosphere of the papal court is wide of the mark.

Fajardo's idle life of dalliance with Barberina is brought to an abrupt end by the cardinal, who in a military-like operation sends him peremptorily packing back to Spain with pertinent and prophetic advice: he ought to marry; and should he enter politics he should be sure to get himself a place among the smart ones ('los que pasan de listos': **II**: 1428). Fajardo is thus admirably equipped, as an intelligent, well-read, sensual, hypersensitive, and independently minded individual, to become an involved commentator on the course of Spanish politics.

After his return to Spain everything comes easily to him at the material level; he obtains a sinecure with the government publication *La Gaceta*; he is cared for by his two brothers Agustín and Gregorio and their wives, and, more significantly, he enjoys the hidden protection of his sister, the nun Catalina de los Desposorios, close associate of Sister Patrocinio[39] at La Latina convent. This secure pattern is at odds with his agitated private life. His passionate affair with *Antoñita la cordonera* (over which he fights a duel) ends with her death, which coincides with the May rising against Narváez in the Plaza Mayor, in which the captain-general José Fulgosio is killed and the young Nicolás Rivero[40] plays a part. He conducts an intrigue with Eufrasia Carrasco and searches frantically and futilely for his idealized Lucila Ansúrez. At Sister Catalina's insistence he agrees, very reluctantly, to marry an ugly and apparently stupid heiress, María Ignacia, daughter of the traditionalist Don Feliciano de Emperán; it is—as Eufrasia puts it (**II**: 1949)—a question of *colocarse*. In this way he renounces his Bohemian independence and acquires a fortune, an established position in society as the Marquis of Beramendi, and a seat in Congress to boot. On making his decision, the narrative voice becomes introspective and passes from the preterite to the present of on-the-spot reflection, including a reference to the 'year of revolutions', A 'hidden polemic' between his new situation and his conscience is thus engaged.

¿Qué organismo social es éste, fundado en la desigualdad y en la injusticia, que ciegamente reparte de tan absurdo modo los bienes de la tierra? Retumba en mi mente, al pensar en esto, el fragor de las tempestades que pavorosas estallan en toda Europa. Mis conocimientos de las teorías o utopías socialistas reviven en mí, y reconozco y declaro la usurpación que efectúo casándome con Mariquita Ignacia. Yo, señorito holgazán, inútil para todo; yo, que no sé trabajar ni aporto la menor cantidad de bienes a la familia humana, ¿con qué derecho me apropio esa inmensa fortuna? (**II**: 1510)

[39] Sister Patrocinio, 1811–91, was born Rafaela Quiroga. Known as *la monja de las llagas*, she became notorious as a result of her claim to have received the stigmata and for her influence over the queen. For the historical and ecclesiastical background see Journeau (*3d*, 1988), who as well as correcting some errors made by Galdós—according to her research, Sister Patrocinio's Order left La Latina convent on 29 Oct. 1845—implausibly denies that the stigmatized nun had any political influence on Isabella II.

[40] 1814–78. One of the founders of the Democratic party and of the newspaper *La Discusión*, Rivero fought in the barricades in the 1866 rising. He was President of the Cortes following the 1868 revolution until the proclamation of the Republic.

He has become, to his amazement, a pillar of the establishment:

A mi nombre va unida, con el flamante título que ostento, la idea de sensatez; pertenezco a *las clases conservadoras*; soy una faceta del inmenso diamante que resplandece en la cimera del Estado y que se llama *principio de autoridad*; en mí se unen felizmente dos naturalezas, pues soy *elemento joven*, que es como decir inteligencia, y *elemento de orden*, que es como decir riqueza, poder, influjo. (II: 1553)

In my view Clara Lida misunderstands the situation when she speaks of Galdós's lack of enthusiasm for 'estos advenidizos sociales' and declares that 'ante la autosatisfacción grandilocuente de Beramendi el tono de burla del autor no pasa inadvertido' (*3d*, 69); Beramendi is far from being self-satisfied and is sarcastic at his own expense. What in fact is important, and at the same time fictionally operative, is that his rise in social status does not prevent his continuing to hold progressive ideas and to be influenced specifically by the Socialist doctrines of Saint-Simon, Fourier, and especially Robert Owen.[41] He thus becomes a dissident from within the Establishment, who comments freely from an independent point of view on the events he describes. Although he chafes at his inaction and idle life, he not infrequently acts on behalf of good causes or friends; he is particularly helpful to Virginia and Leoncio (*Mita* and *Ley*). It is also notable that his wife turns out to be far more intelligent than he has imagined; although she is suspicious enough of the effect of Lucila's beauty on her husband to want her to become a nun (II: 1542), she effectively supports him up to a point in his intellectual pursuits in order to allay his frustrations. Indeed, María Ignacia voices a mild subversion of her own concerning certain traditionalist ideas, for example, on the rosary, on religious images, on Papal secular power, and on frequent confession (1557–8); she even disapproves of convents and of nuns calling themselves the spouses of God (1570). The creation and development of Beramendi's structural role, effectively bridging the gap between *la historia grande* and *la historia chica*, are thus one of the pivotal and most original aspects of the series.

As a result of his rise in society, Beramendi rapidly enters into a surprisingly intimate relationship with the young queen.[42] Galdós had already given his appraisal of the queen's personality in a political article of November 1885, seventeen years after her expulsion and just before Alfonso XII's death:

[41] The rise of Socialist and working-class movements in Spain is alluded to only infrequently in the *episodios*, though there is an overlooked discussion on the subject at Loja, Narváez's home base, with Diego Ansúrez at its centre, in *La vuelta al mundo* (III: 444–7). This lack of coverage corresponds essentially, as Peter Goldman has argued (*3b*, 1971), to the social reality of the time; it is not 'camuflaje' on Galdós's part, as Regalado contends (*3d*, 217).

[42] One might even treat Isabella's favours to Beramendi as an indication of how she picked her lovers. Certainly the queen mother and her husband Riánsares seem to think so, judging by the frigid way they receive the intrusive guest (II: 1609).

Isabel II obedeció siempre a impresiones y sentimientos más o menos pasajeros, y las ideas políticas fueron siempre poco menos que letra muerta. Mujer de corazón y no desprovista ciertamente de arranques generosos, rara vez comprendió los alcances y el sentido intelectual del papel de reina.

Tiene en su carácter el corte acabado de la mujer del pueblo español, así en sus gustos y aficiones. ('La familia real de España': *3a*, *OI*, iii. 94)

Now, as he traces her actions as queen while they unfold in the *episodios* some twenty years later, his assessment remains notably similar.

Thus, at Beramendi's first and extended audience, certain of Isabella's essential characteristics immediately become apparent. One is her unbridled personal generosity,[43] so naïvely declared: 'Soy la dispensadora de mercedes y gracias, soy la Reina que desea serlo haciendo felices a todos los españoles, lo que es un poquito difícil—pero, en fin, se hace lo que se puede' (II: 1607). This trait and its accompanying lack of sound judgement are well illustrated when she mistakes Fajardo for someone else touting for favours. The result is that she calls him back to the palace to apologize and recounts the story of how at the beginning of her reign she had 20,000 *duros*[44] brought to her in coin as a gift to a suppliant.[45] She herself acknowledges how very poorly instructed she has been, as we have noted, about royal duties and obligations: 'Es que no me han enseñado, ni siquiera, el mucho y el poco de las cosas' (II: 1608). This excuse is reiterated in the interview with Galdós in 1902 (VI: 1192).

These characteristics stand out most clearly in the novels *Tormento* and *La de Bringas*, set towards the end of her reign twenty years later, which it is convenient to deal with at this point. Isabella's indiscriminate charitable instincts are here used for a fictional purpose: they serve as the last recourse for financial assistance for those in her service. Thus, in *Tormento* Rosalía de Bringas[46] thinks of the queen as possibly providing a dowry for Amparo to

[43] An article by Pérez de Guzmán (*La Época*, 9 Apr. 1904), quoted by Cambronero (*1b*, 1908: 273), cites a provisional audit, commissioned by Alfonso XII, of the queen's expenditure between 1844 and 1868: 'En estos veinticuatro años, sin contar sus dádivas en trajes, joyas y otros objetos, ni las cantidades esparcidas por su propia mano, había gastado cerca de 100 millones de pesetas en limosnas, pensiones, auxilios de caridad y de protección; hallándose en el número de los agraciados individuos más o menos emparentados con su casa y familia, grandes y títulos arruinados, viejos servidores, monjas, frailes, iglesias y conventos, hospitales y todo género de instituciones benéficas; poetas, novelistas, artistas, periodistas y hombres políticos de todos los partidos . . .'

[44] The *peso duro*, or simply *duro*, was worth about a dollar or five shillings sterling. It was divided into twenty *reales* or five pesetas. (After 1868 the peseta was the official unit of currency, though still little spoken of in ordinary conversation.)

[45] Baroja makes the recipient one of her early lovers: the so-called *Pollo real*, Ruiz de Arena (*4a*, iv. 1148).

[46] Rosalía is linked with the *episodios* by being related to Don Juan Bragas de Pipaón. This is no doubt why in *La de Bringas* Galdós goes out of his way to correct the genealogy of Pez's wife Carolina. She had been previously described, in *La desheredada* (IV: 1032, 1034), as a Pipaón, making her a close relative of Rosalía's. Galdós now converts her instead, most appropriately, into a member of Gloria's fanatical family, the Lantiguas.

enter a convent (a course of action which has little appeal for Amparo herself). 'Es tan caritativa', Rosalía adds, in words similar to those used by Isabella herself in *Narváez*, 'que si estuviera en su mano, todo el dinero de la nación (que no es mucho, no creas) lo emplearía en limosnas.' The same idea of the queen as universal provider is repeated when, having hit on the idea of Agustín Caballero as a benefactor, Rosalía rapidly changes her mind: 'me parece que no necesitamos molestar a la *Señora*, que hartas pretensiones y memoriales recibe cada día, y la pobrecita se aflige por no poder atender a todos' (IV: 1467). In *La de Bringas* Rosalía goes further; when she tells her husband that her new shawl is a gift of the queen's, royal generosity has become a stratagem by which she avoids the consequences of her extravagance. The scene is well built up, with a lively imagined dialogue with the queen, backed up by the narrative voice, to create an authentic impression. *El estilo es mentira* certainly applies here in this Bakhtinian 'parodic stylization':

> 'Ponte ésa, Rosalita . . . ¿Qué tal? Ni pintada.' En efecto, ni con medida estuviera mejor. '¡Qué bien, qué bien! . . . A ver, vuélvete . . . ¿Sabes que me da no sé qué de quitártela? No, no te la quites . . . ' 'Pero, señora, por amor de Dios . . . ' 'No. déjala. Es tuya por derecho de conquista. ¡Es que tienes un cuerpo! Usala en mi nombre, y no se hable más de ello.' De esta manera tan gallarda obsequiaba a sus amigas la graciosa soberana . . . Faltó poco para que a mi buen Thiers [Bringas] se le saltaran las lágrimas oyendo el bien contado relato. (IV: 1590)[47]

And as Refugio indicates when Rosalía turns to her in despair, the queen's absence during the summer at Lequeitio has a disastrous effect: 'Si estuviera aquí la Señora, no pasaría usted esos apurillos, porque con echarse a sus pies y llorarle un poco . . . Dicen que la Señora consuela a todas las amigas que le van con historias y que tienen maridos tacaños o perdularios' (IV: 1663).

Actual examples of royal munificence, dispensed characteristically to worthless figures, are not lacking. One recipient is the mistress of the wardrobe, 'la generala', Doña Tula, who is perennially preoccupied with her offspring. Her late husband, Don Pedro Minio,[48] had enjoyed every possible patronage from the royal family. A purely domestic army officer, 'ganando batallas cortesanas en las antecámaras palatinas' (IV: 582), and a 'nulidad' and a 'bestia condecorada', he was promoted from colonel to general, received

[47] See Bly (*3c*, 1983: 66). The same critic also notes acutely (76) that it cannot have escaped Galdós's notice that the person in charge of the palace administration was the queen's latest lover, Marfori. Bly has two other excellent studies (*3b*, 1980; *3c*, 1981) which throw light on *La de Bringas*.

[48] The origin of the name of the supreme creator of malapropisms may be found in *El amigo Manso*, where Manuel Peña cites, as an example of his ex-girl friend's verbal mistakes, *pedrominio* (no doubt for *predominio*). Galdós seems to have liked the name (and perhaps the word-play), since he later chose the same name for the entirely different title-character of one of his plays. Pedro Minio's career may be based, as is one aspect of Ferdinand VII's favourite the Duke of Alagón, on Sebastián de Miñano's *Cartas del pobrecito holgazán* (1820); see Baquero Goyanes (*3b*; *Memorias de un cortesano de 1815*: I: 1311).

decorations on royal birthdays, and was made a count. His only claim to fame were his malapropisms (*la espada de Demóstenes*,[49] *la tela de Pentecostés*, *el alma de Garibaldi*), so flagrant that they amused a capricious queen not noted for her intellect.

Two other such favourites are earlier incarnations of elderly figures already found in other novels. One is the pretentious and extravagant Doña Cándida de García Grande, who has squandered her late husband's wealth accumulated in the halcyon days of O'Donnell's rule and who, as Galdós reiterates, has, in a curious time-warp, already appeared as an even more decrepit figure (her *segunda manera*) in *El amigo Manso*. She thus fills what Wolfgang Iser terms a 'blank' (*2*, 1974: 279–80; 1978: 182–5) between one work and another. The second is Doña Tula's sister, Milagros, Marchioness of Tellería, who is fighting a valiant battle against increasing years before reaching, eight or ten years later, the 'ocaso en que se nos aparece en la triste historia de su yerno' (IV: 1587), i.e. in *La familia de León Roch*. As befits a member of the parasitic Sánchez Botín tribe, Milagros is insinuating, cunning, and hypocritical. She is Rosalía's role model and her temptress in her spending-sprees after the latter has sampled, almost orgastically, the joys of making the extravagant purchase just mentioned of a fine shawl at a price of 1,700 *reales*.

The one function at which we come near to the queen and her consort in the novel is a set piece, a good example of what Hayden White calls 'emplotment': the dinner served to the poor after the traditional ceremony of washing their feet on Maundy Thursday. The scene is, as Peter Bly has indicated (*3b*, 1980: 22–3; *3c*, 1983: 63), viewed at a distance from a deliberately distorted angle, and through the eyes of the Bringas children, who are watching down the stairwell from the attic skylight; of the dozen men and dozen women participating, only the women are fully visible. In a dream which immediately follows, Isabelita Bringas, impressionable and inclined to epileptic fits, relives a grotesque parody of the ceremony, with the royal part being played by her mother, accompanied by Pez, and her father giving his arm to Doña Milagros; the partial identification discerned by Bly between the queen and Rosalía,[50] each of them adulteresses, is thus re-enacted.

Galdós does not disguise his contempt for this travesty of charity: the grotesque decor and atmosphere of the Sala de Columnas; the conduct of the noisy and impertinent spectators; the arbitrary authority exerted by Doña

[49] Paul Julian Smith (*3b*, 103) incorrectly calls this malapropism 'a slip of the tongue'; it is more than that: it is a sign of ignorance. Smith's dubious Lacanian arguments about the contrast between Demosthenes and Damocles are undermined by the fact that that this error is only the first of three. Torquemada, in his famous speech in *Torquemada en el Purgatorio*, refers to *la espada de Aristóteles*; he immediately justifies himself and proceeds to show the clumsy philistinism he shares with his audience by proclaiming his pride at not knowing anything about Damocles (V: 1100).

[50] Rosalía's chubby Rubens-like looks (IV: 1460) are not dissimilar to Isabella's. See Bly (*3b*, 1980: 8).

Cándida; the bewilderment of the poor women, who hardly eat anything; the haste with which the uneaten food is packed up into baskets to be sold to boarding-houses. There is no argument, declares the narrator, exercising his authorial voice, 'que nos pueda convencer de que esta comedia palaciega tiene nada que ver con el Evangelio'; if he does appear to exonerate to some extent the royal couple from blame, it is no doubt on the basis of one more example of unthinking 'charity', with all the limitations which this implies:[51] 'Formando cadena, las damas y gentiles-hombres los iban pasando hasta las propias manos de los Reyes, quienes los presentaban a los pobres con cierto aire de benevolencia y cortesía, única nota simpática en la farsa de aquel cuadro teatral' (IV: 1585).

In these rapidly sketched incidents the queen appears obliquely as an unconscious participant in a routine farce. Farce is of course the essential quality of Valle-Inclán's portrayal of the queen, first in the play *Farsa y licencia de la reina castiza* (1922) and later in *El ruedo ibérico*. Her nickname, *la reina castiza*, stresses, not too flatteringly, her popular traits, but other unbecoming characteristics are also emphasized: vulgarity, obesity, gluttony, erythema, proneness to superstition, as well as promiscuity (*4a*, Sinclair 28–55). As Valle-Inclán pithily put it, 'La de los Tristes Destinos fue por muchos años Ninfa de los Cuarteles' (*4a*, *Baza* 195).

A second characteristic of the queen's is potentially more favourable: the identification with the *pueblo*, so frequently stressed, which constitutes the basis of her popularity. It is a consistent part of her personality, as is indicated in Fajardo's first physical impression of her, sympathetically and familiarly expressed, but with a hint of polarized attitudes: 'No hay duda que ella ha sabido crearse una realeza suya en perfecta armonía con sus azules ojos picarescos y con su nariz respingada, realeza que toca por un extremo con la dignidad atávica y por otro con no sé qué desgaire plebeyo, todo gracejo y donosura.' She is, he concludes: 'la síntesis del españolismo y el producto de las más brillantes épocas históricas' (II: 1604–5). The irony of the last phrase is clear: Spanish qualities in the nineteenth century are a long way from the brilliance of Spain's past; Isabella's *españolismo* can only be a modern degeneration, *agarbanzado*, to use a favourite term.

The early part of her reign is summed up very cogently by an acute outside observer, the travel writer of Italian origin Antonio C. N. Gallenga, who draws attention to much the same facets as Galdós:

Queen Isabella—heiress of all the sins of her parents, sacrificed in all her natural instincts by a policy which was as short-sighted and stolid as it was execrable, united to a husband who was physically, intellectually, and morally the most contemptible

[51] *The Attaché* (*1b*, 235–41) describes the ceremony (which is taken very seriously for 'its touching and impressive effect', 239) in 1854. According to Bly (*3b*, 1986*b*: 60), it is anachronistic to set the ceremony in the Sala de Columnas in 1868. Gallenga (*1b*, ii. 136–46) has an account of its revised form under Alfonso XII, which he too calls a 'farce'.

of human beings—had, by her sufferings, and also by some undeniably generous impulses, gained the sympathies of the Spanish people, so as to find them indulgent to her juvenile indiscretions. She was, her subjects said, *muy Española*, that is, very patriotic; she wore a mantilla, despised French bonnets and the French nation, was punctual at mass and the bull-fight, fond of pleasure, fond of display, and so *bondadosa* or charitable—indeed, so lavish in her charity, that she never looked whether it was gold, silver, or copper that she flung from her carriage-window among the poor. She had pluck, she had pride, she was good-natured and forgiving; and, for the political blunders of her early years, her subjects were willing to throw the blame on her advisers, and especially on the influence of her mother, Queen Christina, whom they called *La Mala* (the evil one), while on Isabella herself they bestowed the half-endearing epithet of *La Tonta* (the fool). (*Ib*, i. 18–19)

The year 1848 was 'the year of revolutions' throughout Europe. In Spain, it brought no really major upheaval, only even more severe preventative action.[52] When the next year a political crisis of some importance occurs, it is a matter of palace intrigue, not of revolution. The famous *Ministerio Relámpago* is the first incident—typical of many to come—in which the queen, still only 19, is clearly responsible, by commission or omission, for arbitrary political upheavals. In his 1902 interview Galdós, still anxious to learn more about the occurrence, found the ex-queen deliberately reticent: 'Cuatro palabritas acerca del *Ministerio Relámpago* habrían sido el más rico manjar de aquel festín de historia viva; pero no se presentó la narradora en este singular caso tan bien dispuesta a la confianza como en otros' (VI: 1191).[53]

At a time when the powerful Narváez government had just consolidated its position by ruthless oppression and by expelling the interfering British minister Bulwer, Narváez is suddenly replaced by the ultra-traditionalist Count of Cleonard. The event is witnessed through Beramendi's eyes, since he has now become a confidant of Narváez's, as Calpena had been of Mendizábal's in the third series. The sudden change is depicted as the first success of the sinister behind-the-scenes operations of the fictional ultramontane Socobios and the real clerical *camarilla* of Father Fulgencio[54] and Sister Patrocinio.

The inner workings of Patrocinio's convent are revealed in greater detail in

[52] A typical measure was the extended use of the State of Exception, in which constitutional guarantees were suspended. Catalonia, as Prim emphatically pointed out (27 Nov. 1851), was mostly under a State of Exception for eight years. (See *Ib*, Ballbé 159–70; quoted 164.)

[53] In the interview the ex-queen, 'Más generosa que sincera' (VI: 1191), according to Galdós, also defended Sister Patrocinio's actions at this time. Répide comments sardonically that Galdós 'Salió de la entrevista sin averiguar lo que quería . . . salió encantado de la visita . . . cuando tenía que haberse marchado molesto porque un hombre como él hubiese sufrido una especie de toreo o de tomadura de pelo de pura marca borbonesca' (*Ib*, 282–3).

[54] Fulgencio López belonged to the Order of Pious Schools founded by St Joseph of Calasanz; its members were known as Piarists or *escolapios*. Pius IX was educated at a Piarist college at Volterra. Father Fulgencio was the king-consort's confessor and introduced Sister Patrocinio to the queen. In *Narváez* María Ignacia describes him, during a visit to the Emperáns, in a backhanded compliment, as much more urbane and less coarse in manner than is usually thought (II: 1570).

a subsequent *episodio*, *Los duendes de la camarilla*. The ex-nun Domiciana Paredes gives a lively account, confused in chronology but clear in import, of Sister Patrocinio's impact in the old convent of the Caballero de Gracia. She describes the superstitious atmosphere of miracles and demons and Patrocinio's claim to have received the stigmata. Until the Peace of Vergara (1839) the nuns' sympathies were entirely with the Carlists, but once installed in La Latina convent, Sister Patrocinio (*La Madre*) turns towards the Isabelline court and cultivates the infante Don Francisco, the king's father, the king-consort himself, and, through them, the young queen. Domiciana bears witness to the dazzling if disquietening fascination exercised by Patrocinio at the same time as she reveals her own very disturbed personality. The perversities produced by conventual enclosure (an archetypal example is Mauricia *la dura* in Las Micaelas in *Fortunata y Jacinta*) are evident. Domiciana has already advised Lucila to seek *La Madre's* help in saving her revolutionary lover Bartolomé Gracián, at the cost of sacrificing herself by submitting to religious authority. It later emerges that Lucila has been a messenger between the palace and Sor Catalina, and hence Sor Patrocinio (II: 1722), during the *Relámpago* incident.

The historical point that this tale of partly fictional intrigue underscores is that from quite early in her reign the queen was incapable of resisting the *camarilla*,[55] which found even Narváez too liberal for its taste. This debility has the consequence politically of reducing Spain to the level of a serial novel. As Isabella's mother María Cristina—hardly a model of responsibility herself—put it: 'La Historia de España se nos está volviendo folletín', and a bad and foolish one at that (*Narváez* II: 1615). In a similar vein, Narváez is quoted as referring to Spanish politics at this juncture as a 'sainete' (*1b*, Ballesteros Baretta 27) The image of the serial novel or the popular farce in turn relates to the literary activities, with their unjustified flights of fancy, of Alejandro Miquis and José Ido del Sagrario. (See *3b*, Ynduráin; *3b*, Andreu; *3c*, Andreu 1982.)

The conspiracy is excellently conveyed in fictional terms through the general's frank conversations with Beramendi. Narváez is sharply portrayed as a tough, irascible individual with some resemblance to a rooster. He brooks contradiction only from his batman *Borrego* and has the odd habit of addressing anyone around him as *pollo*.[56] He is particularly incensed by Eufrasia, now married to Saturno Socobio, whom he has no compunction in referring to insultingly as the Marqués de Capricornio. Eufrasia, for her part, has emerged

[55] As Eufrasia Carrasco confides to Fajardo, 'Para mí, se apoderaron de un secreto de la Reina, y con ese secreto, cogido como un puñal, la han amenazado' (II: 1624). Llorca (*1b*, 1984: 117–18) attributes responsibility for the incident to her favourite Bedmar.

[56] Payne quotes Balmes as saying: 'What frustrates his work is principally his lack of political thought . . . His instincts, his ideas, his sentiments, and his interests are in perpetual conflict' (*1b*, 25). See also Journeau (*3d*, 1990).

as an arch-conspirer and a venomous adversary: 'una víbora' (II: 1580). She puts forward an alternative view of Narváez, as dominated by envious rivalry towards Espartero, and as a rebel with no respect for authority posing as the champion of law and order (1579). Thus, though he is returned to power within twenty-four hours—hence the nickname given the ministry (according to the *episodio*, by Eufrasia herself)—Narváez has reason at an early stage to feel bitterly disillusioned and to revise his first estimate of the young queen as fundamentally good (1599, 1618).[57] Behaving with a caution he never showed towards any other type of opposition, Narváez manages, with considerable difficulty, to secure the expulsion of Sister Patrocinio, whose departure, with Beramendi's sister Catalina, is witnessed by the marquis himself at the end of the volume: history and fiction remain inextricably intertwined.[58]

One notable incident marks the possibility the young queen still has of capitalizing on the people's affection. This is the attempt on her life on 2 February 1852 by Father Martín Merino, which among other things gave a great boost to the queen's popularity. As Beramendi comments, 'He creído siempre que el pueblo español ama verdaderamente a su Reina. Pero hasta hoy . . . no había yo visto clara la exaltación de ese cariño, que raya en idolatría' (III: 11). It represented one more lost opportunity for her to take advantage of this dwindling asset, as Fajardo acknowledges when he utters to himself an unequivocal warning, very appropriate historically but far too explicit for fictional credibility:

> Sábelo, Isabel; hazte cargo de que este sentimiento lo tienes por ti sola, no por tu padre, que se pasó la vida haciendo lo posible para que le aborreciéramos, ni por tu madre, más admirada que amada; acoge en tu corazón este sentimiento y devuélvelo, como un fiel espejo devuelve la imagen que recibe . . . Esto no es cosa de juego . . . Mira lo que tienes, mira lo que haces y mira con lo que juegas. (III: 11)

Isabella does not possess the intellectual maturity to carry out without adequate guidance secret injunctions like these; like the *pueblo*, in Galdós's view, she is incorrigibly naïve. Eufrasia Carrasco puts it more crudely: 'Es bondadosa, es generosa, pero se diría que nació y la criaron en la calle de Embajadores. Tiene todas las supersticiones de la mujer del pueblo' (III: 578).[59] This is

[57] According to a letter from León y Castillo (*3a*, Casa-Museo), when he was arranging for his old friend to visit the ex-queen, she expressed herself as delighted ('encantada') with what the novelist said about her in *Narváez*. She sent him a signed photograph in Jan. 1903. See Bravo-Villasante (*3b*, 230).

[58] Baroja recounts the expulsion in *Desde el principio hasta el fin*, the last of the *Memorias de un hombre de acción* (*4a*, iv. 1156–7); it is one of the examples Ortiz Armengol gives of Baroja's deliberate effort to emulate Galdós (*4a*, 5–34; cf. *4a*, Longhurst 83). In other cases it is evident that Baroja is concerned, first, to rectify Galdós's low opinion of Aviraneta ('colosal genio de la intriga y un histrión inimitable': II: 316), second, to confront his senior rival's concept of Juan Martín *el empecinado* as 'un pobre patán . . . que no sabe hablar' (IV: 316), and, third, to cover Cabrera's campaign in El Maestrazgo more thoroughly and with greater vehemence against the Carlist general than Galdós had done.

[59] She retained her popularity, however, well into the 1860s. As Carr says (*1b*, 212), 'Like her father she understood the populace and failed to understand respectable politicians'.

confirmed by radical popular literature of the time, which, as Alison Sinclair has shown (*4a*, 28–31), treats her as essentially vulgar in mind and attitude.

To return to Merino's assassination attempt. Not only is it mentioned by both Estupiñá and Isabel Cordero, but it is given quite remarkable coverage in the *episodios*. Merino first appears in *Las tormentas del 48* as the ill-mannered confessor of Fajardo's dying lover Antoñita (II: 1489–90). Then, in *Los duendes de la camarilla*, his life is interwoven into the fictional existence of the Ansúrez family; he is the confessor of Domiciana Paredes, who presents him with the knife Lucila intended to use against her; he fulminates against Narváez; and he performs the marriage ceremony of Lucila and Halconero just before assaulting the queen with that same knife. The historical incident is thus paralleled by the fictional one, and Domiciana's individual treachery is bound up with Merino's crime; they are both the products of the distortion of personality resulting from a perverse religious training.

The attack is viewed from a distinct angle of vision in the first pages of *La revolución de julio*.[60] It is a fine example, as Cardona explains (*3b*), of *historia viva*. Fajardo's brother Agustín, like Estupiñá an 'hermano de la Paz y Caridad',[61] visited Merino in his cell, where he was scandalized by the prisoner's cynicism. Through Fajardo's eyes, we witness another example of an abnormal character led to violence by frustration, caused in this case by something contemporary Spaniards could in part share: his 'aborrecimiento de toda injusticia y del mal gobierno de la Nación' (III: 12). As the governor Ordoñez says, how much better it would have been if Merino had carried out his original intention of killing Ferdinand VII: another intriguing alternative history. Beramendi also expresses his disgust at the stagnation and intransigence of the Church as we see Merino's astonishing composure[62] during his imprisonment, defrocking, and execution, emphatically described as 'la vileza y procederes bajunos del brazo secular' and 'el bárbaro formalismo del brazo eclesiástico' (18).

As for the *pueblo*, it is interesting to compare Beramendi's view with the hostile attitude found in *La Fontana de Oro* and with the patronizing concept held by Juanito Santa Cruz and those around him. Beramendi, overcoming the earlier open hostility, makes the same basic judgement as Juanito, but as a result of his characteristic guilt complex about his parasitic life sees an active and irresistible component in the people's make-up. It is not only that Galdós is more sympathetic to the people in 1902 than he was in 1870 or 1886, but

[60] In the 1902 interview the ex-queen conveyed to Galdós an 'Impresión más de sorpresa que de espanto, y su inconsciencia de la trágica escena por el desvanecimiento que sufrió, efecto, más que de la herida, del griterío que estalló en torno suyo y del terror de los cortesanos' (VI: 1191).

[61] A charitable religious organization which visited and 'comforted' condemned prisoners and accompanied them to the scaffold. Galdós frequently uses the institution, treated with implicit irony, as a means of allowing his characters close access to these events. Thus Serafín de Socobio, also a member, attended Diego de León at his execution (II: 1221).

[62] Baroja admired him for his stoicism; see 'El alma estoica de don Martín Merino' (*4a*, v. 1160–4).

that Beramendi is portrayed as a far more sensitive character than Juanito. The narrator of *Fortunata y Jacinta* pontificates: 'el pueblo, en nuestras sociedades, conserva las ideas y los sentimientos elementales en su tosca plenitud, como la cantera contiene el mármol, materia de la forma. El pueblo posee las verdades grandes y en bloque' (V: III. 7. iii, 407).[63] Beramendi, by contrast, is much more positive: 'Ellos me parecen materia viva, aunque tosca; yo, materia inerte, ociosamente refinada. Ellos marchan . . . son elemento activo . . . Para consolarme de la envidia que me punza el corazón, pienso en la barbarie de ellos . . . pero esto no me vale' (*La revolución de julio* **III**: 108). Earlier, in the same work, he had exclaimed:

¡Infeliz pueblo! Por una noche, por algunas horas no más, le permiten los dioses el uso práctico de su soberanía, de esa realeza ideal que sólo existe en las vanas retóricas de algún tratadista vesánico. *Y en su candidez, en la inexperiencia de su soberanía, es el pueblo como un niño al que entregan un juguete de mecanismo delicado y sutil.* (**III**: 90; my italics)

The italicized sentence could and should be applied to the still youthful Isabella. She cannot utilize her popular resources effectively and her childlike whims constantly undermine the responsibilities she has been entrusted with. The incident is in many ways typical of the qualities of the *episodios*: it is well integrated, but the historical point is a bit too obvious.

Galdós takes great pains to show Isabella from a variety of sources and from a variety of angles at diverse moments of her reign. We see and hear her in imagined conversation, particularly with Beramendi, and listen to judgements of her made by Serrano and Narváez, as well as Eufrasia, among others. From these testimonies it is quickly apparent that Isabella does not learn from experience or—in fiction—from Beramendi's opportune warnings. Youth and inexperience soon cease to provide excuses for her whims. The hidden motif-force behind these sudden changes of direction is often the influence of her lovers. Such capriciousness, moreover, is not exempt from the deceit, cunning, and cruelty associated with her father. By 1854 we overhear the thoughts of General Francisco Serrano,[64] who as her first lover has had ample opportunity for observing Isabella's behaviour:

Conocía muy bien . . . la veleidosa condición de la Reina, sus sarcasmos y disimulos, heredados de Fernando VII y sus preferencias por la política moderada; conocía también, y mejor que nadie, la flaqueza del corazón de Isabel ante las taimadas sugestiones de una beata embaucadora; sabía qué fácilmente se ganaba la Real voluntad, no siendo en aquel nebuloso terreno. (**III**: 150)

The queen's mental immaturity leads to the constant inconsequentiality of her conversation and her actions, demonstrated already in her first discussion with

[63] For the concept of the *pueblo* in *Fortunata y Jacinta* see Sinnigen (*3c*, 1974), Rodríguez-Puértolas (*3c*, 1975), and Ribbans (*3c*, 1977*a*: 67–9).

[64] Francisco Serrano, 1810–85. More amiable, more opportunistic, and with a less defined political direction than his fellow generals, he secured the regency for himself on two occasions.

Beramendi;[65] to her simple belief that she enjoys God's grace and that she is loved by the people; to the extreme ease with which she is influenced; and to her notorious favouritism. The fact is that, as Galdós noted in his 1885 article, she never grows up to a sense of political responsibility. As he asked himself at the time of the interviews with the elderly ex-queen: '¿Verdad, Señora, que en la mente de Vuestra Majestad no entró jamás la idea del Estado? . . . ¿Verdad que le criaron a Vuestra Majestad en la persuasión de que hacer podía cuanto se le antojara, y quitar y poner gobernantes como si cambiase de ropa?' (VI: 1192).

The vulnerability produced by the queen's promiscuous behaviour[66] continues throughout her reign: 'Isabel podía desechar el temor del Infierno por sus personales culpas; pero no por el pecado de consentir que su Pueblo cayese en los abismos del descreimiento y la corrupción masónica' (III: 150). She is the victim, therefore, of what Montesinos (*3c*, iii. 157–60) aptly calls 'el chantaje político' at the hands of her spiritual advisers.[67] Her weaknesses will provide Valle-Inclán with scope for particularly derisory satire. As Sinclair has demonstrated (*4a*, 40–8), he makes discreet use of the scurrilous popular periodical literature of the time. In particular, he draws attention to the notorious papal bull, allegedly negotiated by her confessor Father Claret,[68] which, it is claimed, granted her indulgence for her carnal sins; and he pokes fun, at the beginning of *La corte de los milagros*, at the award of the Golden Rose of virtue to her by the Pope in February 1868.[69]

The most significant of the revolutions which preceded the *gloriosa* of 1868 was the July revolution of 1854, a sort of delayed Spanish contribution to the revolutionary fervour of 1848. The uprising came after the activist authoritarian ministry of Bravo Murillo (1851–2) had been followed by the unprincipled Conservative government led by Luís José Sartorius, Count of San Luis.[70] It

[65] Urey (*3d*, 1989: 109) draws attention to how in her conversation 'she skips from subject to subject like a "pájaro" or a "mariposa"' (II: 1609, 1610). Note the frequency with which words like *antojo* and *veleidosa* are used to describe her actions.

[66] This behaviour had at least the beneficial result of securing the succession to the throne, which would otherwise have passed to Montpensier's line. Of Isabella's numerous offspring, four (Isabel, Alfonso, Paz, and Eulalia) survived to maturity. Beramendi's son Tinito is portrayed as playing with the first three and with Pilar, who died young (*La de los tristes destinos* III: 693).

[67] As V. G. Kiernan remarks, 'a royal sinner who repents on Sundays makes an ideal subject for clerical manipulation' (*1b*, 33).

[68] Antonio María Claret y Clará, 1807–70. He was the author of numerous pietistic tracts in Catalan and Spanish and founded the Order of the Claretians in Vic in 1849. After serving as Archbishop of Santiago de Cuba, he was recalled to Spain in 1859, as titular Archbishop of Trajanopolis, to be the queen's confessor. He left Spain at the recognition of Italy in 1865, but on the Pope's instructions returned shortly after. He accompanied Isabella II into exile. He was canonized in 1950.

[69] Benedetto Croce comments, 'amid universal derision Pius IX had sent the Golden Rose to this queen who was proverbial for her immodesty' (*1a*, 1934: 234).

[70] 1817–71. He was, in Kiernan's words, 'a parvenu, a southerner who had risen by journalism and served as Minister of the Interior, an obediently unscrupulous one, under

was aimed particularly against the corrupt capitalist ventures of María Cristina and Salamanca; the revolution even put Isabella's crown in jeopardy. Three elements were involved in it: first, O'Donnell's bid for power with his indecisive rising at Vicálvaro, witnessed by Fajardo (III: 75), which was followed by the mildly liberal *Manifiesto de Manzanares*, written by Cánovas; second, Espartero's immense prestige; and, third, a genuinely popular uprising lasting three days.

The three appearances on a balcony made by Espartero and O'Donnell[71] in Estupiñá's list correspond essentially to the distinct phases of development of this revolution, which is presented in considerable detail in *O'Donnell*. The first appearance signals the short-lived collaboration between the two generals, Espartero the vain and bombastic 'caudillo de los patriotas', and the cautious and wily O'Donnell: it is 'el célebre abrazo en el balcón' (V: 124). The second indicates Espartero's overwhelming popularity, a popularity he was unable or unwilling to utilize to his political advantage.[72] In the *episodio* the efforts of the historical Calvo Asensio and the fictitious Centurión prove powerless to persuade the leader to take a firm stand (146). The third appearance marks O'Donnell's success in ousting the inexplicably passive Espartero in July 1856 and quelling with a considerable show of force the militia on which the Progressives depended. This brings the active phase of the revolution to an end (*1b*, Kiernan 218–27).

The popular rising brought about one of the most spectacular and pathetic incidents of mob violence: the assault on the house of Francisco (García) Chico, the hated chief of police in Narváez's authoritarian government, and his subsequent murder. It is sufficiently important to be included, as we have seen, in Estupiñá's repertoire of balcony scenes. Moreover, the episode provides the occasion for a direct comparison between Galdós's account at the beginning of *O'Donnell* and the description given by Pío Baroja in *El sabor de la venganza*, part of his 22-volume series of historical novels on Eugenio de Aviraneta, the *Memorias de un hombre de acción* (1913–34).[73] The incident is also related rapidly in Pereda's *Pedro Sánchez* (1883).[74]

Narváez' (*1b*, 38). Beramendi refers to 'la política mohosa, rutinaria' (III: 59) of his so-called 'polacos'.

[71] Leopoldo O'Donnell, 1809–67, Count of Lucena and Duke of Tetuan. Gallenga describes him in 1865 as follows: 'tall and massive, cold, silent, unsympathetic, square-faced, with slightly prominent jaws, white thin hair on lips and chin, bald . . . a conspicuous figure, overtopping by a whole head the minor mortals around him' (*1b*, i. 59).

[72] He is described as uttering 'conceptos de una oquedad retumbante' (III: 124); his empty and ambiguous phrase during this crisis, '¡Cúmplese la voluntad nacional!' (1854), used as the subtitle of Antonio Espina's popular biography, is quoted with disgust by Galdós. Baroja is particularly hostile towards Espartero (*4a*, iv. 707).

[73] It is also described in Baroja's biography *Aviraneta, o la vida de un conspirador* (*4a*, iv. 1327–31).

[74] The three descriptions are very similar and probably depend on a common source, possibly oral transmission. Julio Nombela's contemporary description, in *Impresiones y recuerdos* (*1b*), is mentioned in Baroja's later article on Chico ('La vida de Chico' [1934]: *4b*, v. 1216–20).

Pereda's novel has an autobiographical form, and the events of the *revolución de julio*, in which the protagonist is depicted as playing a prominent part, are described from the perspective of twenty-five years later:[75]

Así murió el famoso don Francisco Chico. Un día se presentó la turbamulta en su casa; le arrancó de la cama en que yacía postrado; le sentó medio desnudo en unas angarillas; cogió después al portero que le servía; echóle a andar junto a su amo; y en ruidosa procesión, calle de Toledo abajo, llegó todo junto, entre oleadas de curiosos y de furias, hasta el último tercio de ella; y allí, a las diez de la mañana, arrimados los reos a una pared, con angarillas y todo . . . ¡cataplum! Esta era ya la tercera justicia que hacían aquellas bondadosísimas gentes. Bajó San Miguel allá; echóles un trepe rudo entre algunos piropos indispensables, y le prometieron la enmienda; pero no se enmendaron cosa mayor. (*4a*, ii. 120)

His account is the most factual and sober, though, as Germán Gullón (*4a*, 82) indicates, his horror at popular justice comes through clearly in the vocabulary chosen ('turbamulta', 'le arrancó', 'furias'); 'bondadosísimas gentes' also has a strongly ironic ring. Pereda comments elsewhere on the rabble which emerges at any crisis and the ease with which riotous conduct can be aroused.

In the previous volume, *Los duendes de la camarilla*, Galdós has prepared the ground by various references to Chico's reputation and appearance. Jerónimo Ansúrez has worked in his house looking after his famous art treasures, but could not stand the hatred Chico aroused among the populace (II: 1691; 1707). After forty years of police service, he has come to be a symbol: 'Era hombre terrible, de sagaz inteligencia para tan ingrato servicio, y a los poderosos inspiraba confianza, como a los débiles espanto. Llegó a ser al modo de institución' (1729). Later, in *La revolución de julio*, Beramendi seeks Chico's assistance in finding Virginia; the police chief there defines himself as '*el testigo presencial* de la Historia de España' (III: 33); he seems to be a figure left over from the eighteenth century, reminiscent of *Coletilla* from *La Fontana de Oro*. His subsequent murder thus takes on a further dimension as a break with an earlier, more arbitrary and authoritarian dispensation.

Galdós's presentation is a judicious blending of 'showing' and 'telling' or 'scene' and 'summary' (*2*, Lubbock 267–72; *2*, Chatman 68–79). As witnesses to the events he provides two familiar characters, Centurión and Telesforo del Portillo (*Sebo*), together with the Hermosilla sisters, known as *las Zorreras*,[76] who are streetwalkers with Progressive leanings: *la historia chica*

[75] In a letter to Pereda of 13 Mar. 1884 (*3a*, '28 Cartas' 33–4) Galdós expresses his unqualified admiration for the novel, and in an article for *La Prensa* (21 Jan. 1888: *3a*, *Prólogos* 300–6) he says, '*La revolución en las calles*, las escenas y altercados en los clubs, la vida periodística, son cuadros de perfecta verdad y hechura' (304; my italics).

[76] Rafaela and Generosa, daughters of a fur-manufacturer (hence the nickname) who took part in the 1854 revolution. Both girls were seduced and abandoned by the fervent conspirator and revolutionary Don Juan Bartolomé Gracián, who is killed by Beramendi with Hermosilla's

casts light on *la historia grande*. The *Zorreras*' bring news of the disturbance, the dialogue lending the scene a spark of immediacy, alternating with passages of summary:

—Véanle, véanle —dijeron—. Desde la plazuela de los Mostenses lo *train*... El *Chico* es el que viene en andas, *y el Cano*, a pie... Que los *afusilen*, que les den garrote... que paguen lo que han hecho.

The real-life popular leader, a bullfighter nicknamed *Pucheta*,[77] who led the revolt, is the lover of one of the *Zorreras*, Generosa. The other, Rafaela, subsequently gives an account in racy popular language of the decision taken by the revolutionaries to kill him, and of the assault on Chico's house. *Pucheta* (whose real name was Muñoz) is thus linked with these picturesque fictional characters. The elderly Chico is dragged half-dressed from his sickbed through the streets seated incongruously on a mattress and pushed along on hand-barrows, as in the other accounts. The whole scene has a grotesque quality reminiscent of Goya; the spontaneous procession created by mob violence includes a large oil portrait carried on one pole, together with a plucked chicken on another. In describing Chico himself, Galdós emphasizes the oxymoronic contrasts: the baseness of the rabble, the grotesqueness of the scene, the serene indifference of the victim:

Seguían las angarillas cargadas por cuatro, de lo más soez entre tan soez patulea; las angarillas sostenían un colchón, en el cual iba el infeliz Chico, sentado, de medio cuerpo abajo cubierto con las propias sábanas de su cama; de medio cuerpo arriba, con un camisón blanco; en la cabeza, un gorro colorado puntiagudo, que le daba aspecto de figura burlesca. Con un abanico se daba aire, pasándose a menudo de una mano a otra, y miraba con rostro sereno a la multitud que le escarnecía, al gentío que en balcones y puertas se asomaba curioso y espantado. (III: 118–19)

To accentuate the frantic aspect of the scene, Chico's mistress runs crazily alongside the improvised tumbrel, still absurdly preparing his morning chocolate and frantically beseeching the mob not to kill him; his henchman *El Cano* is hustled along to his execution by the side of his chief.

Baroja's brief description appears to draw on Galdós for some of the details (the white shirt, the red cap, the action of fanning himself), but, as we would expect from its author, is vivid and direct, with emphasis on physical and sartorial details:

Luego, como un 'paso' de Semana Santa, sentado en un colchón y sostenido en unas parihuelas, apareció en la plaza de Santo Domingo un hombre flaco, amarillo, ictérico, como una momia, ya viejo, con patillas grises.

Iba medio desnudo, cubierto con una camisa blanca y un pañuelo en el cuello, un

assistance (III: 114). Five years later, in 1859, Generosa becomes the mistress of Facundo Risueño, displacing Teresa Villaescusa (*O'Donnell* III: 215). Beramendi and Aransis are, in Galdós's discreetly promiscuous world, occasional clients of the two sisters (III: 583).

[77] *Pucheta* died in the suppression of Espartero's militia in 1856 (*1b*, Kiernan 261).

gorro de color en la cabeza y en la mano un abanico, con el que se abanicaba tranquilamente. Su expresión era fosca, amarga y casi burlona. (*4a*, iii. 1155)

All three accounts clearly reveal fear and hatred of popular violence. Pereda's protagonist refrains from all commentary, except to express some doubt about the ability of the veteran revolutionary general Evaristo San Miguel, appointed Captain-General of Madrid, to control the tumult; Galdós has San Miguel pass majestically by 'con su séquito de militares y patriotas, a trote largo calle abajo' (III: 123) after the shooting. '—A buenas horas, mangas verdes—', Centurión comments cynically, employing one of Galdós's favourite proverbs.

The episode reveals the very different concepts of history held by Galdós and Baroja. Baroja is resolutely opposed to a purely scientific view, claiming that subjectivity plays an essential part in historical judgement and that complete objectivity is impossible ('La historia': *4a*, v. 1124–8). The incompatibility between history and fiction that Manzoni found insoluble, and that Galdós settled pragmatically, was simply not a problem for Baroja. To cite from the Introduction the terms used by Walsh, he not only saw no meaning *of* history, but no meaning *in* history. He disbelieves to a remarkable extent in any causal relationship between events, adopting a purely casual criterion which concedes supreme importance to chance: 'La historia para Baroja es una sucesión de acontecimientos al azar, sin orden ni concierto' (*4a*, Longhurst 143). The issue, essentially, is a personal one. Thus Baroja attributes the murder almost exclusively to the enmity of one Brigadier Castelo, who instigates *Pucheta*'s rousing of the people. He is not concerned with Chico's own record, but takes his execution as a fine example of stoic equanimity in the face of death.[78] His only comment on the motivation of the uprising speaks of 'ese sentimiento simplista de las multitudes', who imagine that by getting rid of Chico they will get rid of injustice. Nor is he preoccupied by the grave questions of political morality involved in the queen mother María Cristina's dubious speculative enterprises. His concern is with individual biography, treated in an informal and unsystematic fashion, without any integration into a broader framework; in this respect, he is not dissimilar to Mérimée, as seen by Lukács. This scene is witnessed entirely through the eyes of Aviraneta, María Cristina's unconditional supporter; and the adversaries *Pucheta* and the queen's morganatic husband are linked on the quite arbitrary grounds that they share the same surname Muñoz: 'Así teníamos un Muñoz arriba (el marido de Cristina) y otro Muñoz abajo (*Pucheta*). La revolución del 54 era un conflicto entre dos Muñoces.'[79]

[78] Compare his article mentioned above (n. 74), where he emphasizes Chico's unusual individual qualities as a 'pajarraco raro y extraordinario' (*4a*, v. 1220). He also has an article on *Pucheta*: 'El torero *Pucheta*' (1933: *4a*, v. 1838–42).

[79] It is worth noting that, ironically echoing the queen mother's consort, the many fictitious characters called Muñoz—Eulalia Muñoz, the marquis of Casa-Muñoz, etc.—are often *nouveaux*

Galdós connects the killing much more closely with the social and political circumstances. We witness the uncontrolled fury of the mob, reflected in the picturesque language of the two *Zorreras*, who none the less give voice to the people's legitimate grievances against the repressive regimes of Narváez and Sartorius and their creature Chico; the vain attempts made by Centurión, now the rather obvious representative of reason and moderation, to calm the situation; and we see the fear, like Aviraneta's, of the former police agent *Sebo*, of being caught up in the frenzy should his identity be discovered. Montesinos (*3c*, iii. 239) remarks, quite rightly, on 'el sabor extraordinario' and 'la mezcla de ordinariez y buen sentido' of this episode. It is one of Galdós's most successful: a historical incident of some importance effectively 'embedded' within a fictitious but authentic social context which allows issues of repression, popular fury, and attempts at temporizing to be ventilated.[80]

The revolution has repercussions too for the fictitious characters of the *episodio*. Beramendi, making a modest contribution to freedom and tolerance by seeking out *Mita* and *Ley*, gets caught up in the uprising, and commits the one really decisive action of his life. He confronts Bartolomé Gracián when he tries to renew his affair with Lucila, and ends up by killing him. Gracián, for his part, is a typical case of a rebel in both the public and private spheres who delights for its own sake and without scruple in the adventure involved, whether it is a challenge to the government or overcoming a woman's resistance to his advances. He has much in common with the later Rodríguez Leal in *Prim*.

An arbitrary intervention by the queen brings what little is left of the revolution[81] to an abrupt end in 1856. O'Donnell's sudden fall, signalled by a capricious breaking of protocol at a state ball on the queen's birthday, is the result of her opposition, urged by the clerical faction, to the key issue of *desamortización*. The important measure of disentailing the church lands had been commenced years before by Juan Álvarez Mendizábal. It was resumed once more, with common lands added to ecclesiastical property, by Pascual Madoz under Espartero and continued under O'Donnell's 'long ministry'. The results were not, as intended, to broaden distribution of the land, but to introduce widespread speculation and, probably, to increase still further the size of the already huge estates, the *latifundios*.

riches; in *El caballero encantado* the tendency of ennobled plebeians to add 'un Casa como una casa a su primer apellido' (VI: 81) is noted. The *Miaus* Milagros and Pura are the daughters of a Muñoz, vaguely related to the Marquis of Casa-Muñoz; though far from being *nouveaux riches*, they share the same aspirations. The saviour of the Real Palace in 1868 is Casimiro Muñoz, just at the time when Santiago Ibero suspects that Teresa has taken as a lover one of the 'Muñoces de Tarancón', i.e. relatives of Riánsares.

[80] The episode may be compared with Scott's treatment of the Porteous riots in *The Heart of Midlothian* (1818), in which both the provocation for the murder/execution of Porteous and the restraint shown in carrying out the lawless act are skilfully delineated.

[81] It is dismissed by Beramendi, in his 'juicio sintético' on the last page of *La revolución de julio* (III: 115), as 'pobre y casera'.

Mendizábal[82] is a statesman whose influence is felt throughout the series. On his funeral in November 1853 Beramendi speaks of the 'tributo tardío y menguado a un hombre que estimo como de los más altos de nuestro siglo por las ideas grandes y la voluntad poderosa' (III: 25). Ironic echoes of his personality occur in the novels. The reactionary porter with the heart of gold and the looks of a gorilla in *Miau* is, in Galdós's puckish way, named after him. It is at the foot of Mendizábal's statue in the Plaza del Progreso that Benina and Almudena have their first discussion in *Misericordia*: 'aquel verdinegro señor de bronce que ella no sabía quién era ni por qué le habían puesto allí' (V: 1887). Bly interprets this as summing up 'Galdós's anguished, almost total anti-historicism at this stage in his career' (*3c*, 1983: 173). I see it rather as an example of how progress is unconsciously assimilated and continued on a personal plane.

The hegemony of the Conservative oligarchy has asserted itself once more, with renewed oppression. It is not long, however, before the queen, with the same arbitrariness as before, grows tired of Narváez.[83] Two short-lived and ineffective governments follow. They are reflected in Beramendi's state of mind; it seemed that 'el alma de la Nación, como la de él, sufría un acceso de pesada somnolencia' (III: 188). Not long afterwards the queen calls on O'Donnell, late at night, to take office again. Galdós gives the scene a cosy domestic setting which corresponds to the domestically inclined leader.[84] O'Donnell's 'long ministry' is thus ushered into existence. This will be the subject of the next chapter.

[82] A sympathetic portrayal of him as a prudent but frustrated reformer is given in the *episodio* which bears his name.

[83] Both Répide (*1b*, 191) and Llorca (*1b*, 1984: 171) attribute his fall to a dispute over her current favourite Puig Moltó. It occurred shortly after a mysterious confrontation between Narváez and the king-consort in which Narváez's aide, the Marquis of los Arenales, and General Uriztondo, Don Francisco's assistant, were killed. (Répide 189–90; Llorca 167–70.)

[84] As has been frequently noted, he is portrayed as reading a serial novel to his wife in bed (*la historia chica*) when he is summoned to form a government (III: 190–1).

3

The Reign of Isabella II

Military Rule and Revolution, 1858–1868

WHEN he came to power in 1858, O'Donnell was extremely conscious of the dangers of the previous institutional instability. Through the movement he founded known as the 'Liberal Union' (*Unión Liberal*) he attempted with some success to produce a stable new order through the reconciliation of the opposing political forces; his ministry was destined to last longer than any other (nearly five years: 1858–63) during Isabella's reign. The inception of this policy is dealt with at length in *O'Donnell* and continues, with certain deviations towards other topics, in the following *episodios* (*Aita Tettauen and Carlos VII, en la Rápita*).

Exceptionally, in O'Donnell's case Galdós thrusts a leading historical character into the limelight. To lay bare his objectives, skilled use is made of the rhetorical device which will become known as the stream of consciousness; typically, it is called up as part of a semi-conscious dream process. As a result of his abrupt dismissal in 1856, O'Donnell is shown as suffering from insomnia, provoked by the tingling ('hormigueo') of his nerves. Thus, ideas which his conscious self suppressed or rationalized during the daytime are unlocked as 'libre y atrevido pensamiento' (III: 165) at night as he passes, in Becquerian fashion,[1] from waking to somnolence. The image used is the spool or 'devanadera' of his thought, from which '*con no poco trabajo* estos fraccionados pensamientos' (163; my italics) are extracted: the narrator is fully conscious of the incongruity of his recounting such entirely private musings. The substantial passage is expressed in direct speech, interrupted by two commentaries from the narrator. The first occurs as the long-limbed general turns over in bed, the second as he begins to slip into unconsciousness, the sign of which is a growing separation from reality as he presumes to address his monarch as 'tú', only to revert once again to semi-conscious meditation. Finally, after the unwonted unequivocal frankness of again using the familiar form of address, he falls asleep. The expression of these utterances is

[1] Cf. *Rima* LXXI: 'No dormía; vagaba en ese limbo | en que cambian de forma los objetos, | misteriosos espacios que separan | la vigilia del sueño' (*4a*, 103). Because of the poet's early death, it is easy to forget that he and Galdós were almost contemporaries. Galdós wrote a substantial article on Bécquer in *El Debate* (13 Sept. 1871), which was overlooked until recently resuscitated (*3b*, Cazottes).

colloquial ('la pícara Desamortización', 'menos mal', 'Cuidado, ¿eh?') and it contains direct addresses, rhetorical questions, and familiar repetitions; as it reaches its conclusion, on the advantages of disentailment and Isabella's deep responsibility, these qualities are accentuated, but without any break in continuity or severing of the formal grammatical restraints such as we find in James Joyce or Virginia Woolf.

These ruminations give the positive side of the movement: the folly of the clerical autocracy fostered by the queen and tolerated by Narváez, the opportunity to forge a sensible, intelligent, élitist coalition ('Con esta flor de los partidos amaso mi pan nuevo . . . *Unión Liberal*': **III**: 163), the key issue of disentailment to provide prosperity; and, to accomplish all this, the impelling need to curb the queen's blind caprices. It is in some respects a first attempt to occupy the political middle ground which was later essayed, slightly more to the left, by Amadeo's democratic monarchy and then, further to the right, by the Restoration; as might be expected, the young Cánovas played an important role in the Liberal Union.

The more negative side comes out in the action of the *episodio*. The narrator euphorically hails the new movement as the reign of 'todos los hombres que al saber de cosas de gobierno reúnen la distinción y el buen empaque social' (**III**: 191), but the last two phrases have an ominously snobbish ring. Its relativist and flexible ideas, subordinating principles to tolerance and opportunity, gloss over its cynical desire to be all things to all men:

> Estábamos en tiempo de tolerancia, de transacción, pues la Unión Liberal, ¿qué era más que el triunfo de la relación y de la oportunidad sobre la rigidez de los principios abstractos? Se transigía en todo; se aceptaba un mal relativo por evitar el mal absoluto, y la moral, el honor y hasta los dogmas sucumbían a la epidemia reinante, al *aire de flexibilidad* que infestaba todo el ambiente. (**III**: 193)[2]

As V. G. Kiernan puts it,

> Liberal Unionism would never acquire a soul. Manipulation and bribery and jobbery were the breath of its life; not philosophers like Ríos Rosas but wirepullers like Posada Herrera were its tutors. Parliamentary life was the camouflage for *caciquismo*, buying and selling of votes by local bosses in return for government favours. (*1b*, 249–50)

One overriding concern is apparent: an intense preoccupation with the most pressing material considerations, the means of earning a living, and, in

[2] The *tertulia* surrounding Jenara Baraona in 1834 is a clear anticipation of both the Liberal Union and the Restoration: 'En el seno de este partido, que en un tiempo se llamó de *los sabios* y en sus albores se llamó de *los anilleros*, había gente de gran mérito, aleccionados los unos en la práctica del liberalismo, otros algo amaestrados en el arte político que faltaba a los liberales. Ellos fueron los primeros *maquiavélicos* ante quienes sucumbió la inocencia angélica de aquellos candorosos doceañistas que principiaban a no servir para nada. A falta de principios, tenían un sistema, compuesto de engaño y energía' (**II**: 313–14; *4a*, quoted in Fletcher 23; Jenara is there referred to by her alternative name of 'Generosa').

extreme cases, of securing enough food: the synecdoche 'garbanzo' in Galdós's popular language. Hence the agitation by crowds of hungry supporters of Narváez or O'Donnell to get jobs. Starvation, as Centurión eloquently bears witness, is what links the unemployed with the peasant insurrectionists savagely executed at El Arahal in Andalusia in 1857. The lot of *Mita* and *Ley*, and with them Santiuste, is equally grim.

The Liberal Union is, for the moment, enthusiastically supported by the Marquis of Beramendi and accepted by Eufrasia, now Marchioness of Villares de Tajo. Beramendi, appropriately, strives to conciliate all interests; he manages to keep *Sebo* in post and he persuades the cultured Cándido Nocedal, despite his extreme clerical persuasion, to pardon Leoncio Ansúrez for his elopement with Virginia. But the scope of the Liberal Union goes much beyond this, ending up as a hedonistic feast, expressed in the person of the suitably named José (de la) Riva (y) Guisando. As Diane Urey observes, the banquet Brizard offers Teresa at which Guisando is present 'displays in detail how the three major codes of activities in the novel—politics, sex, and eating—are interconnected' (*3d*, 1989: 132). Guisando, for whom the two things which redeemed sin are 'el buen gusto y la opulencia', is 'una imagen sintética de la sociedad de aquel tiempo' (III: 177). It is a society which is based on false or optimistic expectations of material success: 'Era don José de la Riva algo nuevo y grande en nuestra sociedad: la esperanza del bienestar y de la alegría destronando a la miseria triste' (179). Hedonism thus formed the outward show of a sybaritic community bent on an investment spree for which disentailment[3] was an essential stimulant.

This important subject yields one of the finest examples of the interaction between *la historia grande* and *la historia chica*. Its protagonist is one of Galdós's most distinctive creations, Teresa Villaescusa, the most prominent of the extraordinary group of women characters[4] depicted in the fourth series, as Urey has shown (*3d*, 1989: 102–5), in more or less direct comparison with the queen. The ironic term used to link them is 'soberana'.[5]

Teresa raises, moreover, the whole question of sexual morality and the status of women.[6] Evidently dissatisfied with the usual options available to women, she read widely and attended church assiduously, so that her mother

[3] For the early stages of disentailment see Marichal (*1b*, 66–9, 156–65).

[4] Marie-Claire Petit is evidently mistaken when she states that the *episodios*, 'consacrés le plus souvent au récit d'événements historiques et politiques, appartiennent aux hommes et les femmes y ont peu de relief' (*3c*, 10).

[5] Lucila is perhaps less closely identified with Isabella than Urey suggests; in my view her relation with Beramendi is more important. Urey also claims, more controversially, that this link reflects the narrative process: 'the inter-penetration of these two characters describes the process of the historical novel . . . What results from the merger of Lucila and Isabel is a theory of narrative, of the role of point of view in history and fiction' (*3d*, 1989: 103).

[6] It would be a worthwhile task—for another occasion and for someone better qualified—to relate the portrayal of Teresa with the now voluminous feminist studies of the 19th-cent. novel. For a sound discussion of some aspects of the problem in Spain consult Scanlon (*1b*).

fleetingly suspected she might wish to become a nun. She is not disposed to conform by accepting under pressure any of her countless uninspiring *novios*, the last of them the unsavoury Alejandro Sánchez Botín, brought in from novels like *La desheredada* and *La de Bringas*. Instead, she is attracted by one of the most radical of the democrats, a 'demagogo', according to her mother, Sixto Cámara. Cámara was involved in the El Arahal revolts and died some years later (1862) as the result of a Republican revolt at Olivenza. He was, in Galdós's words, 'el más loco de nuestros demócratas . . . aquel visionario que padeció la generosa demencia de querer implantar la República con tres republicanos' (*Aita Tettauen* III: 244). Then Teresa is suddenly left unprovided for by the death of her father, a colonel suffering cruelly from cancer, who kills himself having done his duty under Serrano in suppressing Espartero's militia. Misdirected by the persistent irresponsibility of her mother Manuela Pez, known as the *sutil tramposa*, and disgusted by the hypocritical actions of her friend Valeria de Socobio, Teresa makes a conscious decision, when she takes over Guillermo de Aransis[7] from Valeria, to lead what counted as an immoral life.

Galdós takes care not to portray her as a romantic victim in the manner of *La Dame aux camélias*.[8] Within the well-known type of the kept woman or *cocotte*, Teresa is conspicuous for her self-awareness; given the painfully limited choices open to women, her decision is a rational and practical one, however risky and liable to harsh exploitation it may be. As Aransis indicates when instructing Teresa on the unexpected restrictions her new life places on her freedom in public (III: 160), within respectable society only a married woman has a certain freedom of movement, which may allow her to be discreetly unfaithful:[9] the flagrant and immediate example is Valeria.[10] This perspective is also clearly reflected in Virginia Socobio's attitude to Teresa: given Teresa's independence and her generosity, Virginia concludes that she must be married—a countess perhaps, and possibly Beramendi's mistress (203; 205). Teresa has chosen a more honest course, even though it is condemned (and at the same time condoned) by society (*3d*, Rodríguez 145); her situation is in some ways similar to Fortunata's at the time of her liaison with Feijoo. Moreover, Teresa does not lose her straightforward reactions or

[7] The name may associate the dissolute aristocrat with the haughty Marquesa de Aransis in *La desheredada* and with the Catalan family of that name in *La loca de la casa*.

[8] The popularity of the famous novel by Alexandre Dumas the younger (1848, and performed as a play in 1852) was greatly augmented by Verdi's extremely successful operatic version *La traviata* (1853). It is constantly used as a metonymic euphemism for an 'immoral' woman with a 'heart of gold'.

[9] Rosalía de Bringas is the case of someone who takes advantage of this limited freedom; Fortunata, when living with Maxi, is an example of someone incapable of doing so successfully.

[10] It should be noted that, like practically all the women in Galdós's novels, Valeria and Virginia are very poorly educated, in this case as part of a deliberate policy by their mother (III: 19).

her warm generous instincts; in this she is like Leonor in *Realidad*. Beramendi is immediately impressed by her magnanimity in releasing Arensis from any obligation towards her.

Yet she has an agonizing personal predicament to face, which is part and parcel of the subordinate status of women. When she falls head over heels in love with Juan Santiuste, she has to make the very difficult decision about whether or not to opt finally for a life of honourable poverty with him, as Leoncio and Virginia have done so heroically. At their crucial meeting their dialogue is direct, without tags to indicate speakers or even pseudo-dramatic directions. When they separate, Juan takes from her some blades of grass Teresa has been sucking, which in a fine metonymic image represent oxymoronically their dilemma: 'Son amargas. Toda la vida es amarga, pero contigo el amargo es dulzura' (III: 209). Near to a final decision to join Santiuste and *Mita* and *Ley* in a *huerta* on the outskirts of Madrid, she successfully avoids her mother, but bumps into Manuel Tarfe, who advises her, as one who is characterized by her 'pasión del buen vivir y la pasión de repartir el bien humano' (224), against a life of deprivation. As she walks slowly towards her destination, she engages in a long meditation, expressed in fine free indirect style, on what she should do.

Her discourse, punctuated with passages of summary, consists of fluent colloquial utterance, with frequent questions and exclamations, as she speculates about the new and brilliant lovers her mother, Tarfe, and Serafina have each lined up for her; the diegetic preterite contrasts with the imperfect of the mimetic episodes, though the latter tense also serves to reflect her continuous mental and physical indecision ('se paraba . . .', 'volvía los ojos hacia Madrid', 'Poníase de nuevo en marcha lenta'). Much of the discourse is double-voiced, in Bakhtin's terminology, as expressions characteristic of the narrator like 'sanos impulsos de moralidad' are intermingled with the reflection of her speech: 'Loca la habían vuelto entre todos.' She sees some poor families, who mark what would be her lot if she decided for Santiuste: 'Groseros le parecieron los hombres, desgarbadas las mujeres, flacuchos y pálidos los niños' (III: 225). She turns momentarily towards Madrid, but then instinctively, like birds returning to their nest, continues towards the *huerta*. The summary reflects her sentiments, but is not expressed in her own words: 'sintió la urgencia de resolver y ordenar en su mente un aluvión de ideas que en ella entraron como huéspedes alborotadores.'

In a long series of questions in free indirect style, she asks herself how she would manage to look after Santiuste properly; the narrative voice breaks through at times, somewhat clumsily, in expressions like 'antes de lanzarse resueltamente en la vida pobre'. And how, too, would she be able to help the poor around her? As she reaches the other side of the *tapia* where she was to meet her friends, she asks herself whether she should speak directly with Juan. Better not, she concludes, and vividly pictures to herself the impatience of those awaiting her. Fearing that Juan might set out to look for her, she makes off back towards Madrid, buying from some old and impoverished street vendors water and doughnuts for herself and swarms of wretched children, for all of whom she ingenuously expects the Liberal Union to offer some relief.

It is now dark and she loses her way, with the result that, when she approaches the centre, she finds herself gazing on the fine house of the most famous entrepreneur of them all, the Marquis of Salamanca,[11] whom she was popularly thought to belong to already: a clear proleptic hint that he may be her next lover (*3d*, Urey 1989: 247 n. 20). She witnesses the procession of carriages containing the accustomed groups of ostentatiously wealthy ladies, aristocrats mingled with *nouveaux riches* like Eufrasia, who is escorted by Riva Guisando. Such stalwarts of the Liberal Union as the fictional Tarfe and the real-life Alboreda accompany them on horseback; finally comes the figure on whom all hopes depend: O'Donnell, travelling unpretentiously with his wife 'en una vulgar berlina'. The book ends, therefore, on a note of expectancy. The fate of these expectations and of Teresa's parallel decision is left to a subsequent volume.

Teresa thus embarks on a life of sexual mobility and commercial free enterprise which corresponds to the financial and hedonistic mobility of the new governing coalition, making her, in Galdós's ironic perspective, the numen of the Liberal Union[12] (III: 222). No matter that this means that she is treated as a commercial object—her lovers are called *contratistas de amor* and she is compared to animals, such as the highly bred horses Risueño deals in. As a commodity, for the time being, she is a very highly desirable one. In another respect Galdós separates himself from conventional prejudices: he spares us any heavy moral censure at her expense.[13] She is no more reprehensible (though, as we shall see, she is much more vulnerable) than the exploiters of other commercial assets are.

In all this there is a complex and sustained parallel with the queen.[14] Similar to Isabella in her popular instincts, her generosity, and even her

[11] José Salamanca, Marquis of Salamanca (1811–83), was the entrepreneur, banker, and politician initially associated with María Cristina in railway development and commemorated in the 'barrio de Salamanca'. Galdós comments: 'En aquel tiempo, el vulgo señalaba como de Salamanca todo lo superior: las poderosas empresas mercantiles, los cuadros selectos y las estatuas, las mujeres hermosas, los libros raros y curiosos . . . Homenaje era éste que tributaba la opinión a uno de los españoles más grandes del siglo XIX' (III: 193).

[12] So called by Manuel Tarfe, himself nicknamed 'O'Donnell el chico' because of his own allegiance to the Union.

[13] Galdós is far less inhibited about expressing women's sexuality than has usually been assumed. Compare, for example, Jenara Baraona, in whom Benítez (*3d*) finds the prime example of a change to a more open attitude from 1875 onwards: a result, he claims, of the influence of *Pepita Jiménez*.

[14] Urey (*3d*, 1989: 129). Once more, I disagree about the self-reflexive functions she sees in Teresa: 'Teresa is like the Clío who is both a creator and a creation of the narrative' (143).

atrocious spelling, as well as in her sexual promiscuity, Teresa is in sharp opposition to the monarch over the Liberal Union policies. As its numen, Teresa instinctively and ironically does more for disentailment and the redistribution of resources than any of the actions of the queen or her government: 'Pues, siguiendo paso a paso la historia integral, dígase ahora que al tiempo que Isabel de Borbón decía con desgarrada voz de maja: "Yo no desamortizo", la otra maja, Teresa Villaescusa, gritaba: "Juro por las Tres Gracias que a mí nadie me gana en el desamortizar"' (III: 169). Not only does she help redistribute wealth by subtracting it from her lovers ('Aligerar a los demasiado ricos es obra meritoria') but she displays many examples of great personal generosity; she is even mistaken for one 'de esas damas de la Beneficencia' (203), like Doña Guillermina, another accepted if eccentric role open to ladies well connected enough and independent enough to undertake it. Through Tarfe, she secures jobs for two of her relatives: for Centurión, thereby converting him from radicalism to the Liberal Union, and, despite his betrayal of her, for Leovigildo Rodríguez, hard pressed as a result of his growing family. She collaborates fruitfully with Beramendi in bringing about a reconciliation between Virginia and her father.

Her lovers are examples of representative politico-social types. Aransis, Marquis of Loarre, is the impoverished aristocrat, unable to resist luxurious pursuits; Beramendi has to use his powerful influence with O'Donnell to have him sent off on a diplomatic mission overseas. Isaac Brizard is a French capitalist who has invested heavily in one of the major economic developments of the time, the railways;[15] he takes her to Paris, providing a necessary part of her new sophistication. Her return is marked, significantly, by the birth of the infante Alfonso: her fate is thus linked with a key event in nineteenth-century history. After being protected by a shadowy old marquis, she becomes attached to a representative Andalusian *cacique*, Facundo Risueño, who among other things secures Beramendi's election to parliament with the arch-fixer José Posada Herrera. We shall follow her subsequent development later.

The practical consequences of *desamortización* are not passed over. It is evident that, in ironical contrast with their support for the clerical faction, much of the wealth of both Saturno de Socobio (*Las tormentas del 48* **II**: 1456) and Fajardo's future father-in-law Feliciano de Emperán (1508) derives from the earlier sale of church lands. In addition, in the scramble to invest which followed disentailment there was a premium on financial liquidity, with the

[15] Carr (*1b*, 265) points out that half of the capital for the Spanish railway network was French and that the centre of control of the system was in Paris. For the importance of railway development see Vicens Vives (*1b*, 1972: 202–7) and Fontana (*1b*). It is a central theme of Torquemada's speech at the banquet in his honour (V: 1101). Galdós waxes lyrical about the freedom the railway opens up when Santiago Ibero makes his escape to France: 'Oh, Ferrocarril del Norte, venturoso escape hacia el mundo europeo, divina brecha para la civilización' (III: 675). Compare 'un tren que parte es la cosa del mundo más semejante a un libro que se acaba' (*Tormento* IV: 1568). See also Casalduero (*3b*, 1970).

result that Fajardo's brother Gregorio and his ambitious and unscrupulous wife Segismunda Rodríguez, nicknamed *Medusa*, were able to profit from the sale by devoting themselves to moneylending; and in her new prosperity Segismunda, like Socobio, aspires to the social prestige of the titles which proliferate at this time, even if only a papal one in her case. The link between disentailment and usury, no longer tied to its purely political context, receives full treatment, of course, in the Torquemada novels. This will be considered in Chapter 6.

In much the same way, an abiding social evil such as *cesantía*, so fundamental a part of the politics of rapid turnover and personal allegiances, has a considerable role in the *episodios*.[16] Typical examples, as we have seen, are the recurring figures of Centurión and *Sebo*, desperately hoping for positions after the 1854 revolution and subsequently seeking to survive changes of government. The most tragic case, however, is the earlier example of Faustino Cuadrado, in *Las tormentas del 48*. Cuadrado is dismissed from his post as a consequence of justly rebuking his subordinate, the young and influential Fajardo. The hidden clerical influence is all-powerful: Fajardo's sister, Catalina, has direct access to Patrocinio at the La Latina convent and hence to the palace. Once the process is under way, even Fajardo can do nothing to remedy the situation. The unfortunate Cuadrado,[17] having resorted to political conspiracy, subsequently becomes a victim of Narváez's propensity (the famous *cuerdas de Leganés*)[18] for deporting potential undesirables to the Philippines (II: 1515).

O'Donnell also had another card to play: a completely unnecessary nationalist war in North Africa (1859–60)[19] to arouse popular patriotic fervour, after the style of Napoleon III (III: 242). In the comprehensive coverage in *Aita Tettauen*[20] Galdós draws for much of the fighting on Pedro Antonio de Alarcón's extremely popular *Diario de un testigo de la guerra de África* (1860), but his outlook is very different. As José Schraibman has

[16] In the contemporary novels the same type of situation holds but is viewed from a different angle, for in the fictional context individual reactions to adverse circumstances take on greater importance. The analogous case of Villaamil in *Miau* will be discussed in Chapter 6.

[17] In *O'Donnell* Galdós gets him confused with another redundant civil servant, *Segundo* Cuadrado, also protected by Beramendi (Sainz de Robles, in III [*Censo*] 1526).

[18] So called because suspects were brought together for deportation at Leganés, on the outskirts of Madrid (see II: 1681–8).

[19] A war in which both O'Donnell and Prim consolidated their military reputations, becoming Duke of Tetuan and Marquis of los Castillejos respectively; Prim, particularly, distinguished himself for his reckless bravery. In Mar. 1861, invited by Pedro Santana, Spain annexed the Dominican Republic, threatened by Haiti; it relinquished sovereignty in July 1865. The war of the Pacific in 1865 and 1866, recounted in *La vuelta al mundo en la 'Numancia'*, was a similar military distraction.

[20] For a careful and detailed study of *Aita Tettauen* and the African part of *Carlos VII, en la Rápita* see Torres Nebrera (*3d*). He also points out very pertinently that Morocco was still a festering problem at the time of the composition of the *episodio*, and that a few years later the military intervention in Morocco sparked off the *Semana trágica* (401).

indicated, 'No cabe duda de que Galdós, entre otras cosas, está dando una respuesta a lo escrito por Alarcón' (*3d*, 542).[21] Initially, in Ansúrez-Halconero's family group, a typical war mania prevails which encompasses old Halconero himself, the young Vicente, Santiuste, Jerónimo (with reservations), and even Lucila. Alarcón himself appears in the novel, his unabashed nationalism—'el poeta de la guerra' (III: 263), Santiuste calls him—being the starting-point for Santiuste's evolution from an uncritical acceptance of patriotic fervour to a Tolstoyan pacifism (*3d*, Collins). Santiuste ends up by becoming aware—reflecting Galdós's own awareness—that 'en el fondo de todo esto no hay más que un plan político: dar sonoridad, empaque y fuerza al partido de O'Donnell' (263).[22]

There is also a radical contrast in racial attitudes. Within his heroic exaltation of war, Alarcón displays respect for the Moors and bitter contempt for the Jews.[23] In contrast, Galdós presents a composite perspective of all three peoples,[24] with an implicit vision of racial harmony between them. Through the remarkable chronicle of the pseudo-renegade Christian El Nasiry (Gonzalo Ansúrez), in which our awareness that we are dealing with a false document prepared for political purposes is deferred until later, we are given ironic insight into Muslim mentality, while Jewish attitudes are revealed by the Sephardite speech of Mazaltob.[25] This somewhat intuitive ideal harmony is expressed, as is typical of Galdós, in terms of sexual relations, by means of Santiuste and his love affairs. The general impact of Galdós's treatment of the African war is unequivocal: by indirect implication he appears resolutely opposed to imperialist military adventures and in favour of the equitable reconciliation of races, however difficult and complex this may be (*3b*, Schyfter 101–16; *3c*, Gilman 25–36).

None of the contemporary novels is set as early as O'Donnell's long ministry, so that the references that are made to it are necessarily of the second or third type listed in the Introduction. In *El amigo Manso* Manso projects back to the time of his mother's friendship with Doña Cándida, summing up rapidly (second type) key events—political, economic, cultural—of the time: 'allá en los días, no sé si dichosos o adversos, del consolidado a 50, de la guerra de Africa, del *no* de Negrete, de las millonadas por venta de

[21] An 'anti-*Diario de la Guerra de África*', Torres Nebrera calls it (*3d*, 403).

[22] In an obituary article in *La Prensa* (31 July 1891; *3a*, *Prólogos* 450–5) Galdós praises the style of the *Diario*, emphasizes the importance of *El escándalo*, and deplores Alarcón's later intolerance and unadaptability.

[23] '¡Cuánta dignidad en el Agareno! ¡Qué miserable abyección en el Israelita!' (*4a*, 1000). See Schraibman (*3d*, 540–1).

[24] Galdós received copious documentation from a correspondent in Tangier, Ricardo Ruiz Orsatti. Some of it arrived too late to be included in *Aita Tettauen* but was utilized in the early chapters of *Carlos VII, en la Rápita*. Galdós visited Tangier in Oct. 1904, but was unable to reach Tetuan because of bad weather. See Ricard (*3b*, 1968).

[25] Martínez Ruiz (*3d*) carefully studies Galdós's treatment of Sephardite customs and language, noting that he is slapdash about obvious phonic characteristics while taking great pains with morphological and lexical aspects.

bienes nacionales, del ensanche de la Puerta del Sol, de Mariano y la Grissi, de la omnipotencia de O'Donnell y del Ministerio largo' (IV: 1176). García Grande, for his part, was an extremely undistinguished member of the politico-economic set surrounding O'Donnell, and ended up by losing his fortune. His widow, Doña Cándida, continues the profligate spending and concealed indigence of the *Unión Liberal* era in *La de Bringas* (set in 1868) and *El amigo Manso* (set in 1878–81).

The most interesting references in the novels to the African war are found in *Torquemada en la cruz* (1893). When the veteran Hipólito Valiente describes its heroic actions at some length to the blind Rafael, the description provides at once a vicarious pleasure and an exaltation of Rafael's patriotic zeal:

> Para Rafael, en el aislamiento que le imponía su ceguera, incapaz de desempeñar en el mundo ningún papel airoso conforme a los impulsos de su corazón hidalgo y su temple caballeresco, era un consuelo y un solaz irremplazables oir relatar aventuras heroicas, empeños sublimes de nuestro Ejército, batallas sangrientas en que las vidas se inmolaban por el honor. (V: 990)

This enthusiasm fits very well with his aristocratic, military, and patriotic views, inflexibly frozen in the past; but Galdós slyly indicates in passing that Rafael's aunt had married the daughter of a saddle-maker, whose living no doubt depended on military contracts (V: 995). It should be noted too that with an ironic twist Torquemada is described as resembling O'Donnell: 'Dice *Rumalda* que me parezco algo a O'Donnell cuando volvía de Africa' (968): it is another fixed reference to a no longer directly relevant past situation, the third type of historical reference.

The failure of O'Donnell's efforts to stabilize the system brought about from 1863 onwards an alternation between repression and attempted revolution and a drift towards the overthrow of the dynasty. The period following the fall of his long ministry in February 1863 is chronicled in the last two *episodios* of the series, *Prim* and *La de los tristes destinos*. Three of the contemporary novels, the *El doctor Centeno–Tormento–La de Bringas* trilogy, are set in the period. Moreover, Galdós's first direct experience of Madrid life dates from this time.[26] He was therefore in a stronger position to make *la historia grande* and *la historia chica* coalesce.

The action of *El doctor Centeno* (1883)[27] takes place in that year, 1863.[28] Its

[26] For an anecdotal account of his life in Madrid at this time see his late *Memorias de un desmemoriado* (1915–16; VI: 1430–73), on which consult Percival (*3b*, 1982) and Berkowitz (*3b*, 1948: 74–84).

[27] Germán Gullón has argued for the unity of this loosely knit novel (*3c*). Cardona (*3c*) has discerned an influence of Goethe's *Bildungsroman Wilhelm Meister*, linking it with his change from a Romantic dramatist to a Realist novelist. Hazel Gold has a fine study of its structure and historical implications (*3c*, 1989); for further discussion see Chapter 6.

[28] Its specific time is underlined by the narrator: 'A 10 de febrero de 1863, entre las diez y las once de la mañana' (IV: 1296). The date is further emphasized by the reference to the issue of *La Iberia*—Calvo Asensio's paper—Cienfuegos is carrying: 'fijaos bien en la fecha, que era por febrero de 1863' (1297).

'discourse time' is the earliest in which any contemporary novel is set; this fact separates it to some extent from other novels and makes the interchange of characters more difficult. Presumably because of this protracted time-scale, Galdós makes sure to write from the narrator's present ('story time') twenty years later. At the beginning of part II he deliberately links the two periods as he meditates on the changes which have occurred—the disappearance of Doña Virginia's boarding-house—and the fate of the characters. Most are listed as survivors, with varying degrees of prosperity, like Zalamero, Poleró, Arias Ortiz, Cienfuegos; Sánchez de Guevara has died in a military insurrection. About others, including Miquis, the narrator evokes the *ubi sunt?* topos, raising an expectation which is not for the moment satisfied: '¿Estos y otros que no nombro, ¿dó están? ¿Viven? ¿Se salvaron o se sumergieron para siempre?' (IV: 1367).

Apart from the natural development of a few characters in the sequel *Tormento*—Polo, Amparo, Refugio—there is little follow-up in later novels. It is surprising, perhaps, that Celipín (Felipe) Centeno should disappear after *Tormento*, when the subject of his education played such a central role in the earlier novel. Alejandro Miquis's story is fittingly concluded, and his brothers Augusto and Constantino thrive in other spheres. The characters who do turn up again in later novels and *episodios* are essentially minor ones: stock figures like José Ido del Sagrario, the most enduring (*3c*, Shoemaker 1951), Federico Ruiz, Leopoldo Montes, and Basilio de la Caña.

Quite an amount of political discussion takes place among the fictitious characters of the novel. The most notable figure involved in political comment is the ageing and pompous concierge of the Observatory Florencio Morales y Temprado. Characterized by a pretentious *muletilla*, 'entre paréntesis', he is full of high-flown concepts he can never manage to get quite right; an enthusiast for mineral water, he becomes intoxicated with rhetoric instead of alcohol. His leg is mercilessly pulled by Alejandro Miquis and Cienfuegos, and he voices an astonishingly incoherent mixture of patriotism, liberalism, and religion: 'yo he sido siempre hombre de orden, muy español, muy enemigo de lo extranjero y de la tiranía', he declares, '¡Libertad, religión!' (IV: 1306–7). One side of him is opposed to freedom of religious practices and to foreign democracy, thinks Spanish products (including its generals) superior to those of any other nation, and denies the 'obstáculos tradicionales' which Olózaga has defined. The other side is 'un progresista platónico y vergonzante que se iba callandito a la Tertulia alguna vez' (1307), who admires Olózaga and shares his oratorical mannerisms. As Federico Ruiz said, he represents 'una petrificación, en la cual se veían hasta tres ideas perfectamente conservadas, duras e inmutables como las formas fósiles que en un tiempo fueron seres vivos'. A throw-back therefore to the days of Espartero, when these beliefs were not incompatible, he is a more extreme case of the fossilization of progressive ideas found in Don Baldomero Santa Cruz, an old-fashioned

Progressive who 'pensaba el 73 lo mismo que había pensado el 45; es decir, que debe haber mucha libertad y mucho palo' (V: I. 8. ii, 85),[29] but who disliked the conspiratorial *tertulia*. Just as incongruously eclectic is Morales's choice of the three great men of the century: Antonelli, the advocate of papal spiritual and temporal supremacy, the imperial Napoleon III, and the old-fashioned Progressive Olózaga, all of whom conspicuously fail in the long run to achieve their objectives.

Public opinion is astir at this time as never before, according to the narrator, an anonymous participant who describes Doña Virginia's boarding-house, in which he and other students live: 'En pocas épocas históricas se ha hablado tanto de política como en aquélla, y en ninguna con tanta pasión. Jamás tuvieron parte tan principal en las conversaciones populares los chismes palaciegos y las anécdotas domésticas de altas personas' (IV: 1370). For this he blames the censorship, which produces *hablillas* which poison the atsmosphere and bring about the fall of empires. Desultory political remarks among the residents, Sánchez Guevara, Poleró, Arias Ortiz, as well as Miquis and Cienfuegos, are remembered by Felipe in a sketch which still retains something of a *costumbrista* tone. The names of O'Donnell and Narváez are bandied about and predictions are made about the fall of the former—which is imminent—and the return of the latter; clerical influence, in the figures of Sister Patrocinio and Father Claret, is also indicated. The somewhat unruly young people spend much time and energy making fun of two figures: Basilio de la Caña, with his pompous statements about economics and the budget, and Leopoldo Montes, with his *muletilla* 'bajo el prisma de . . .', his dubious conquests, and his imagined political conversations. Not surprisingly, most of this talk is about the Progressives, who are both the up-and-coming force and an unknown quantity.

At this juncture comes Calvo Asensio's sudden death, described as follows:

> En tal día enterraban con gran aparato de gente y público luto al atleta de las rudas polémicas, al luchador que había caído en lo más recio del combate, herido de mortal cansancio y de fiebre; hombre tosco y valiente, inteligencia ruda, que no servía para esclarecer, sino para empujar; voluntad de acero, sin temple de espada, pero con fortaleza de palanca; palabra áspera y macerante; temperamento organizador de la demolición. Reventó como culebrina atacada con excesiva carga, y su muerte fue una prórroga de las catástrofes que la Historia preparaba. (IV: 1346)

The event has been prepared for by Morales, who is depicted as a friend originating from the same village—Mota del Marqués—as the political leader (IV: 1306), and by quick references to *La Iberia*, the journal he edited. At his

[29] Don Baldomero Santa Cruz's name recalls that of his namesake Espartero; still more remarkable, the name of Espartero's wife is Jacinta Santa Cruz. Obviously, Galdós had the idolized first Progressive general in mind at the formative stage of creating his well-heeled middle-class family.

funeral Morales makes a passing comment on his former profession: '—Entre paréntesis, si no hubiera cambiado su farmacia por la condenada política, todavía viviría. Era un mocetón' (1346). Yet the 'emplotment' seems inadequate for such a chunk of direct unassimilated narrated history, even though it does at the same time bear witness to Calvo's importance. Another sign of Galdós's esteem for the dead man's integrity and intransigence is the fact that he is the only politician saved from general denunciation of less resolute left-wing leaders by José Izquierdo in *Fortunata y Jacinta*: '¡Si viviera Calvo Asensio! Aquél sí era un individo que sabía las comenencias, y el tratamiento de las presonas verídicas. ¡Vaya un amigo que me perdí!' (V: I. 9. v, 110).

The funeral is shown not only as a grand occasion,[30] but as the event which for a time prevents Felipe Centeno from delivering Miquis's important message to Doña Isabel Godoy; held up in the Puerta del Sol, he climbs a lamppost to get a better sight of the huge procession led by 'un señor alto y gordo, de presencia majestuosa'. His resourceful friend Juanito del Socorro is carrying errands for this same gentleman, whose qualities now seem even more impressive—'señor grueso, persona de tan admirable presencia que a Felipe le parecía, si no rey, un dedito menos' (IV: 1348); he calls him Don Salustiano. The politician alluded to in this oblique way is once more Olózaga, the senior and still (with Calvo Asensio gone) the most important civilian politician among the Progressive forces.

Too much political symbolism should not be read into the characters themselves; Bly offers such an interpretation of Doña Isabel Godoy and Pedro Polo y Cortés, both identified with the Bourbon monarchy, and of Amparo Sánchez Emperador, 'the embodiment of ideal Spain' (*3c*, 1983: 42), which I find dubious. Doña Isabel is evidently a reflection of an antiquated style of life[31] going back to the *ancien régime* (hence the relation with Godoy, no less strong because it is denied); and she esteems her namesake as 'esa perla de las reinas' (IV: 1356). For his part, Polo's surnames are certainly not casual. His first denotes extremes and his second links him, by name, origin, and temperament, with Cortés; it implies, as I see it, an autocratic imperialism reminiscent of the *conquistadores* both as a pedagogue and as a sensual man who is a reluctant priest. Amparo, whose story is worked out some four years later in *Tormento*, is, for her part, more of a representation of the extremely limited choices offered a poor unprotected lower-middle-class girl than an embodiment of Spain. Eamonn Rodgers sums up these choices as 'successful marriage, prostitution, and the convent', adding that in Amparo's case 'The first seems initially unlikely and the second unthinkable, so the third seems

[30] 'Los progresistas hicieron alarde de su fuerza en el entierro de Calvo Asensio, director de *La Iberia*' (*1b*, Ballesteros Baretta 67).

[31] Geraldine Scanlon (*3c*, 1978) makes a good case for the novel as a study of 'obsolescent values' represented by Federico Ruiz, Florencio Morales, Pedro Polo, and Doña Isabel Godoy.

the only realistic choice' (*3c*, 1987: 96).[32] But under the rather misleading label of prostitution many gradations are to be found, and Galdós has a special interest in exploring them.

In his portrayal of nineteenth-century society the kept woman, the discreet mistress, the hidden adulteress, all play an important role, and these choices, however undesirable, are often the least disadvantageous available to many grievously under-privileged women. Amparo, having been seduced by Polo, is already highly vulnerable (her two familiar names record the effect she has on each of the two men involved with her: to one she is *Tormento*, to the other 'Amparo'). I do not share, therefore, the rather hostile opinion of her held by some critics.[33] Her horizons are certainly narrow and conventional, and she is irresolute and ineffectual, but when she seeks to make the most of her one opportunity, this is surely only to be expected from a defenceless girl. It seems out of proportion to demand a sacrifice from someone as undervalued by society as Amparo, in order to redeem a passionately frustrated seducer, whose whole attitude, if deserving of compassion, is none the less aggressive and uncompromising. Of the three options just mentioned, Amparo has the most attractive—a wealthy marriage[34]—tantalizingly close, but the final outcome is the next best alternative available, quite different from Refugio's more flagrant and precarious level of prostitution. It represents in my view Galdós's ironic comment on society, with little or no moral reflection on Amparo—or, for that matter, on Refugio—and is, tacitly, vastly preferable to the convent to which Rosalía wanted to dispatch her. Her situation, as I see it, is similar to that of the much more emancipated and socially alert Teresa Villaescusa, already discussed.

To return to Calvo Asensio: in the *episodios* he appears several times. In *O'Donnell* he is a supporter, with Sagasta and the fictional Centurión, of Espartero against O'Donnell in the July revolution of 1854. Like the vast majority of his fellow citizens, he succumbs, together with Olózaga, to the patriotic rhetoric of the African war, which is explicitly denounced by the narrator:

> Los partidos de oposición, deslumbrados por el espejismo histórico, cayeron en el artificio. Olózaga y Calvo Asensio cantaron en el Congreso las mismas odas que en sus púlpitos entonaban los obispos . . . Decía Calvo Asensio que 'el dedo de Dios nos marcaba el camino que debíamos seguir para aniquilar al agereno'. Estas y otras elocuentes pamplinas arrebataban al auditorio y encendían más la hoguera patriótica. (III: 243)

[32] Needlework as such was such sweated labour as hardly to be an option. Refugio waxes eloquent on the subject (III: 1489).

[33] Rodgers (*3c*, 105–10), Rodney Rodríguez (*3c*), and Charnon Deutsch, who is even more censorious: 'morally Amparo is as ill as her decrepit apartment' (*3c*, 40).

[34] This is the solution, as Andreu points out (*3b*, 143), of the virtue-rewarded *folletín* novels. I do not agree, however, that the conclusion is a 'castigo —atenuado si se quiere— por querer salir de su condición socio-económico a través del matrimonio con un indiano rico' (149).

By this time Prim[35] has emerged as the acknowledged military leader of the Progressives (apart from the withdrawn symbolic figure of Espartero). As anticipated to some extent in earlier volumes,[36] his star begins to rise decisively in the *episodio* which bears his name. His ascendancy is skilfully interwoven into the lives of the fictitious characters. The first and most important of these is the young Santiago Ibero, whose adventures mark the beginning of the narration and mirror the diverse attitudes towards the general. These are subject to a series of rectifications. Bursting with enthusiasm, Santiaguito leaves home hoping to join Prim's Mexican expedition. Influenced in the provincial atmosphere of Samaniego by Don Tadeo Baranda, he first thinks of the expedition as a heroic reconquest of part of the old Spanish empire, with Prim as a new *adelantado*. Then the unpolished patrons he meets at an inn at Almazán allege sordid motives of lucre and opportunism by everyone in public life, including Prim. Ibero reacts violently to such insinuations. In this he is supported by Milmarcos, an ex-sergeant of the African war representing 'la historia viva', who demonstrates from first-hand experience a personal allegiance to his hero. From his encounter with the up-and-coming Juanito Maltrana emerges a more sceptical realism about Prim's expedition, which has already departed; he explains the Mexican debt and the three-power intervention; he holds personally 'una idea muy mediana' of Prim. From Maltrana and his friends (among them a very impudent Segismundo Fajardo) Ibero assimilates information about the political scene; he learns for the first time about *neos*, or clerical absolutists, and about the euphemistic *obstáculos tradicionales* (III: 554) to the Progressives gaining power, as well as absorbing the atmosphere of the capital and catching a glimpse of the royal family, received in a very cool fashion by the populace.

With Ibero's disappearance and the frustrating search for him which ensues, Galdós is able to bring Beramendi's intellectual circle (Manuel Tarfe, Jesús Clavería, and Juan Santiuste) into the action. The course of the Mexican intervention and Napoleon III's scheme to place Archduke Maximilian on the throne are elucidated. Prim's promising future is discussed. Both Beramendi and Tarfe are now moving away from the Liberal Union towards the Progressive leader. Beramendi, in particular, views him as more than just another

[35] Juan Prim y Prats, 1814–70, Count of Reus and Marquis of los Castillejos. A Catalan who worked his way up from the ranks, he owed his military prestige to his exploits in the African war and his diplomatic reputation to his withdrawal from the Mexican expedition when Napoleon's imperialist aims became clear.

[36] For example, his first appearance in *Montes de Oca* (II: 1162), when he encounters Santiago Ibero, with prophetic or proleptic irony typical of Galdós, on the corner of the Calle del Turco, scene of his eventual assassination. In *Bodas reales* (II: 1314–15) we see him raising forces in Reus against Espartero and Mendizábal, for both of whom, as Galdós sarcastically remarks, he will later support the erection of statues. In *Los duendes de la camarilla* Narváez sends him to Puerto Rico (for the second time) as captain-general (II: 1717, 1724); the revolutionary Gracián is scheduled to accompany him.

political general: 'Yo veo en la ambición de Prim lejanías que tú no ves' (**III**: 564).

The chance to reverse the traditional policy comes immediately after Prim has made the prudent but courageous decision to withdraw the Spanish forces from Mexico in 1862 in the light of French imperial expansion.[37] This course of action, though strongly disapproved of by the O'Donnell government, won the queen's enthusiastic support, albeit for the subjective motive of spiting Napoleon III. O'Donnell's eventual fall after his long administration is presented as typically sudden, representing one more example of the queen's lack of political consistency. The opportunity then afforded her of bringing Prim and the Progressives into the government is skilfully portrayed; its adroit advocate is Eufrasia. Prim delivers a statesmanlike speech in the Senate and visits the queen. Further evidence of the importance of the period is given as Galdós notes a constant buzz of dissatisfaction in the country in early 1863, at just the time when young Santiago Ibero reappears: 'un zumbido de inmenso moscardón . . . nunca en lo que llevábamos de siglo, había expresado cosas tan feas ni tanto desprecio a los altos poderes.' The ineffective attempt at electoral reform by a group of 'historic', i.e. old-fashioned, Conservatives is described vividly, but the queen, typically, nods ('La bondadosa y antojadiza Reina no veía ni oía nada de esto. Descuidada dormía': **III**: 571). The attempt is finally frustrated by the lurch to the right produced by the king-consort's *camarilla* during one of her pregnancies. 'Prim estaba volado. Dicen que, cerrando el puño, gritó a sus amigos: "Caballeros, a conspirar"' (574). 'Los obstáculos tradicionales' have reasserted themselves once more, obstacles which in a famous speech (3 May 1864) Prim promises to eliminate within two years and a day. This entails the alternative policy of *retraimiento*: withdrawal from the political process as a tacit prelude to an armed uprising. Conspiracy, not just to gain power, but to topple the monarchy, becomes the order of the day.

Calvo Asensio's death occurs immediately afterwards. It is recounted in terms which recall and expand the previous description; his destructive polemic power is reiterated, but his potential for constructive labour is now added; his harsh, tough qualities are seen as linked with his Castilian roots; his old and new professions—pharmacy, journalism—are metonymically linked together:

Triste fue aquel verano. Murió Calvo Asensio de traidora enfermedad, que hubo de rendirle y acabarle en pocos días, dando con su vigor físico y mental en la sepultura. Era un hombre de gran empuje para la destrucción política; para el construir habría sido seguramente un hombre útil, pues en su voluntad existían seguramente las dos caras de la acción. Su talento no era florido, sino adusto, genuinamente castellano; su

[37] Gallenga describes him as follows: 'It was in Mexico, in 1861, that he gave proof of a political wisdom infinitely more valuable than any military ability . . . I never had any doubt that Prim was the only statesman who had a future in Spain' (*Ib*, i. 141–2). John Hay describes him as a 'soldier, conspirator, diplomatist, and born ruler; a Cromwell without convictions; a dictator who hides his power; a Warwick who mars kings as tranquilly as he makes them' (*Ib*, 315).

palabra, de secano, sin verdor ni lozanía; pero sabía, como pocos, imprimir a las ideas el germen fecundo y sembrarlas luego en millares de entendimientos. No había venido, como casi todos los políticos, de los campos abogaciles; era un farmacéutico que administró a su país enérgicas tónicas y estimulantes. Su farmacia se llamaba *La Iberia*. (III: 572)

The subject, though it is 'told' rather than 'shown', has ample justification as part of the political spectrum, for Calvo's disappearance deprived the Progressive forces of a much-needed civilian leader in the 1860s and 1870s; his honest toughness contrasts sharply with the well-known malleability of his close associate Sagasta, who succeeded him as editor of the influential Progressive newspaper *La Iberia*, which Calvo Asensio had owned and directed since 1854.

The illness and death are also explicitly interwoven into the private events of the volume: '*Como no hay manera de separar aquí lo público de lo privado*, digamos que la hermosa y desenvuelta Teresita Villaescusa fue atacada de la misma enfermedad [*typhoid*] que dio con Calvo Asensio en la sepultura' (III: 572; my italics). Teresa does in fact recover, after a delirious outpouring of confused political references[38] worthy of *Confusio* and taking the last rites. Her illness and recovery mark the beginning of a parallel emergence from the selfish moneymaking policies of the Liberal Union into a more compassionate role, as a kind of symbol of unorthodox and progressive values; she is by this time a firm supporter of Prim.

Narváez quickly returns to power, with the incompatible charge of resisting recognition of the government of Italy and of reconciling the Progressives. Galdós gives a vivid portrayal of how the queen, dissatisfied with him and as capricious as ever, charges Istúriz, the elderly Progressive turned Conservative, with forming a government, only to change her mind by the time the ministers attend to take the oath of office next day. The necessary reconciliation with Prim does not take place. The stage is set for the coming revolution.

The energy Prim displayed in pursuing his aims is extraordinary. He was quite undaunted in promoting risings and in conspiring in various European capitals. As Galdós writes evasively in his 'Crónica de Madrid' (23 April 1865; VI: 1288), skirting clear no doubt of the censorship: 'La curiosidad pública continúa huroneando en busca de cierto simpático general, que tan pronto está en Bayona como en Suiza, tan pronto se pasea a las orillas del sombrío Rhin como del alegre Arno.' In *Prim* the narrator writes in familiar tone: 'Por Dios, si era valiente el hombre, a perseverante y cabezudo no había quien le ganase, pues apenas fracasado en una tentativa de pronunciamiento, ya estaba

[38] Among the many 'absurdos trueques de nombres' is a Prim–Cándido Nocedal government and Ruiz Zorrilla and González Bravo as alternative and antithetical candidates as minister. Prim is warned about the *neos* and the king-consort, and his assassination is foreshadowed: '¡Eh, tú, dile a Prim que le van a matar!' (III: 572).

metido en otra, sin perder su brío ni la ciega confianza en estas arriesgadas aventuras' (III: 594).

The next stage, however, belongs not so much to Prim as to Emilio Castelar.[39] The queen, faced with a financial crisis, decided, in an apparent gesture of personal generosity, a *rasgo*, to hand over the national assets to the nation, retaining only 25 per cent for her personal use. In an article entitled 'El rasgo'[40] in his journal *La Democracia*[41] Castelar denounced the deal, pointing out that the property was not hers to give; it constituted an alienation of public funds and a huge donation to the queen. As a result, the word 'rasgo', as Gilman has shown (*3c*, 287–8), has a special resonance in certain works, especially *Fortunata y Jacinta*.

Narváez's government ordered the rector of the University of Madrid, Pérez Montalbán—briefly glimpsed as a bespectacled clean-shaven elderly gentleman in the Ateneo with another respectable real-life character, Father Miguel Sánchez (III: 582)—to remove Castelar from his chair. When the rector refused, he was dismissed also. The result was *La noche de San Daniel* (10 April 1865), 'the first effective student agitation in Spain' (*1b*, Carr 296). Galdós in real life and Juanito Santa Cruz in fiction had a part in the event;[42] Galdós by his own account received some blows at the hands of the Veteran Guard (VI: 1430), while Juanito escapes lightly—a night in jail—thanks to his father's influence.

The incident which throws ironic light on the superficial character of the protagonist in *Fortunata y Jacinta* is tackled with striking immediacy in *Prim*. Appropriately enough, the scene is set with an extensive description of the intellectual club, the Ateneo,[43] with its library and its *habitués*, the significance of which, to Galdós and his contemporaries, is undisputed. Before matters become serious, the situation is commented on, from diverse viewpoints, by various intellectuals—some fictional, some real[44]—from the Ateneo, together with other people on the street, like the *Zorreras*, who reappear more or less unchanged from the time of the 1854 revolution eleven years before. A few, like Beramendi's feckless friend Aransis, vaguely support the government

[39] 1832–99. A university professor famous for his oratory, he was the most conservative and unitarian of the Republican leaders.

[40] 'Véase pues, si tenemos razón; véase si tenemos derecho para protestar contra esta proyectada ley que desde el punto de vista político es un engaño; desde el punto de vista jurídico, una usurpación; desde el punto de vista legal, un gran desacato a la ley; desde el punto de vista popular, una amenaza a los intereses del pueblo' (*1b*, quoted in Cambronero 1908).

[41] 25 Feb. 1865 (*1b*, Llorca 1966: 88). Galdós mistakenly says it was published in *La Discusión*, edited first by Rivero, then by Pi y Margall.

[42] For Galdós's part in *La noche* see *Memorias* (VI: 1655) and Pérez Vidal (*3b*). Galdós also speaks of the disturbance with bitter sarcasm in his contemporary 'Crónica de Madrid' (23 Apr. 1865; VI: 1285) as 'una descomunal batalla que convirtió en campo de Agramante la Puerta del Sol, liza desigual entre el inofensivo pito y la bayoneta'. Twelve people were killed in the incident.

[43] For the importance of the Ateneo see Ruiz Salvador (*1b*) and Azaña (*1b*).

[44] Among the many real people listed as present are Antonio Fabié, Laureano Figuerola, and Urbano González Serrano.

of Narváez and González Bravo; others, like the Hermosilla sisters, are resolutely opposed. For his part, Beramendi's brother Gregorio denounces Prim as the instigator and advocates a tough line with the demonstrators; Father Sánchez, who is no revolutionary, finds the Conservatives inefficient as well as despotic, and engages in a not unfriendly dispute with the Democrat Rivero. Colonel Manuel Pavía,[45] evidently engaged in conspiring with Prim, observes that hose-pipes would have been quite effective in controlling the disturbance; another suggestion made is to support the radical direction of the uprising by singing the *himno de Riego*. As the *pueblo* joins in, the disturbance assumes greater proportions. Beramendi himself secretly rejoices at the bloodshed, which will bring forward the revolution he thinks inevitable.

Some incidental effects involving real people are described subsequently. In the confusion the Conservative minister Antonio Benavides is assaulted by the police, and his fellow minister, responsible for prosecuting Castelar, the famous writer Antonio Alcalá Galiano,[46] has an apopleptic stroke and dies shortly afterwards. The veteran politician Antonio de los Ríos Rosas, respected for his probity, delivers a tremendous denunciation in Congress. The blending of real events and fiction helps create an effect of plausibility.

The *Noche de San Daniel* incident contributed to Narváez's fall some months later.[47] The succeeding O'Donnell ministry finally recognizes Italy and promulgates some liberal measures, causing Beramendi to speculate as to whether O'Donnell might be capable of preventing the coming revolution. He doubts it, and he is of course right: Prim's determined conspiracies against the dynasty continue.[48] In the fiction they are interwoven into the story of Teresa Villaescusa. It is she who describes to Beramendi the failure of Prim's first insurrection at Valencia in April 1865. She is also present at the second attempt on the city on 10 June, and there interrogates Santiago Ibero, who during his long absence from the action has been protected and instructed by Captain Ramón Lagier.

Impatient and constantly frustrated by excessive caution and occasional treachery from his subordinates, Prim continues to attract unconditional

[45] There were two General Manuel Pavías (confused by Butler Clarke [*1b*, 175] and Sainz de Robles in III [*Censo*] 1739) in the Spanish army of the 19th cent. The one referred to here is Manuel Pavía Rodríguez de Albuquerque (1827–95), who despite strong Liberal antecedents dissolved Congress by force after the defeat of Castelar in Jan. 1874. The other is the Conservative Manuel Pavía y Lacy, Marquis of Novaliches (1814–96), who supported Isabella II to the last and was defeated at the battle of Alcolea.

[46] In his 'Crónica de Madrid' (23 Apr. 1865; VI: 1285) Galdós also records the death of the former Romantic and liberal hero whose fiery oratory had been evoked in the final days of the 'trienio constitucional' in *Los cien mil hijos de San Luís* (I: 1687–90). Despite his friendship with a member of the family (Pepe Alcalá Galiano), Galdós treats the deceased politician severely as an apostate, and does not associate his death in any way with the *Noche de San Daniel* disturbances.

[47] The evacuation of Santo Domingo (*1b*, Répide 211) was also a factor.

[48] The queen's failure to return to Madrid for fear of cholera caused much resentment. Gallenga speaks of the unresponsive, indeed 'glacial', reception she received on her return on 27 Dec. 1865 to open Parliament after five months' absence (*1b*, i. 54).

loyalty from a variety of individuals, including such humble figures as Vicente Jiménez. He is the 'Espejo y norte de todos... Lagier veía en él como un enviado de Dios; Ibero, la encarnación de un pueblo que lucha por desatarse de ligaturas, cuyos nudos estaban endurecidos por los siglos' (III: 598). Teresa's latest lover, Jacinto Rodríguez Leal, is a tireless conspirator on Prim's behalf, but she suffers the consequences of Leal's profligate spending, his alternation of violence and affection towards her, and her mother's callous intrigues against him. After he is killed by the Civil Guard through Manuela's treachery, Teresa, completely exhausted, is taken by Ibero and Clavería to the action at Villarejo (January 1866), where the indomitable leader is to suffer yet another rebuff. Exemplifying his grandiose spirit, Prim offers, despite the reverse, a magnificent reception for his followers at his estate at Urda, which provides the occasion for Teresa and Santiago to hold a long discussion on love and on the spiritualism Santiago has learnt from Lagier, before he suddenly departs on a secret mission to Madrid for Clavería. Teresa is left to search for him frantically.

Despite the leader's determination and the enthusiasm of his supporters, exemplified by Santiago Ibero, with their rousing cry of *Prim*, *Libertad*, the *episodio* is imbued with a certain theatrical and rhetorical quality.[49] The real history turns out to be no less implausible than *Confusio's* idealistic history, which alternates with it. The associations with such figures as Hernán Cortés give Prim the legendary character appropriate to a dead hero viewed from the perspective of a generation later. The Cervantine influence Ricard has rightly discerned (*3d*, 1970–1: 344–5), with its evocation of knight-errantry, is by no means casual: 'Resuscitaba en nuestro tiempo la andante caballería, desnudándola del arnés mohoso y vistiéndola de las nuevas armas resplandecientes que han forjado los siglos' (III: 601). Here, more than anywhere else, the question of an *a posteriori* judgement or a 'necessary anachronism' comes into play. All readers know that the dynamic victor of the coming revolution will be killed and the revolution itself largely frustrated as a consequence. Yet to give an authentic impression of the time, the enthusiastic hopes that the moderate Progressives (like Galdós himself) pin on Prim have to be depicted with fervour as more than vain illusions. For these reasons a more distant, more ironic narrative approach is appropriate.

The most bloody of the sequence of uprisings is the revolt of the sergeants of the San Gil barracks on 22 June 1866 and their subsequent execution. The revolt was planned as part of another general rising of Prim's supporters. The other barracks in Madrid—El Retiro, La Montaña—were also to have taken part, as well as civilian units led by eminent Progressives. In Galdós's account in *Prim* the enthusiastic organizer of the uprising is the bearded idealist José

[49] Sinclair notes that 'Both conservative and liberal sections of the popular press drew on the same aspects of Prim for their satire: his theatrical manner, his oaths, his rhetoric' (*4a*, 83).

Rivas Chaves. The novelist pays tribute to him as a fine example of *candor progresista*, without whose sacrifice the country would never have emerged from its hereditary stagnation; and the aspirations of men like Chaves are focused on Prim:

Chaves fue de los más esclarecidos patriotas, de los más candorosos mártires por la idea, que martirio y candor parecen la misma cosa, y el hombre se dejó ir a su ruina y descrédito por secundar valerosamente las ideas de libertad y justicia que sintetizaba en cuatro letras el sugestivo nombre de Prim. Prim era la luz de la Patria, la dignidad del Estado, la igualdad ante la Ley, la paz y la cultura de la Nación. Y tal maña se habían dado la España caduca y el dinamismo servil, que Prim, condenado a muerte después de la sublevación del 3 de enero, personificaba todo lo que la raza poseía de virilidad, juventud y ansia de vivir. (III: 641)

In the event, owing to a leak of the plan to O'Donnell, only the San Gil barracks rose up. The killing of some of their officers by the sergeants[50] was unpremeditated and strictly against orders. The leader, Blas Pierrad, though valiant, was a poor substitute for the charismatic presence of the Catalan general. The uprising was ably and ruthlessly suppressed by O'Donnell, his brother Enrique, Serrano, and Narváez (who was wounded in the fighting); troops from one barracks—San Gil—fought against their comrades from others—El Retiro, then La Montaña. Thus, as Galdós notes, friends become enemies and only fortuitous circumstances determine who can be called loyal and who treacherous (III: 650). The military leaders (Pierrad, Captain Hidalgo) and the civilian conspirators, who included Castelar, Rivero, Sagasta, Becerra, and Martos, all managed to escape, but the other ranks—the so-called 'sergeants'—paid the penalty. The grim suppression calls forth one of the most heartfelt condemnations of futile bloodshed in support of a despotic regime by a narrator who feels himself personally involved and expresses himself in emphatic, immediate style:

¡Lástima del brío militar empleado sin fruto y perdido en el torrente político más espumoso! Creyérase que el morir hombres y más hombres era necesario, por ley fatal, para la consolidación de nuestros altares y tronos, de perfecta índole asiática. ¡Viva Dios que ningún poder se asentó jamás sobre tan ancha y alta pila de cadáveres! (III: 651)[51]

The searing permanent effect that the rising and its aftermath left on Galdós comes out clearly in his three reconstructions of the incident, separated by a very long time-scale: twenty-four, forty-one, and fifty years respectively after the actual event and at a distance of over twenty-five years between them

[50] The sergeants had grievances of their own regarding promotion prospects. See Headrick (*1b*, 152).

[51] Galdós had a fascinated horror of executions. Apart from Merino, see e.g. the executions of Juan Martín *el Empecinado*, Ulibarri, Montes de Oca, Diego de León, Jaime Ortega, and especially the degrading execution of Riego and the heroic one of Patricio Sarmiento, in *El terror de 1824*.

(*Ángel Guerra*, 1890; *La de los tristes destinos*, 1907; and *Memorias de un desmemoriado*, 1916). It is not narrated in the contemporary novels as a current event.

In their essentials the material facts are similar in all three accounts and doubtless correspond to the historical data. The description of the forced serenity of the condemned men is also comparable. Where the accounts differ substantially is in their narrative and emotional context. In the late *Memorias*, these events are recalled personally and directly in the most emotional and apocalyptic terms:

> Los cañonazos atronaban el aire; venían de las calles próximas gemidos de víctimas, imprecaciones rabiosas, vapores de sangre, acentos de odio . . . Madrid era un infierno. A la caída de la tarde, cuando pudimos salir de casa, vimos los despojos de la hecatombe y el rastro sangriento de la revolución vencida. Como espectáculo tristísimo, el más trágico y siniestro que he visto en mi vida, mencionaré el paso de los sargentos de Artillería llevados al patíbulo en coche, de dos en dos, por la Calle de Alcalá arriba, para fusilarlos en las tapias de la antigua Plaza de Toros.
>
> Transido de dolor, les vi pasar en compañía de otros amigos. No tuve valor para seguir la fúnebre traílla hasta el lugar del suplicio . . . (VI: 1430–1)

In *Ángel Guerra* the hero suffers, at a stressful moment when he dreads presenting himself to his sick and domineering mother, from two successive nightmares (*3c*, Ruiz Ramón 33–8). The first, not linked to a specific incident, is a dream of falling from the fifth storey of an unfinished building. The second draws from his childhood the distant but harrowing personal experience of the San Gil executions. The actual nightmare is not evoked; surprisingly perhaps, no attempt is made to convey the twelve- or thirteen-year-old's unmediated impressions. Instead, the narrator's account starts largely from outside, with some interrogative sentences reflecting use of free indirect style. He recounts the vote taken by Guerra's schoolmates (he is described as 'intrépido') in favour of witnessing the most unusual spectacle; at this stage they are 'saltando de gozo'. The direct telling, after young Guerra has climbed a tree next to a rubbish dump, is gradually infused with Ángel's feelings as his bravado and curiosity combine with fear and horror: 'le amargaba horrorosamente la boca; sentía dolorosa opresión en el pecho; pero la curiosidad pudo más que el instintivo terror . . .' The bare and staccato narrative, instead of becoming free indirect discourse, takes on the hybrid double-voice aspect defined by Bakhtin; without reflecting the boy's utterances, it defines his reactions: the 'espectáculo muy nuevo' of a few lines back has become an 'espectáculo horrible'; the 'silencio de agonía' is succeeded by an 'horroroso estrépito'; and his actual recollections come to be evoked: 'bien lo recordaba Guerra' (V: I. 3. v, 1239).

In *La de los tristes destinos*, on the other hand, the prevailing emotional atmosphere of the time is much more developed. Whereas in *Ángel Guerra* the women who offer cigars and water to the prisoners are unnamed and do not form part of the story, and the tragic situation is personalized in the prisoner's truculent riposte '¿Para qué . . . quiero cigarros ahora?', in the *episodio* the most important among the distraught friends of the prisoners turn out to be none other than the same *Zorreras* we have seen before. Rafaela is depicted as the lover of one of the victims, Simón Paternina, who had previously helped Teresa to locate Santiago Ibero. Occasion is now provided for describing him as a fine young man, an enthusiastic partisan of Prim's and at the same time a devout Catholic. Rafaela devotes half a *duro*—earned in honest work, not in prostitution—in masses for his soul and pictures him as informing God of the wickedness of the Conservatives.

The fortitude and dignity of the condemned men are pathetically maintained as we learn they enjoyed an opulent last supper and—in contrast with the sarcastic prisoner in *Ángel Guerra*—smoked the finest cigars on their way to execution. Through the *Zorreras* and their friend Pepa *Jumos* the grave social questions of the responsibility for the savage sentences and the evident class discrimination against the underprivileged are raised in racy dialogue, full of anacolutha:

> confiábamos en que *la Isabel* perdonaría . . . [*declares Rafaela*] Para perdonar la tenemos . . . ¡Bien la perdonamos a ella, Cristo! . . . A un general sublevado le das cruces, y a un pobre sargento, ¡pum! . . . Tu justicia me da asco.
>
> —No hables mal de ella —dijo *la Pepa* con alarde de sensatez— que si no perdona es porque no la deja el zancarrón de O'Donnell, o porque la Patrocinio, que es como culebra, se le enrosca en el corazón . . . (III: 657)

How directly is the queen responsible? Or is she simply ill-advised, by Sister Patrocinio or by O'Donnell? These are shown to be questions which concern the populace, which still retains some trust in her. It is clear, however, that whatever sensitivity towards the people she once possessed has now deserted her completely. The historical evidence seems strong that, against the wishes of O'Donnell, the queen herself desired the harshest of punishments. Baroja quotes the general as stating: '¿Pues no ve esa señora que si se fusila a todos los paisanos y soldados prisioneros va a derramarse tanta sangre que llegará hasta su alcoba y la ahogará en ella?'[52] Valle-Inclán also reflects her naïve anger, when she incongruously demands '¡Pegar fuerte, a ver si enmiendan!' (*4a*, *Corte* 10).[53]

In *Ángel Guerra*, though Galdós remains interested in drawing attention

[52] 'Isabel II estuvo en esta ocasión frenética; quería el exterminio completo del pueblo', declares Baroja (*4a*, iv. 1171). According to Répide (*1b*, 222), 'la sangrienta sed de Isabel no se saciaba', and he notes that some of those shot were completely innocent. He also quotes O'Donnell's comment, with minor differences, as does the historian Ballesteros Baretta (*1b*, 75); it is no doubt authentic. Llorca (*1b*, 1984: 208) denies the queen's severity and points out that she allowed Castelar, who was hiding in Carolina Coronado's house, to escape, but she entirely overlooks the unprotected sergeants.

[53] For further details see Schiavo (*4a*, 360) and Sinclair (*4a*, 53).

once more to a deplorable political atrocity, the direct consequences are no longer important. The incident, in Bly's words, is 'a pertinent measuring stick by which to gauge an individual character's reaction and motivation', though I do not think it is 'also an escape-valve for his [Galdós's] own conflicting views' (*3c*, 1983: 163). The emphasis lies entirely on the personal anguish the executions created at the time (Ángel suffers from vomiting and is confined to bed for three days) and still create seventeen or eighteen years later in one self-willed boy who is momentarily to take up the cause of revolution. Hence the attention accorded to the gruesome details:[54] the varying stances of the men as they were shot; their incongruous postures as they fell; the shouts, smoke, and dust; the finishing off of those still alive at point-blank range; the even more horrifying effect of the bodies wrapped in blankets; and the haunting impact of the frenzied unidentified figure by his side who had the appearance of a Greek mask and whose hair literally stood on end: this is the image that persists most vividly in his dreams at moments of stress, as when he thinks he has killed Arístides (**V**: II. 4. vii, 1380). History is 'embedded' in a masterly (if still conventional) fashion as a subtle temporal mode in fiction in one case, and systematized, vitalized by fictional accretions, with no loss of authenticity or personal involvement, in the other.

In the upshot, the bloodshed has been futile and even self-defeating. The narrator's words in the *episodio* offer a clear case of what Genette calls iterative narrative (*2*, 116–17) to make a generalization which is at once ironical and distancing: 'heroica medicina contra las enfermedades de autoridad que por aquellos días y en otros muchos días de la historia patria, padecía crónicos achaques y terribles accesos agudos' (**III**: 655).

Immediately following on and interwoven into the action of *La de los tristes destinos* is the interview Manuel Tarfe has with the queen to seek a pardon for Leoncio Ansúrez and Santiago Ibero, who are in imminent danger of deportation. In granting the request, Isabella, now obese, more bourgeoise in appearance, and quite disillusioned, indirectly answers the criticism of excessive harshness: 'Eso para que digan que no perdono, que no soy generosa' (**III**: 662). She also refers to Virginia's unhappy marriage in terms which apply strikingly to herself: 'Y no hay que culpar a Virginia, sino a sus padres, que la casaron con un hombre afeminado y bobalicón, sin maldita la gracia para el matrimonio'; her 'sonrisa picaresca' indicates that she is thinking of her husband. By recommending haste in obtaining their release and protesting unnecessarily in a cynical change of subject that 'Yo quiero mucho a O'Donnell', she forecasts the minister's imminent fall.[55] This peremptory

[54] Compare the scene in *War and Peace* when Pierre, expecting to be the next victim, witnesses the execution of five prisoners by the French (XIII, ch. LXVI).

[55] See Iser, commenting on Jane Austen: '[The reader] is drawn into the events and made to supply what is meant from what is not said. What *is* said only appears to take significance as a reference to what is not said' (*2*, 1978: 168).

dismissal gives rise to his famous exclamation of disgust (unconsciously throwing light on the treatment of the lower classes at the time): 'Me han despedido como despedirán ustedes al último de sus criados' (663).

In short, O'Donnell was allowed by the *camarilla* to reap the opprobrium of the executions before Narváez was summoned to take over. The conclusion attributed to Adelardo López de Ayala (later author of the famous manifesto of the revolution, 'España con honra'), 'Esa señora es imposible' (III: 664), points directly towards the revolutionary alliance of Progressives, Democrats, and dissident Unionists which is soon to bring down the monarchy. In a short conversation with Narváez, Beramendi finds the prime minister desperate, hounded as he is from right as well as left—'Yo me encuentro con la revolución enfrente y con la reacción detrás' (689)—and prophesying, rather facilely, that it will be the death of him. In effect, once O'Donnell and Narváez, who, for all the monarch's capriciousness, would not countenance an uprising directed against her, are dead,[56] her downfall quickly becomes inevitable. They were, as Ibero says later, 'Los dos puntales de la Monarquía'. González Bravo, appointed in place of Narváez, was foolhardy enough to promote two traditionalist generals, Manuel Pavía y Lacy, Marquis of Novaliches, and José de la Concha, Marquis of La Habana, and exile the more dominant Unionist generals, including Serrano, together with Montpensier. In the narrator's words: 'Fue un ademán de suicidio' (732).

Meanwhile the fictional story has continued. Teresa Villaescusa and Santiago Ibero, still completely involved in the anti-dynastic conspiracy, meet as Ibero is escaping to France. Teresa eventually joins him there. As a result of her love Teresa is completely reformed: 'No es la juiciosa que se equivoca, sino la equivocada que rectifica, la fatigada que se sienta y se adormece en la tardía enmienda de sus errores' (III: 706). It is noteworthy that there is no conventional dwelling on the shameful nature of her actions; for Galdós they are 'errores', not 'pecados'.[57] By comparison, even a compassionate writer like Thomas Hardy, though intent on stressing that Tess is 'A pure woman', cannot avoid rubbing in the social fact of her involuntary loss of virginity: 'Maiden no More', an 'immeasurable social chasm was to divide our heroine's personality thereafter from that previous self of hers' (*4b*, 103).

Teresa, for her part, encourages Ibero to follow his bent in supporting revolutionary activities, with the result that he takes part in the hopeless invasion of Aragon in which the general Manso de Zúñiga is killed. She also investigates his idealistic love for Salomita Galán, who has become a

[56] O'Donnell died on 5 Nov. 1867 at Biarritz, Narváez on 20 Apr. 1868, while still in office.

[57] Compare the solidarity with Amparo expressed by her fellow workers in *La tribuna*. As Geraldine Scanlon comments: 'Although Amparo is abandoned pregnant, she is not made to suffer social ostracism or a devastating loss of status, nor is there any suggestion that illicit sex has set her on the slippery slope to total degradation—a common fate for the morally lax fictional heroine' (*4a*, 141).

nun, spurring him to overcome it. On account of her past life and her age, she offers him the opportunity to give her up. Explicitly rejected is the false Romantic image of Marguerite Gautier's self-sacrifice in *La Dame aux camélias* when Armand's father comes to rescue him as Ibero's father is about to do on Santiago's behalf: 'Aquí no hay *damas de camelias*, ni Cristo que lo fundó' (III: 735).[58] Lastly, under the guidance of her French mentor Ursula Plessis she learns a useful and satisfying trade—fine embroidery—an opportunity denied women in Spain.[59]

In Paris Ibero joins his fictional fellow conspirators Jesús Clavería and Manuel Santa María. He gets to know and assists Sagasta, even in common domestic tasks; a vehement figure called Don José also plays some part (information on his actual identity is deferred: it is Paúl y Angulo). Ibero is sent to London to join Ruiz Zorrilla and Prim; justified worries about a Liberal Union/Montpensier conspiracy[60] cause him to be suspected at first. Prim's voyage in disguise on the *Zaragoza* and the turning of the treachery of his Italian aide Antoni to his advantage are recounted, with Ibero as an eyewitness. The grandiose effect of Prim's famous speech of 18 September on national sovereignty is evoked: 'Oído por la marinería el grito del General, ya no sonaron más los fríos clamores de ordenanza, sino que estalló un "¡Viva Prim!" inmenso, ardoroso, y confundido con el estruendo de la artillería.'[61]

On landing at Cadiz, Ibero recalls Ferdinand VII, forcibly taken to the customs house by the Liberal Government in 1823 and, in *Confusio's* imagination, executed there (III: 755); 'real' history and imagined history are again confronted. Ibero has an emotional reunion with his old mentor Lagier, who gives him sage advice on the priorities of heroism and peace: '*Nacemos como un libro en blanco . . . Reconstruid vuestras personas con actos buenos*' (758): precisely what Santiago and Teresa will do at the end. Informed by Tarfe that Teresa has taken to her old ways again with one of the Muñoz family, the distraught Santiago returns to Madrid, where he gives a lively description of the revolution in the streets before ending up in the royal palace, where, among portraits

[58] Galdós similarly avoids another stereotype of the fallen woman's fate: Zola's *Nana*.

[59] See Miller (*3b*, 109–10). Similarly, Aurora's entrepreneurial attitude in *Fortunata y Jacinta* is a result of her French contacts. Compare also Goldman's discussion of the liberating effect, in certain contexts in 19th-cent. Spain, of Feijoo's gift of a Singer sewing-machine to Fortunata (*3c*, 1982). Isidora, of course, will have no truck with the Relimpios' sewing-machine.

[60] At the same time Galdós cannot resist indulging in quite lengthy tourist descriptions of Paris and, especially, of London: the result, evidently, of his travels.

[61] As Llorens notes (*3b*, 144–5), Prim's words are habitually given a purely negative tenor: 'Destruir en medio del estruendo lo existente', as Unamuno put it, affirming that 'Aquel bullanguero llevaba en el alma el amor al ruido de la historia' (*4a*, i. 794). In fact, Prim's speech is much more constructive. The complete text reads: 'Destruir en medio del estruendo los obstáculos que sistemáticamente se oponen a la felicidad de los pueblos, es la misión de las revoluciones armadas; pero edificar en medio de la calma y la reflexión es el fin que deben proponerse las naciones que quieren conquistar con su valor su soberanía, y saben hacerse dignas de ella conservándola con su prudencia'.

of monarchs, including Isabella II's, he dreams of Teresa. The parallel between the two women thus continues. As Bly says, 'The Palace scene has thus established not the similarity but rather the *superiority* of Teresa to her royal counterpart—because of her fidelity and sincerity' (*3b*, 1980: 12).

When, in the last stages of the monarchy, Beramendi and his wife visit the queen with their children, she inordinately praises María Ignacia's late father, provoking in Beramendi a silent comment about her error in supporting absolutism. He also muses about deficient religious education, ridiculous court etiquette, and the arbitrary patronage involved in her insistence on bestowing new titles on them. Her unawareness of why she has lost the allegiance of her subjects and her empty confidence in God's grace cause a summing up and a farewell in which he creates intertextually the title of the *episodio*:

No invoques al Dios verdadero mientras vivas prosternada ante el falso. Ese Dios tuyo, ese ídolo fabricado por la superstición y vestido con los trapos de la lisonja, ese comodín de tu espiritualidad grosera, no vendrá en tu ayuda, porque no es Dios ni nada . . . Adiós, reina Isabel. Has torcido tu sino . . . *Adiós mujer de York, la de los tristes destinos* . . . Dios salve a tu descendencia, ya que a ti no te salve. (III: 704)[62]

Here we have an example of the inevitable priority given to historical events within the *episodios*, especially when what is occurring is of crucial importance. From the perspective of the fictional story, this is presented in far too explicit a manner; yet it is obviously urgent to give a considered, if fictional and oversimplified, contemporary opinion of the key event taking place.

After the defeat of the loyalist forces under Novaliches at Alcolea, Beramendi goes to see the queen again in order to put before her one of her rapidly narrowing options: an immediate abdication in favour of Alfonso,[63] to which Eufrasia adds the idea of seeking Espartero's protection for Alfonso at Logroño—a possible reconciliation of divergent interests which Beramendi enthusiastically endorses. Carlos Marfori,[64] the queen's latest lover, arrogantly scotches this suggestion as beneath the monarch's dignity. We are regaled with a final meditation by Beramendi on 'las 100,000 víctimas inmoladas por Isabel desde su cuna hasta su sepultura [*i.e. her exile*]. Véase la tragedia de este

[62] For the reference to *Richard III* and a contemporary political catch-phrase, drawn from a speech by the influential traditionalist politician Antonio Aparisi y Guijarro, see Shoemaker (*3d*). Galdós also applies the phrase to Amadeo during his reign (*3a*, *Artículos políticos* 22).

[63] An idea attributed to Salamanca (*1b*, Romanones 112), which might have obtained the acquiescence of Serrano (III: 775).

[64] Marquis of Loja, 1818–92. He is described in *La vuelta al mundo* as 'uno de los principales mantenedores del feudalismo narvaísta' (III: 447). Through the protection of his uncle, Narváez, he became successively deputy, governor of Madrid, and minister, but it was his position as chief executive of the royal palace which brought him the queen's favour and unlimited power. Galdós, through Beramendi, contrasts his 'caballerosidad' and 'firmeza de sus ideas retrógradas que sostenía con modestia y sin ofender a nadie' in his youth with his changed personality ('ya era otra cosa': III: 776) when he reached the palace.

reinado, toda muertes, toda querellas y disputas violentísimas, desenlazadas con esta vulgar salida por la puerta de Bidasoa' (III: 778); the passage sums up, perhaps necessarily, the course of the reign, even though it is very much at the expense of fictional verisimilitude. Everyone—Eufrasia, the people of San Sebastián, Teresa—is portrayed as sorry for the queen personally.

Shadows of the forthcoming upheaval are everywhere in evidence in the novels from *Tormento* onwards. The disruption caused by Amparo's disgrace is paralleled by social and political turmoil at the end of 1867 and beginning of 1868. Equally symptomatic are the theft of Bringas's new cloak at the palace ball and the radical agitation picturesquely exaggerated by Bringas, welcomed by the bitter Caballero, and played down by Rosalía: 'Esos horrores sólo están en el entendimiento de mi pobrecito Bringas' (IV: 1564). Caballero's 'immoral' decision to take Amparo with him to Bordeaux as his mistress is a confirmation for Bringas of the widespread moral and political disorder: 'La revolución no tarda; vendrá el despojo de los ricos, el ateísmo, el amor libre' (1568).

Bringas thus stands as the exemplar of inflexible obscurantist views, opinions which are at times reinforced by Rosalía. When they moved into their new apartment, Rosalía suggested putting the portrait of O'Donnell somewhere else (the lavatory is understood) 'por indecente' (IV: 1462), while Narváez's retained its place of honour.[65] In *La de Bringas* the political situation has become more sharply focused; the whole action of the novel, confined geographically and emotionally to the royal palace, is now used in metonymic fashion as a microcosm of Spanish society on the verge of revolution.[66]

Bringas himself, ironically identified in every physical detail with Louis Adolphe Thiers, is manifestly inferior in all respects to the French economic historian, so prominent in another revolutionary situation, the collapse of the Third Empire. Rather than the proponent of productive work like Thiers (*3c*, Varey 65–9), Bringas is merely 'el ratoncito Pérez', his frantic activity being devoted to the preposterous hair-picture in the lowest and most derivative taste (*3c*, Gold 1986). His obsession with hoarding recalls the classical miser and is the reverse of the dynamics of capitalism; in highly ironic fashion, his savings are, for free spenders like Milagros, another example of useless entailed property, 'dinero de manos muertas' (IV: 1605).

The main source of political commentary is the ubiquitous Manuel de Pez, in conversation with Bringas and others. When the Unionist generals are

[65] For a comment on this apartment see Charnon Deutsch (*3c*).

[66] Galdós demonstrates, as Bly says (*3b*, 1980: 3), a 'very intimate knowledge of the Palace interior'. Note particularly the description: 'Es que durante un siglo no se ha hecho más que modificar a troche y moche la distribución primitiva' (IV: 1578), quoted by Varey (*3c*, 63). As Shoemaker notes, 'Rosalía's fall parallels the course of the larger national *ambiente*' (*3c*, 1959: 423).

exiled, Pez at first takes a studiously neutral line, in phrases characterized by pedantry and anaphora: 'La revolución, de que tanto nos hemos reído, de que tanto nos hemos burlado, de que tanto nos hemos mofado, va avanzando, va minando, va labrando su camino' (IV: 1594). While Bringas is ill, he is careful not to alarm him too much by 'ponderar los peligros de Trono ni el mal curso que tomaban las cosas' (1618). In a remarkable passage the voluble eloquence of the bureaucrat reduces the narrator to a semi-conscious state (*coma vigil*) with the comings and goings of Prim, Cabrera, and the queen. The natural medium is free indirect discourse, with Bakhtinian double-voiced effect; this is not a narrator to be relied upon! What is conveyed, in renewed anaphoric phrases, is the deepest pessimism: 'Este era el país de perdición, un país de aventuras, un país dividido entre la conspiración y la resistencia. Así no podía haber progreso, ni adelanto, ni mejoras, ni tampoco administración' (1624)—this last, evidently, the essential thing for the archetypal bureaucrat.

Later, in conversation with Bringas he criticizes 'aquella obcecada política de González Bravo . . . La prisión de los generales y del duque de Montpensier era una torpeza' (IV: 1641), and, to Bringas' indignation, takes slight consolation in the involvement of the Liberal Union. Francisco and Rosalía, for their part, rejoice in the expulsions as a sign of forceful government (1629). Even in one of Isabelita's nightmares, among indications that money is scarce and a diabolic vision of Pez, there are, in Genette's terminology, proleptic forebodings, from Bringas, of bloody revolutions: 'Vargas y su papá dijeron también que iban a correr ríos de sangre, y que *la llamada* revolución iba a venir sin remedio' (1638). Paquito de Asís (named, obviously, after the king-consort, as his sister is named after the queen) is exposed, to his father's horror, to Krausist and revolutionary ideas at the university (1656, 1658). Finally, when Bringas learns that the navy under Topete has revolted, he is mentally and physically overwhelmed: it is 'la de acabóse' (1666). Pez declares that 'tal derrumbamiento [no] se podía componer, pues la reina estaba perdida y no tenía más remedio que meterse en Francia', even though Galdós, typically, gives voice to other opinions, such as Doña Cándida's—hardly a reliable authority—that the queen had only to appear in Madrid to quell the disturbances (1667).[67] For Bringas, Prim is the monster who has brought about the dreaded revolution, and the loyal traditionalist rejoices at the false rumour (a pseudo-prolepsis) that the revolutionary leader has died at the barricades; his actual death two years later is thus anticipated by authorial hindsight: to a reading public of 1883, his eventual disappearance is more significant than his living presence.

[67] This course of action was supported initially by Concha, the Marquis of La Habana, but he quickly changed his mind (III: 775, and *1b*, Ballesteros Baretta 103). Cambronero, dismissing the report that the railway line has been cut at Burgos as 'un ardid de los enemigos de la dinastía', goes on to affirm that 'no es conjetura infundada suponer que la presencia de Isabel en Madrid y su abdicación en el Príncipe de Asturias hubieran conseguido parar por completo el movimiento revolucionario' (*1b*, 1908: 310). Salamanca, according to Cambronero, advocated this course.

The consistent and meticulous use of historical detail essential for an *episodio* is not required. Instead the novel offers an intense concentration on the effects of the crumbling of a style of life. It focuses on the end, or apparent end, of a dynasty and a system, with two divergent reactions presented. On the one hand, Bringas's apocalyptic conviction that all traditional values (such as they were) are irreparably lost: 'Era el acabamiento del mundo'; moreover, the narrator has already told us in *Tormento* (IV: 1458), in a long-lasting analepsis, what a shadow of his former self Bringas had become some fifteen years later: the revolution has clearly been catastrophic to him personally. On the other hand, we witness the cynical, accommodating view of Rosalía and Pez and their confidence in facing the unknown: Rosalía, 'que no miraba la revolución con ojos tan implacables como su marido; que confiaba en la vuelta de la reina . . . y que, en tanto, había que esperar los sucesos para juzgarlos' (III: 1669–70), and Pez, who 'con reposado lenguaje y juicioso sentido, se defendía enalteciendo la teoría de los hechos consumados, que son la clave de la Política y de la Historia' (1670); among his (unremunerated) *faits accomplis* was, of course, his affair with Rosalía. Each in his/her own fashion now seeks, and will no doubt find, ways of adapting him/herself to the new circumstances,[68] in the assurance that things will not change too much. About Rosalía, despite the narrator's prudent refusal to get any further involved with her, we are informed that 'en lo sucesivo supo la de Bringas triunfar fácilmente y con cierto donaire de las situaciones penosas que le creaban sus irregularidades'. With such a promise of new endeavours, it is surprising that Rosalía does not appear in any subsequent novel.

The take-over of the palace in *La de Bringas* and *La de los tristes destinos* offers an interesting comparison.[69] In the first Galdós gives his anonymous narrator, already involved in shady dealings with Bringas and Pez and over-friendly with Rosalía, this responsibility; there is no great extremist menace here, for the revolutionaries who entered the building, though they horrify Bringas, are remarkably meek, and the new administrator is very accommodating to his friends. External evidence confirms the surprisingly peaceful transfer of the building, and the prominent figures who were involved, such as Nicolás Rivero and Pascual Madoz, take pride in this achievement.[70] With

[68] Pez, naturally, has friends and relatives in the revolutionary junta: 'dondequiera que volvía mi amigo sus ojos, veía caras pisciformes' (IV: 1670). See also an 1885 article by Galdós: 'Tras de una perturbación más o menos grande, según las localidades, volverán las cosas al estado antiguo, y todo seguirá lo mismo, los capitalistas siempre explotando, los obreros trabajando siempre y viviendo al día' (*3b*, quoted in Goldman 1975: 7).

[69] Again Bly's comparison is excellent. In my opinion he has the proportions exactly right when he says (*3b*, 1980: 14) that *La de Bringas* is three-quarters a socio-psychological novel and one quarter historical novel and that *La de los tristes destinos* is three-quarters historical novel and one quarter socio-psychological. It seems to me rash, however, to treat *La de Bringas* as an *episodio nacional* (*3c*, 1981: 62–9) without considerable qualification.

[70] The subject was discussed by the *Cortes constituyentes* on 20 Mar. and 8 May 1869. Gallenga (*1b*, i. 214), independently or drawing on the Cortes sessions, also bears witness to the peaceful transition. The intendent appointed to take charge of the royal palaces was José Abascal.

his characteristic distrust of popular disturbances, Galdós provides a special explanation in the *episodio*. The palace is saved by the presence of mind of a typesetter, one Casimiro Muñoz, who, dressed in his most formal clothes, takes charge of the building on behalf of the *Junta revolucionaria*, clears out the rabble who are already there, and brings in as many respectable citizens as he can find. An inventory is made and nothing is missing. In this way the peaceful occupation that Pascual Madoz proclaimed is accomplished: 'Fueron a Palacio y se apoderaron de Palacio casi voluntariamente más de 200 paisanos sin organización, y dicho sea en honra suya, salvaron la propiedad; no faltó nada, absolutamente nada de las grandes riquezas que aquel palacio contenía' (*1b*, Cambronero n.d.: 13).

What does the 1868 revolution augur for the future? Beramendi's conclusion is that the queen herself will not return, and rather obvious hints of the eventual restoration are given. Has it therefore all been in vain? Bly emphasizes the pessimistic side (*3b*, 1980: 10), and not without justification. As has been pointed out,[71] the two characters who evidently represent the best of Spain, Teresa Villaescusa and Santiago Ibero, tested in difficult circumstances and now newly reunited and conscious of their unorthodox life, leave Spain along with the exiled queen. Unlike Leoncio, Ibero refuses Tarfe's offer of a job or money, and turns his back on the opportunistic values which link the so-called 'Glorious' Revolution with the Liberal Union: 'Desde aquel momento, el infeliz Ibero, solo, errante, sin calificación ni jerarquía en la gran familia hispana, miró desde la altura de su independencia espiritual la pequeñez enana del prócer, hacendado y unionista' (III: 766).

The pseudo-dramatic dialogue between the two lovers with which the *episodio* concludes offers a sustained oxymoronic contrast emphasized by chiasmus between the departing queen and themselves; one fleeing from Freedom, the others seeking Freedom; one suffering from *tristes destinos*, the others pursuing *alegres destinos*:

> IBERO—Doña Isabel no volverá, ni nosotros tampoco... Ella, destronada, sale huyendo de la Libertad, y hacia la Libertad corremos nosotros. A ella la despiden con lástima; a nosotros nadie nos despide; nos despedimos nosotros mismos diciéndonos: Corred, jóvenes, en persecución de vuestros alegres destinos. (III: 781)

And when they have crossed into France at Hendaye, Ibero takes up the 'España con honra' slogan and inverts it as, in the last lines of the narration, they seek an unknown European destiny:

> IBERO—... Adiós, *España con honra*.... Somos la *España sin honra* y huimos, desaparecemos, pobres gotas perdidas en el torrente europeo. (III: 781)

Individual freedom to develop, as advocated by Lagier, is preferred over the

[71] e.g. Montesinos (*3c*, iii. 149) and Lida (*3d*, 71), as well as Bly.

new but very fragile opportunity of overcoming the cynical transactions of the past. Yet it would be wrong, perhaps, to dwell too much on the pessimism, since Galdós, in accordance with his normal tactics, offers no conclusion for the moment; the opportunity for change is there if only it can be seized effectively. The narrative is to that extent open-ended.

Critics have frequently remarked on the considerable personal sympathy Galdós shows towards Queen Isabella. In *La de Bringas* there is no direct assessment of her actions: *la señora* is simply the invisible presence at the hub of the vicious circle of patronage and intrigue, destined soon to disappear. Sympathy there certainly is for her in both the *episodios* and the 1904 piece,[72] but this is not to be taken as a complacent or reprehensible exoneration of her conduct.[73] As Diane Urey sensibly observes, 'The picture that emerges of her, although not devoid of sympathy, is too replete with irony to be deemed favorable' (*3d*, 1989: 109). What is important to bear in mind is the steady development within the *episodios*: the queen's behaviour emerges gradually before the reader's eyes. She is presented most favourably at the beginning of her reign, when her spontaneous and popular qualities offered (in novelistic, not historical, terms) some hope of prevailing. She progressively loses her popularity, however, so that by the late 1860s she no longer commands any respect or esteem: her fresh sensuality has degenerated into obesity, her naïvety into cynicism, her popular charms into vulgarity, and her capricious generosity into cruelty. Her promiscuity has become an international scandal. The free-living Teresa, who leaves Spain at the same time, has shown herself her superior. Isabella herself reveals a complete *desencanto* with political life and on her departure makes a historically authentic statement (*1b*, Clarke 305) which demonstrates how out of touch she is from reality: 'Francamente, creí tener más raíces en este país' (III: 777).[74]

In the 1904 article there is an attempt to sum up her career. Galdós shows great understanding of the effects of her environment, the complete lack of preparation for her duties, the responsibilities thrust upon her in her adolescence, the disastrous marriage foreign intrigue imposed on her; and, no

[72] Seco Serrano greatly exaggerates the 'vínculos de cordial amistad personal con Don Benito' (*3d*, 267) and appears to believe that incidents previous to *Narváez* derive from their personal contact.

[73] 'La simpatía por Isabel le hace a Galdós cometer la injusticia de ser condescendente con los matones que la sostenían en el trono' (*3d*, Regalado García 426). Compare Carr (*3d*, 185), who, while admitting 'a great deal of "sentimentalismo ñoño"' in Galdós's attitude, claims that 'Galdós was surely partly correct in taking the view that she *was*, to a large extent, a victim of circumstances.'

[74] Gallenga comments on her departure as follows: 'the infatuated Queen crossed the frontier . . . with Marfori—her latest and most worthless favourite—on one side, and her confessor, Father Claret, on the other. On the right, her besetting sin; on the left, the easy atonement' (*1b*, i. 134).

less, he appreciates the fatal consequences of her inalterably childish and superficial mentality and of her ingrained favouritism. He is fully aware, moreover, of how subject she is, like the political generals who dominate her reign, to the constraints of historical tradition and circumstance. In many respects she is, in her virtues and her vices, a representative figure. Despite this degree of understanding, Galdós is none the less unequivocal in his assessment of her political legacy. He is in agreement with modern historians (e.g. *1b*, Carr 214) that the great political error of the time was the disinheriting of the Progressives. Her personal laxity forms an incongruous and oxymoronic contrast with her political subordination to clerical reaction and intrigue. Caprice, with its attendant qualities of frivolity and hypocrisy, is erected into a norm of conduct. *Caciquismo*,[75] an offshoot of patronage, is given free rein. In short, the balance sheet is not a favourable one. It does, however, raise in Galdós's mind the fascinating question of what it might have been, and his speculations on this point, in the 1904 article and later exemplified in the figure of Santiuste-*Confusio*, are of great interest. They deserve separate treatment in Chapter 7.

[75] In an article in *La Prensa* (25 Jan. 1891) Galdós is extremely critical of what he calls 'la verdadera lepra del régimen constitucional . . . la planta venenosa de nuestro organismo actual. Los gobiernos conservadores la han fomentado con descarada audacia, y los liberales no han podido o no han querido extirparla' (*3a*, *Cartas desconocidas* 446–7).

4

Prim and the Glorious Revolution of 1868

THE September revolution of 1868, called 'la Gloriosa', was the major political upheaval in Spain since the Peninsular war.[1] It set out to make a fresh, if ill-defined, start by eliminating a corrupt dynasty and launching a new *España con honra*. It thus initiated a six-year period of revolutionary fervour, experimentation, and turmoil ('el sexenio revolucionario').

The revolution had an indelible effect at the time on the twenty-five-year-old Galdós.[2] During his return from a second visit to Paris, the revolution found him in Barcelona. Instead of proceeding to Las Palmas, he managed to reach Madrid in time for Serrano's triumphant entry, followed by Prim's. In an 1885 newspaper article he spoke of the 'hecho más grave que España ha escrito en los anales del siglo XIX: el destronamiento de doña Isabel II... Aún están frescos en la memoria de todos los sucesos de aquellos días en que Serrano compartía con don Juan Prim la popularidad más grande de que han gozado los hombres políticos en España' (*3a, OI* vi. 294–5).[3]

Emilia Pardo Bazán is aware of its potentially momentous nature when she declares in *La tribuna* (1883):

> Ocurrió poco después en España un suceso que entretuvo a la nación siete años cabales, y aún la está entreteniendo de rechazo y en sus consecuencias, a saber: que en vez de los pronunciamientos chicos acostumbrados, se realizó otro muy grande, llamado Revolución de septiembre de 1868.
>
> Quedóse España al pronto sin saber lo que pasaba y como quien ve visiones. No era

[1] Josep Fontana raises the question of whether the 1868 revolution was, in Marxist terms, a 'social' revolution or merely a political one. His forthright answer is that it was a revolution of business interests claiming to be a social revolution, containing 'trazas de un gigantesco engaño: de un golpe de estado disfrazado de revolución' (*1b*, 126–7). See also Nicolás de Albornoz, 'El trasfondo económico de la revolución', and other studies in Lida and Zavala (*1b*).

[2] For a thorough account of the revolution and Galdós's reaction to it see Bly (*3b*, 1980: 1–8). Gallenga has a good description of the event (*1b*, i. 129–99). Narcís Oller recalls the revolution in a small Catalan town—Valls—in 'Records de noi: La Revolució de Setembre,' (*4a*, 1125–33).

[3] Bly (*3b*, 1980: 3) points out that this was written only just over a year later than *La de Bringas*. Another example occurs in *Miau* (1888), when the narrator refers to the revolution as 'el mayor trastorno político de España en el presente siglo' (V: 588). In an article of 1886 Galdós declared that 'los derechos políticos se conquistaron de un modo definitivo en la revolución de 1868': a significant though limited achievement (*3b*, quoted in Goldman 1971: 122–3). For the literary consequences see López-Morillas (*4a*).

para menos. ¡Un pronunciamiento de veras, que derrocaba una dinastía! (*4a*, ch. IX, 'La Gloriosa', 104)[4]

Moreover, post-revolutionary politics and love life are inextricably bound together in the existence of the heroine, Amparo, as Germán Gullón has pointed out (*4a*, 50). Amparo, a worker in a tobacco factory, enthusiastically embraces Federalist ideas and, naïvely believing that equality will now prevail, allows herself to be seduced by a middle-class youth, who of course does not fulfil his promise of marriage. The novel portrays a strong sense of solidarity among working-class women, as well as an ingenuous Messianic belief in the Federal Republic. Amparo's son is born, at the end of the book, at the same time as the Republic is proclaimed, with the implication that the fate of the child and the system will be equally uncertain.[5]

Although the rising had three titular leaders (Serrano, Topete, and Prim) and Serrano was the nominal head of the movement, the key figure on whom the success of the revolution depended was Prim. By July 1869 a monarchist constitution had been approved and Serrano, imprisoned in what Castelar called a 'jaula de oro', had been effectively neutralized as regent. Prim, now prime minister, set about the difficult task of finding a suitable constitutional candidate for the vacant throne. His firm rule ended with his assassination in December 1870, just before the arrival of the newly elected monarch Amadeo I of Savoy,[6] who reigned from January 1871 to February 1873. The first phase of the 'sexenio revolucionario' is completed.

Prim is the last figure on Estupiñá's extensive list of historical celebrities witnessed 'en un balcón', though he is not mentioned by name: 'había visto, en fecha cercana, a otro personaje diciendo a gritos que se habían acabado los reyes' (V: I. 1. ii, 35). His search for a monarch has relatively little impact on the novels, for which it was already past history. Instead, he counts as an imposing figure of the immediate past: the third type of historical reference outlined in Chapter 1. Thus Don Baldomero laments, at the time of Amadeo's abdication (V: I. 8. iii, 83), that Prim is no longer alive, just as Bringas had deplored Narváez's recent death from a very different political viewpoint ('Si viviera don Ramón', IV: 1667) when the 1868 revolution broke out. One occurrence which is specially recalled is Prim's famous refusal to countenance any restoration of the Bourbons in the so-called *sesión de los*

[4] For the historical background of the novel, the first to deal with working-class political activities, see Henn (*4a*), and for an acute examination of the gender issue in the novel, Scanlon (*4a*). Francisca González Arias indicates some points of contact with Isidora and Fortunata (*4a*, 137–8).

[5] The otherwise incongruous change of direction ('peregrina idea': *4a*, Gullón 52) in *Memorias de un solterón* (1896), in which the son persuades his widowed father to marry Amparo, corresponds to a completely different concern: the impelling desire for an heir (cf. the end of *Fortunata*), as well as the eager availability of Sobrado to fulfil that role.

[6] Amadeo of Savoy, Duke of Aosta, 1845–90, was the second son of Victor Emmanuel II, first constitutional monarch of reunited Italy.

jamases of 20 February 1869.[7] It is frequently cited as a sign of the fallibility of human judgements. So Don Baldomero, after the discussion with Juanito quoted above, contradicted his son 'y sacaba a relucir los *jamases* de Prim' (V: I. 8. ii, 86) against the Restoration. Doña Lupe also quotes Prim, with a certain amusing incongruity, as if he were still alive, when her nephew Maxi adamantly refuses to contemplate a restoration of Fortunata to the domestic fold: 'Ya te he dicho que no es prudente soltar *jamases* tan a boca llena sobre ningún punto que se refiera a las cosas humanas. Ya ves el bueno de don Juan Prim qué lucido ha quedado con sus *jamases*' (V: III. 5. ii, 361). The remark also serves to mark the political time-scale: the Bourbon restoration, like Maxi's domestic one, though Doña Lupe does not know it, is at hand. It is curious that this famous parliamentary session is not dealt with in the *episodio*.

For boys playing war-games in *La desheredada*, in that dangerous transition from boys to men, Prim is the hero: '¡Soy *Plim*!' (IV: I. 6. i, 1000). His charisma and the visual effect of his gallant posture, with his characteristic *ros* serving as a synecdoche symbolizing his valour,[8] remain strong; they are still recent enough at the time of the action of the novel—some fifteen months have elapsed since his death—to command respect, and they last, with diminishing force, until the time of writing, 1881.[9]

This theatricality, not devoid of vanity, is the quality which Valle-Inclán exploits in his adverse and brilliantly caricaturesque portrayal of the Catalan general. Far more concentrated than Galdós or Baroja in his historical interests, Valle is fascinated by the potential for satire, especially visual or scenic satire, of the last year of *la reina castiza*'s reign. In *La corte de los milagros* Prim is portrayed as a hero of playing-card histrionics: 'El General Prim caracoleaba su caballo de naipes en todos los baratillos de estampas litográficas' (*4a*, 9). Later, in *Baza de espadas*, he is depicted as false, hollow, and vainglorious in appearance and utterance:

Don Juan Prim, verdoso, cosméticas la barba y la guedeja, levita de fuelles y botas de charol con falsos tacones, que le aumentaban la estatura, sacaba el tórax. Pisando

[7] 'Se aventuró [Prim] a decir que la dinastía de los Borbones quedaba hecha trizas y que había desaparecido para siempre de España, y aunque es indiscreto aplicar el adverbio *siempre* tratándose de acontecimientos políticos, siempre inseguros y mudables, en aquel caso especial tenía la convicción de que los Borbones no volverían *jamás*, *jamás*, *jamás*; frase que produjo un efecto sorprendente, que se recordó cientos de veces y que hasta dio nombre a la sesión en que fue pronunciada, pues se la designó en aquella época por la "sesión de los jamases"' (*1b*, Cambronero, n.d.: 3). Prim repeated the same sentiment and the same phrase in the Cortes with even more emphasis on 11 June 1870 (*1b*, Ballesteros Baretta 158–9).

[8] These characteristics are visible in his portraits, one of which Galdós describes and criticizes in his 'Crónica de Madrid' of 22 June 1865 (VI: 1289). It is by Román Sanz and depicts the battle of los Castellejos. Another heroic portrait is by the French painter Henri Regnault. Surprisingly, Bly does not mention the iconography of Prim in his excellent book on Galdós and the visual arts (*3b*, 1986).

[9] 'Esta ilusión, que era entonces común en las turbas infantiles, a pesar de la reciente trágica muerte del héroe, se va extinguiendo ya conforme se desvanece aquella enérgica figura. Pero aún hoy persiste algo de tan bella ilusión' (IV: I. 6. i, 1000).

fuerte y abriendo vocales catalanas, hacía temblar el Trono de Isabel II. Decoraba sus jaquetones propósitos con la retórica progresista que resplandece en los himnos nacionales. (*4a*, 164).

In another equally savage *estampa* he is shown as an example of arrogant theatricality:

Soldado de aventura, con una fe mesiánica en su estrella, no dejaba de mirarse sobre un bélico corcel de tiovio, bordando los campos hispánicos, como otro patrón Matamoros. Con albures de cuartel y arrogancias de matante, presumía que, puesto el ros sobre una ceja, tosiendo fuerte y echando roncas, podía ser el salvador de España (*4a*, *Baza* 181).[10]

But to return to *La desheredada*: the boys are a juvenile, folk-hero version of contemporary adult violence, in the spirit of the popular strip cartoons or *aleluyas*; their play, too, like its model, will culminate in death: 'Era una página de la historia contemporánea de España, puesta en aleluyas en un olvidado rincón de la capital. Fueran los niños hombres y las calles provincias, y la aleluya habría sido una página seria, demasiado seria' (IV: I. 6. 1, 1001).[11] They seem to declare: 'España, somos tus polluelos, y, cansados de jugar a los toros, jugamos a la guerra civil' (1002). Conversely, those who resort to actual civil war—Cantonalists, Carlists—are behaving like irresponsible children. Finally, we should remember that for the protagonist, Mariano, the fight and *Zarapicos*'s sudden end are a foretaste of his attempted regicide and subsequent execution.

What is retained in the novels, then, of an essential figure in Spanish history—almost comparable to Lincoln in the impact his life and assassination had on his country—is a rather blurred heroic memory and the image of civil strife. His career gives rise to occasional reflections on the absence of a strong man or on dogmatic prophecies, but no longer forms any part of the historical process. Thus no attempt is made to fill in biographical facts, to assess his character or the quality of his political decisions, since these factors have no validity for the fictional plot.

Considerations of this sort belong to the *episodios*. We have a new series, the fifth, undertaken at a time (October 1907) when Galdós had already embarked on his new political venture as a Republican *diputado*. He speedily took an active part in forging an alliance between Republicans and Socialists (*3b*, Fuentes). Initially, he had apparently intended to give up at the end of the fourth series,[12] obviously a good stopping-point, but one which would leave

[10] See Sinclair, who speaks of the 'devastating condemnation' and 'exceptionally harsh treatment' afforded Prim (*4a*, 84).

[11] Carmen Bravo Villasante (*3c*) claims, not improbably, that the episode is based on a popular *aleluya* called *La pedrea*.

[12] At the time of the serial publication of *España sin rey* in *El Liberal* (16 Jan. 1908), it is stated that 'Al publicarse el año pasado *La de los tristes destinos* experimentaron honda pena los

unsaid much which profoundly preoccupied him as he re-entered the political arena. Next he proposed to limit the series to only two volumes, *España sin rey* and *España trágica*. The inference is that, given his political activities and his work as a dramatist, continuing in this genre is not now Galdós's top priority. Certainly, as Gilman (*3d*, 1986: 50) has argued, the pain of striking a sensitive nerve in regard to near contemporary events (this was the reason which he earlier alleged for abandoning the subgenre in 1879) prevails once more with increased force; the death of his admired Prim affected him deeply. At the same time he can hardly have experienced much satisfaction in deciding to conclude at this tragic and inconclusive point.

Be that as it may, the first two volumes, basically similar to the previous series, deal exclusively with Prim. Even here political attitudes tend to be embodied in the fictitious characters rather than described in direct narrative. Failure to appreciate this fully causes some critics[13] to underplay the historical content. Thus, although there is no doubt about the impact Prim exerts over events, the characteristic method employed is to record comments on his character or actions drawn indirectly from various sources.

Even the most adverse judgements of his personality (and Galdós, characteristically, gives a wide gamut of opinions) acknowledge his power and significance. In *España sin rey*, for example, the wayward traditionalist Celestino Tapia declares that the Progressive forces would disappear into thin air without 'ese hombrecillo desmedrado y lívido . . . ese Prim, monstruo que parece un arrapiezo, saco de malicias, vaso de bilis. Su perversidad es tan grande como su inteligencia. Y ahí le tiene usted: es el amo . . . Prim es el estorbo insuperable, la rémora, el atasco' (III: 813). The Carlist Trapinedo, shortly to adhere to Cánovas, is more equitably impressed: 'Justo será decir que le agradaron la persona enjuta y el amarillo rostro del general de los Castillejos, así como su oratoria, ceñida, clara, de genuino estilo militar' (843). Ángel Cordero,[14] in *España trágica*, an impassioned partisan of Montpensier's, criticizes Prim's lack of the administrative and economic virtues attributed to his leader, together with his extravagance, insatiable ambition, and plebeian mentality. Prim's chief advocate and confidant is Colonel Santiago Ibero,

admiradores del gran novelista, entendiendo que éste, después de treinta y tantos años, ponía término definitivo al monumento nacional de los episodios.' Berkowitz seems to allude to this decision when he states that '[in] 1905 Galdós announced that with the publication of two more historical novels—*La vuelta al mundo en la 'Numancia'* and *Prim*—he would abandon the genre for ever' (*3b*, 1948: 341), but he mistakenly omits the last of the series, *La de los tristes destinos*. The same note in *El Liberal* indicates that 'el insigne Pérez Galdós ha terminado el *Episodio Nacional* con que inaugura una nueva y última serie, la cual constará solamente de dos tomos.'

[13] e.g. Dendle: 'Historical recreation was often accessory to the exigencies of a fictional narrative rooted in a moral vision of Spain' (*3d*, 1980: 183); and Regalado García: 'En *España sin rey* los elementos novelescos desbordan a los históricos' (*3d*, 445).

[14] No relation, apparently, of Benigno Cordero. In sharp contrast with his homonym he represents the most grossly materialist of bourgeois values, incarnated in the Duke of Montpensier. See Llorens (*3b*, 113–14).

whose initial allegiance to Espartero (who declined the throne offered to him) has now turned, in a way typical of many progressive Spaniards, to Prim. Vicente Halconero, consorting at this time with a group of Federalist sympathizers, admires him 'por su energía, por su buen tino como pastor de pueblos y por su habilidad o astucia política; que en él se manifiestan reunidos el león y el zorro' (945), but he is not sure of his moral principles; he thinks him Machiavellian and rejects what he calls 'ese positivismo [que] será siempre mirado como una ignominia en esta nación romántica'. These doubts arise from the widespread accusations that Prim was planning to sell Cuba to the United States,[15] something which Ibero staunchly denies. As usual, the historical issue is integrated into the fiction; Prim's dispatch on the Cuban question to the American ambassador Daniel E. Sickles is passed to Halconero by Pilar Calpena and then nearly destroyed by his mistress Eloísa. Finally it is shown to Beramendi's nephew, Segismundo Fajardo. It reveals that Prim was ready to contemplate devolution or even independence for Cuba and willing to use the United States as a third party. If this were made public, it would, as Segismundo says, be denounced as 'una atrocidad, una vergüenza con taparrabo, una ignominia sobredorada'. Galdós's hindsight enables him to make 'una previsión profética' (956) or proleptic leap forward in time, as does Prim himself; 'Ya lo llorarán,' he declares, after the projected scheme has been vehemently rejected by the supreme council of war. Naturally, in the light of the Spanish–American war and its outcome Prim's explorations appeared by 1909 far-sighted and prudent rather than shameful or antipatriotic. In the novels another aspect of Cuba is touched on in *El amigo Manso*, with the return of the *indiano* José María Manso, complete with fortune, political ambitions, Cuban wife from Sagua la Grande, mother-in-law, sister-in-law, and a *mulata* and a black boy as servants.

On rare occasions, however, Prim is presented directly. When he reaches a low point in his popularity and stones are thrown at him in the street (March 1870; **III**: 931), he is able to impose himself by dint of personality. More extensively, he is shown, accompanied by two of his real-life military counsellors, Sánchez Bregua and Azcárraga (later to enjoy successful military and political careers), as giving a detailed account to Santiago Ibero—in colloquial direct speech, full of exclamations, repeated phrases, and rhetorical questions—of his aims and policies. Extremely concerned at the high level of political passion in Spain ('El agua hierve, hierve'), Prim claims to have attempted to combine energy with patience. He emphasizes the need for long-term considerations: 'El único filósofo que puede dar obras duraderas es el Tiempo, y nosotros, plantados en un *hoy* apremiante, tenemos la misión de resolver el problema de un solo día . . . Este día puede ser de veinte, de cincuenta, de cien años.' Unwilling to contemplate a personal tyranny, he aims

[15] On Prim's attitude to Cuba see Santovenia (*1b*, 198–214).

through a new monarchy at 'la fusión perfecta del principio monárquico y del principio democrático'. He singles out three Republican politicians whose opinions are disquieting to him: Estanislao Figueras,[16] who had declared that he preferred 'la república más loca a la monarquía más cuerda y liberal' (947); Francisco Pi y Margall,[17] because of his rigid theoretical position; and his former friend and supporter José Paúl y Angulo,[18] whose insane zeal had led him to plan, to Mazzini's alarm, a march on Rome to dissolve the reactionary Ecumenical Council.[19]

It is interesting to note in passing a different case of insertion of history into literature. In the short novel by the Catalan novelist Narcís Oller entitled *La bogeria* [Madness] (1899), Prim is the idol of the deranged revolutionary protagonist Daniel Serrallonga, who has assembled a remarkably extensive collection of pictures of his hero. The narrator lists Prim's accomplishments as displayed in the prints: as a heroic soldier, tribune of the people, seasoned politican, and thoughtful diplomat, adding two qualities which were missing to Catalan eyes:

> ... les dues toques que desllustren la brillant biografia del gran fill de Reus: l'ajuda que per ambició personal dispensà un dia als moderats fent-se a canonades contra els fills de sa pròpia regió i aquell assentiment, bon xic provincià burguès, que solia dar, davant dels castellans, a la nefasta sinonímia d'Espanya i Castella, sens dubte per adular-los. (*4a*, 710)
>
> [... the two traits which tarnish the brilliant career of Reus's great son are the support which through personal ambition he at one time offered the Conservatives by shelling the children of his own region and his acquiescing in front of Castilians, no doubt to ingratiate himself with them, very much in the style of a provincial bourgeois, in the pernicious identification of Spain and Castile.]

The first criticism is of the same opportunism in opposing Espartero about which Galdós waxed sarcastic in *Bodas reales* (II: 1315), the second is of his acceptance of centralist orthodoxy rather than allegiance to the distinctive characteristics of his native Catalonia.

The most complex problem facing Prim was the question of succession. All that is required in a novel is a rapid synopsis of options in accordance with the second category established in Chapter 1. In *Fortunata y Jacinta* such a

[16] 1819–82, a prominent Federalist from Catalonia and the first president of the Federal Republic.

[17] 1824–1901. He was the most intellectual, austere, and legalist of the Federalist leaders. He gives his own account of the reign of Amadeo and of the First Republic (*1b*, Pi y Margall [1874] 1980; see also *1b*, Jutglar).

[18] 1840?-92, a fervent Federal Republican from Jerez, active in the September revolution. After Prim's assassination he lived abroad, vehemently denying his involvement in the murder.

[19] Nicolás Estévanez describes this hare-brained scheme in his *Fragmento de memorias* (*1b*, 328–9). In Galdós's copy in the Casa-Museo in Las Palmas (*3a*) the passage is marked in pencil in the margin.

telescoping technique is associated with Juanito Santa Cruz. It serves two purposes: to characterize Juanito's fickleness as the narrator displays his changes of political affiliation (we have already witnessed his short-lived radicalism in the *Noche de San Daniel* incident), and to provide his audience in the process with a short list of possibilities:

Juan era la inconsecuencia misma. En tiempos de Prim, manifestóse entusiasta por la candidatura del duque de Montpensier.

—Es el hombre que conviene, desengañaos, un hombre que lleva al dedillo las cuentas de su casa, un modelo de padres de familia.

Vino Don Amadeo y el *Delfín* se hizo tan republicano que daba miedo oirle.

—La Monarquía es imposible; hay que convencerse de ello. Dicen que el país no está preparado para la República; pues que lo preparen . . .

Pues señor, vino el 11 de febrero [*Amadeo's abdication and the proclamation of the Republic*] y al principio le pareció a Juan que todo iba a qué quieres boca.

—Es admirable. La Europa está atónita. Digan lo que quieran, el pueblo español tiene un gran sentido.

Pero a los dos meses, las ideas pesimistas habían ganado ya por completo su ánimo:

—Esto es una pillería, esto es una vergüenza . . .

Por graduaciones lentas, Juanito llegó a defender con calor la idea alfonsina . . . (**V**: I. 8. ii, 85–6)

This enumeration covers the essential contenders—Montpensier, Amadeo, the Republic, Alfonso—while representing a very low order of slick political comment by a superficial playboy.

The problems are systematically rehearsed[20] in the two *episodios*. Indeed, the confusion of the entire political situation is vividly and humorously captured in the erroneous way Don Wifredo's servant Filiberta uses the word *caos*, applying a false singular to the leading political tendencies:

¿Sabes tú, Filiberta, lo que es el caos? [*asks Don Wifredo*]

Señor, como saberlo, no lo sé . . . pero ello debe ser algo parecido a la República federal, porque ésta no se les cae a la boca . . . Pero el otro *cao*, el de Carlos VII, también tiene pelos . . . Y para que estemos más divertidos, *cao* de Montpensier, *cao* de Espartero, y del demonio coronado. Digo, señor, que no ganamos para caos. (**III**: 883)

España sin rey encompasses what is frequently called 'la maldita Interinidad', the dangerous period dominated by Prim before the choice of a new monarch is finally made. The traditionalist priest Cristóbal de Pipaón puts it strongly, in his own idiosyncratic idiom, indicating that it serves the Montpensier cause, which he links, through the Orleanist family, with the French Revolution:

[20] For a concise account of the various candidates ('comidilla de todas las bocas en aquellos días': *España trágica* **III**: 903) and the problems raised, see Fernández Almagro (*1b*, 61–81).

[21] Carlos de Borbón y de Este ('Carlos VII'), 1848–1909, gave new vigour to Carlist claims after the expulsion of the rival branch of the family. He was known to his adversaries as *el Niño Terso*.

El intríngulis de esto bien claro se ve: que España se aburra, que España se desespere y a gritos pida la conclusión de esto que llaman *interinidad*. España padece este grave mal, y es forzoso curarla, *desinterinizarla*: el *desinterinizador* que la *desinterinice* no puede ser otro que ese franchute avariento y ruin a quien yo llamo *Antonio Igualdad*, amamantado, como su padre y su abuelo a los pechos de la Revolución francesa. (III: 828)

Much of the book is centred on the absurd Carlist figure Don Wifredo de Romarete, Bailío de Nueve Villas and Knight of Saint John. Obsessed with his Catholic dignities and hierarchical pretensions, he arrives in Madrid from ultra-traditionalist Vitoria reflecting the naïve, anachronistic, and optimistic claims of Carlist legitimacy. Through him we learn in passing that Juan de Borbón has abdicated (3 October 1868) in favour of his son Carlos,[21] who is to prove a formidable adversary in the second Carlist war. The partial identification that is made between Don Wifredo and Don Quixote adds a dimension of idealistic integrity and sympathy to the portrait as well as underlining his unpractical nature. At the same time his political activities are intricately bound up with the fictional story of Fernanda Ibero's love for Juan de Urríes.

Introduced to the Assembly by Urríes, Don Wifredo serves as a vehicle by which we witness the Cortes at work during the debates on the constitution.[22] Through his innocent but prejudiced eyes and interspersed with his harmless gallantries, we are presented to a wide range of political figures under the presidency of Rivero. Among them are the elegant and persuasive *pollo antequerano*, Romero Robledo, the verbose Carlist Vinader, the extremely radical José María Orense, Marquis of Albaida, and the up-and-coming young Liberal politician Segismundo Moret y Prendergast,[23] admired for his handsome appearance and his eloquence. Later Romarate is bemused by the relentless if innocuous atheism of the Federalist Francisco Suñer y Capdevila.[24]

The set piece is the debate between Father Vicente Manterola and Castelar (12 April 1869), in which the latter delivers his speech[25] on the question of religious freedom ending with the famous *Dios en el Sinaí* passage. Castelar's

[22] Galdós himself attended meetings of the Cortes at this time as correspondent of the magazine *Las Cortes* (*3b*, Berkowitz 1948: 68; *Memorias*, VI: 1434).

[23] Moret is singled out no doubt because he is the only politician of this period who survived up to the time of writing (Oct. 1907–Jan. 1909) when Galdós was a *diputado*. Moret was leader of the Liberal Party at that time and would become prime minister less than two years later, in Oct. 1909, after the *Semana trágica*.

[24] A later appearance in the Cortes is José Echegaray, renowned for the incorporation of scientific vocabulary into his speeches. The narrator allows himself an ironic remark about the dramatist who was to share the Nobel Prize for Literature in 1904 (Galdós will be an unsuccessful candidate himself a few years later): 'Lástima que no sea usted dramaturgo' (III: 833).

[25] For the full text see Castelar (*1b*, 1873: i. 257–84).

oratory is dazzling and overwhelms even his opponents, among them Don Wifredo, whose simple solutions now no longer seem adequate. A similar disillusion on the personal plane occurs with his discovery of the clandestine love of Urríes and Céfora. From then on his initial idealism suffers a constant decline, until his pristine innocence and integrity are wholly shattered during the drunken orgy with *Paca la africana* into which he is led by Tapia. In his inebriated speech Castelar's rhetorical phrases and other clichés of the time become hilariously contorted in Bakhtinian parodic stylization:

—Grande, grandísimo es el Dios en el Sinaí . . . El trueno le precede; la chispa le acompaña . . . la tierra se echa a temblar; los montes se ríen a carcajadas . . . Pero en mí tenéis un dios más grande, más bonito . . . ¿No me declaráis el más bonito de los dioses? Yo soy el amador de Paquita; yo bebo en sus ojos la idea espiritual de Chinchón, y vengo a predicaros la libertad de aquellos cultos que practicaron caldeos y macabeos, fenicios, egipcios y estropipcios . . . ¡Viva mi *africana* con honra! (**III**: 822)

A mental derangement follows, during which he pursues the idea of redeeming *Paca la africana*, whom he sees again at the wedding of a companion. He now utters a highly confused mixture of overheard conversation, political conviction, private concerns, and prophecy, in which the most important item is an anticipation of Prim's murder: 'Después viene la noticia del día, la más sonada, la más gorda . . . Que han matado a Prim . . . Se cree que haya sido Tapia el matador . . . Conste que el tal Tapia no es carca, sino *montpensierista*' (**III**: 839).

On his partial recovery, after a period of contrition under the guidance of the priest Don Pedro Vela, he launches a Quixotic campaign to compel Urríes to marry Fernanda or face a challenge from him. The whole episode provides an imaginative example of good integration of public and personal motifs; despite Urey's arguments (*3d*, 1989: 180), his logical or intuitive premonitions, only partly fulfilled, do not, in my opinion, go beyond the limits of deduction or realist presentation.

Another representative of old-fashioned ultramontane views is Doña Carolina Lecuona, Marchioness of Subijana. Although she is so prolific in her sentimental reminiscences of thirty years before concerning 'the queen', Doña Francisca, the original pretender's first consort, and her unacknowledged court (**III**: 795–8), Doña Carolina's ideological stance shifts with her circumstances. Having had her property rights restored by Serrano's government, she comes to adopt the flexible position of following the tide of events: 'Yo veo la procesión histórica . . . ando con ella, tras ella' (838). Thus she assumes a series of equivocal attitudes: she praises the intellectual skills of Republican politicians, frequents the court which the regent Serrano's ambitious wife is setting up, and, finally, accepts an eventual Bourbon restoration in the figure of Alfonso. Typically, she does not at this time commit herself to either of the two groups of society ladies—the *provisionales*, who support the initiative of the regent's wife, or the old aristocracy known by the unflattering nickname of *de la Carrera* (de San Jerónimo); the term sardonically foreshadows the identification of these same ladies with the prostitutes of Madrid in the later *mantillas blancas* episode.[26]

Unyielding supporters of the ex-queen Isabella, like Don Pedro Vela, still exist, but once she has yielded to the pressure to abdicate, which she does on 25 June 1870, the allegiance of this faction passes to her son. Although in his presentation Galdós neither eschews hindsight nor allows it to get out of hand, the wearing away of the revolutionary impulse and the gradual but inexorable progress towards the Restoration are quite apparent. So it is that we find Cánovas,[27] leader of the *alfonsinos*, exercising a powerful influence behind the scenes. The inevitable Eufrasia, now living in a house in the fashionable 'barrio de Salamanca', where the most opulent thoroughfare has just had its name changed—definitively as it happened—from 'Boulevard Narváez' to 'Calle Serrano', offers a magnificent dinner which is 'un taller de historia' (III: 846). It is attended by the real-life Marquis of Orovio, a reactionary ex-minister who has put at Cánovas's service 'su honradez, su experiencia de covachuelista y su ardiente devoción borbónica' (852) and who will have a controversial position in his first Restoration cabinet. Also present are fictitious representatives of the traditionalist Basque country, Don Luis Trapinedo, the future Count of Gauna, and his wife María Erro. Their ancestral Carlism now becomes 'platonic' and Gauna, as an old friend of Cánovas, goes over to tacit support of the Restoration. On learning the news of the Liberal demands made by the erstwhile Carlist general Ramón Cabrera,[28] Cánovas indicates (849), as Romero Robledo had earlier, that these demands, whether accepted or not, render the failure of the Carlist cause inevitable. This too will accrue to Alfonso's advantage. Even Don Wifredo becomes convinced that a Carlist monarch would not last long and that the support enjoyed by Alfonso from 'estos ricachones y las damas bonitas vestidas a la última moda de París [que] son la fuerza social efectiva' (875) is the decisive factor.

Cánovas, for his part, is prepared to play a waiting game; as he declares in

[26] 'La de Subijana, por la *promiscuidad* de sus relaciones, era tan pronto *de la Carrera* como *provisional*' (III: 841; my italics).

[27] Antonio Cánovas del Castillo, 1828–97, historian and architect of the Restoration. Earlier he had been an active participant in O'Donnell's Liberal Union. Galdós first presents him in *La revolución de julio*, when Beramendi, who has met him at the house of his uncle Serafín Estebánez Calderón, describes him as follows: 'Es malagueño, cecea un poco; su talento, duro y poco flexible, me cautiva precisamente por eso, por la dureza y rigidez... Este no dice más que la mitad de lo que piensa, y hará, creo yo, el doble de lo que dice. Así me gustan los hombres' (III: 30).

[28] 1816–77, Count of Morella. Known as 'el tigre del Maestrazgo', Cabrera was renowned for his cruelty and ruthlessness, both before and after his mother was vindictively shot by Liberal forces. During his long exile he married an English heiress and finally abandoned the Carlists. See Chant (*1b*) for details of his life at Wentworth House, Virginia Water.

a famous phrase, when asked what he did in the Cortes Constituyentes: 'Esperamos, y esperando hacemos la Historia de España' (**III**: 846). His view is that 'La política de los últimos años había producido, por errores de todos, una gran fuerza expansiva y revolucionaria. No era prudente ni práctico oponerse al empuje de esa enorme fuerza desencadenada. No había más remedio que dejarla correr hasta que por el continuo roce se gastara' (848). It is precisely the same policy as Doña Carolina's, as Galdós explicitly indicates: 'Cánovas, conforme en esto con la ingeniosa marquesa de Subijana, no pensó en andar a contrapelo de la procesión política: iba con ella muy a retaguardia, esperando la madurez y oportunidad de los fines que perseguía' (840). At the same time the basic weakness of monarchist sentiment, which, as Cánovas irately points out (851), makes the introduction of a foreign dynasty especially precarious, is revealed by the lack of respect accorded to the disinterred remains of Charles V. Cánovas's hopes for Conservatism lie not in the old aristocracy, but in self-made men of humble stock like Juan Antonio Iranzo, who is proud of his origins and of having seized the chance opportunity of becoming a financial agent and making a fortune: 'La vida de Iranzo era en esa historia uno de los pasajes de mayor potencia documental' (849). For the moment, however, Iranzo hedges his bets: he too attends the *tertulia* offered by the regent's wife, and later votes in favour of Amadeo.

From within the ranks of those who supported the revolution, the claims of the Duke of Montpensier, the candidate of the old Unionists like Serrano and Topete, are pressed with vigour and upheld by abundant money and unscrupulous intrigue.[29] All his resources do not succeed in making the duke popular, and he is widely accused of meanness; hence the nicknames disdainfully given him: *naranjero* (through his exploitation of the orange-crops from his San Telmo estate in Seville) and *Monsieur Combien*, from his propensity for hard bargaining. As Louis Philippe's son, he has the disadvantage of facing the hostility of Napoleon III. What he needs for his success, as Don Wifredo points out in conversation with Cristóbal de Pipaón, is Prim's support: 'El es rico, y ricos sus partidarios. Si Prim por él decide, ten por cierto que será rey' (**III**: 827), but he adds, with prophetic foresight, that if the general is not amenable (which he is not: he rejects Urríes's proffered bribes, 877), Montpensier's supporters are capable of murdering him.

Montpensier's hopes are irrevocably dashed when he kills the liberal infante Don Enrique de Borbón in a duel fought in March 1870. The treatment of the subject displays both strengths and weaknesses. Galdós skilfully integrates the incident into the fictional story through Halconero and a member of his group, the real-life politician Emigdio Santamaría, described as 'un perfecto

[29] See Gallenga (*1b*, i. 254–64) for Montpensier himself and for his San Telmo estate. Roger Utt (*3b*) produces fascinating indications that Galdós stopped writing for *Las Novedades* in Jan. 1869, when that newspaper went over to supporting Montpensier.

modelo del tipo arábigo levantino' (III: 919), who is the infante's second in the episode; and he utilizes the opportunity to raise the question of how authentic the liberalism of a Bourbon prince can be. Don Enrique's radical fervour is put in doubt by Segismundo Fajardo. He brings up the pertinent example of the transformation of the revolutionary Louis Napoleon into the emperor Napoleon III, while an unidentified 'voz de bronce' emphatically denounces all the Bourbon family. But, as Montesinos (*3c*, iii. 288) indicates, the attention accorded to the incident seems disproportionate: the duel, with its unexpectedly tragic consequences, is described in rather tiresome detail, after which the infante's life and his Masonic affiliations are needlessly rehearsed, first by the narrator and then by Felipe Ducazcal; a letter to him from Espartero is reproduced; and the event is integrated into the fiction only to the extent that at the funeral Vicente meets his ex-mistress Eloísa again. This excessive treatment may be the result of a personal experience of Galdós himself; significantly, he refers to the *tertulia* of the Canary Islanders at the Café Universal, where the incident was discussed, as 'de las más amenas de Madrid' (930). At the same time, as an incorrigible spinner of alternative histories (see Chapter 7), he was evidently fascinated by an opportunity lost: Don Enrique, truly liberal or not, was clearly a far more suitable suitor for Isabella II's hand than his brother Francisco;[30] Ducazcal declares that 'la reina le quiso siempre' (937).

Among fictitious characters supporting Montpensier, the Count of Ben Alí manipulates his district and associates in the manner of a typical Andalusian *cacique*; but it is his opportunist brother, Don Juan de Urríes, in his double role of Don Juan and political intriguer, who is the more active agent. He is sent off on missions of subversion or bribery all over the country, even stirring up trouble for the government by fomenting risings on either the Carlist or the federal side. He travels to Tarragona, for example, at the time of the Federalist revolt, 'con el objeto de ver si en el revuelto río federal era fácil pescar alguna trucha que pudiese comer el señor Duque' (III: 882). He claims to have bought Sagasta's allegiance, but has no further success to report with other leaders. Finally, he is ordered by his brother to make a suitable marriage of convenience to a wealthy widow, regardless of his obligations to Fernanda and his affair with Céfora. No attempt is made to give him any deep characterization or to present his personality except from outside.

Fernanda Ibero, for her part, is a carefully and slowly delineated character; she has the tough, uncompromising, loyal traits of her family, and Urríes's

[30] See *Bodas reales* (II: 1372–3) for Enrique as an aspirant to Isabella's hand. In *España trágica* Galdós reports Demetria as saying that Isabella encouraged her brother-in-law to thwart Montpensier's pretensions in every way possible (III: 930). The infante certainly succeeded, but the method adopted had rather drastic consequences. John Hay has an account of the duel in *Castilian Days* (*1b*, 371–88).

betrayal of her is a betrayal of the unspoilt values of the country. In her distress at this treachery she feels a warm sympathy for the unorthodox couple Santiago, her brother, and Teresa in France, about whom Jesús Clavería brings such favourable reports to Santiago's parents (III: 870). She is also closely linked with the deranged Don Wifredo, whose platonic love for her causes him to attempt quixotically to take responsibility for killing Céfora. The *episodio* ends, as the Civil Guards approach, with a piece of pseudo-dramatic dialogue between the two and their servants Filiberta and Marciana.

Céfora, too, is a curiously ambiguous character. She remains mysterious for much of the narrative, then Eufrasia explains her parentage, which Céfora herself reaffirms to Urríes. Her Jewish mother Mesooda, 'del tipo de Ruth', is vehement in her passion and her faith; she threw a jug of boiling oil at her lover and inveterate philanderer, Miguel de Nanclares, Doña Carolina's husband. The external racial conflict of *Gloria* has become internalized in Céfora, who alternates between the extremes of sensuality—love for Urríes—and mysticism: desire to enter the convent. She may owe something to Concha-Ruth Morell, with whom Galdós had a long-drawn-out relationship not long before.[31]

The second part of Fernanda's story, her love for Vicente Halconero in *España trágica*, which ends with her sudden death, also has an independent development; it may betoken the impossibility of a clear-cut, direct solution to Spain's problems and may be related to Prim's disappearance, as has been claimed,[32] but I find the link a tenuous one. At all events, her fiancé Halconero is somewhat reminiscent, in his role of narrator-reflector in *España trágica*, of Beramendi's in the fourth series. His childhood has been traced in *Aita Tettauen*. Conditioned by a physical handicap which left him lame, he was, not unlike Maxi Rubín, fascinated by soldiers and military parades. At the time of the African war he was perplexed by the part his renegade uncle Gonzalo (El Nasiry) would play and eagerly sought information about the course of the war from Santiuste. In *La de los tristes destinos* he offers a theoretical, bookish version of history, in contrast with young Santiago Ibero, the incipient man of action: 'La simpatía cordial que entre ambos se estableció

[31] For details see Lambert (*3b*). The affair may have started in 1881. Galdós supported Concha in her not very successful acting career. In 1897 she converted to Judaism and adopted, significantly, the name Ruth. She accompanied him to the Basque provinces in 1898 and proofread some of the third series of *Episodios nacionales*. Through the influence of his family Galdós finally broke with her in 1900. Her approximately 160 letters in the Casa-Museo in Las Palmas reveal strong, often incoherent, emotions, ranging from love to bitter complaints and despair. As Gilbert Smith has shown (*3c*; see also *3b*, Madariaga 76–7), some of the text of her letters was utilized in *Tristana*. It is more than possible that the vehemence attributed to Céfora's Jewish blood derives from Concha-Ruth Morell. At all events, it is one more example of Galdós's keen interest in the Jews. See Schyfter (*3b*) and Chamberlin (*3b*, 1981).

[32] e.g. Brian Dendle (*3d*, 1980: 160), who sees 'the brittle Fernanda' as symbolizing 'the disappointed hope for a possible happier future for Spain'; and, even less acceptably, Don Juan de Urríes as 'the representative of the Revolution of 1868'.

al primer trato, se explica por el estrecho parentesco de sus almas. *El uno era la Historia libresca; el otro, la Historia vivida*, ambas incipientes, balbucientes, en la época de la dentición' (III: 668; my italics). Now, in *España trágica* the well-read Halconero clearly conveys much of Galdós's own ideological position, its hopes and its disappointments; and the revolution is seen, like Halconero himself, as crippled:

> De cuanto pudiera decirse acerca de Vicente Halconero, lo más fundamental es que provenía espiritualmente de la Revolución del 68. Esta y las ideas precursoras le engendraron a él y a otros muchos, y como los frutos y criaturas de aquella Revolución fueron algo abortivos, también Vicente llevaba en sí los caracteres de un nacido a media vida. (III: 900)

His theoretical bent continues, as he flirts with Federalist ideas; in this he is contrasted with a real person who is given as an example of an unreflecting cult of action: one Felipe Fernández, known as *el Carbonerín*. He does have his potential moment of action, when he is provoked to challenge such a consummate fencer as Paúl y Angulo by his gratuitous insults to his beloved mother. (The duel in real life was between Paúl and Ducazcal, who was wounded.) Instead Halconero is unheroically beaten up by Ducazcal's *partido de la porra* in a brawl with Paúl's gang. After the tragedy of Fernanda's death, he eschews violence and conflict and settles down to a more subdued and humdrum existence with Pilar Calpena, whose inconsequential chatter ventilates many current events. Having taken his personal stand against Paúl, Vicente now evolves, like Spain, from revolution towards calculated moderation and reconciliation. His friend Enrique Bravo makes a similar transformation, as he passes from being an aggressive Federalist to a frenzied claimant for a sinecure in Cuba.

Another relevant case is Segismundo Fajardo, initially as unrestrained as his namesake in *La vida es sueño*. He is a character who provides, by means of an individual evolution, a shorthand review of possible political attitudes and disruptive social actions. Politically, he develops in an opposite direction from Halconero, from conformist opportunism to radical republicanism; next he makes a leap to Carlist autocracy and then a rebound to Federal anti-centralism:

> Había comenzado su vista política alistándose en la Unión Liberal; al estallar la Revolución del 68 se pasó a los demócratas de Rivero; poco después, por no sabemos qué piques y despechos, dio un salto tan grande que fue a caer junto a Don Carlos; del carlismo se vino a la República y seducido por las doctrinas de Pi, abrazó el federalismo con fervor delirante. (III: 918)

Socially, he devotes himself to a hand-to-mouth Bohemian existence from which he is periodically rescued by Halconero. He will reappear again during Amadeo's reign and with the Restoration.

Through Halconero and Fajardo Galdós demonstrates how powerful the new surge of Federalist sentiment was, but it is portrayed as speedily outgrowing its capacity for moderation and as developing theoretical doctrines which remain out of touch with reality (III: 881–2). The upshot is the uncontrolled and savage uprising in Tarragona in the autumn of 1869, when the acting governor, one Raimundo Reyes García (882), is killed; the incident is compared to a childish game, which equates it with the tragedy in which Mariano is involved in *La desheredada*.[33]

Galdós is quite emphatic, if too explicit for fictional purposes, about blaming the Federalists for failing to support Prim, much closer to them ideologically than to the Unionists and the Conservatives. This proneness to division he sees as an endemic defect of the progressive parties:

Porque los federales de aquel tiempo, como todo partido español avanzado, padecían ya de mal de miopía, o sea, el ver de cerca mejor que de lejos. Jamás apoyaba a sus afines; en éstos veían el enemigo próximo, y cerraban contra él, descuidados del enemigo lejano, que era en verdad el más temible. (III: 943–4)

As a result, in a crucial vote at the session known as de San José, 18–19 March 1870, Prim is barely saved by his own eloquence and by the votes of the so-called *Perlinos*, a disgruntled group of individuals without fixed allegiance.[34]

Prim, for his part, remains steadfast in his monarchist, though antidynastic, policies. Among other candidates for the Spanish throne momentarily favoured are the retired Portuguese king-consort, Ferdinand of Coburg,[35] who was most reluctant to be considered, and Espartero, to whom Prim sent Ibero as an emissary. Espartero also refused, thus dashing Pilar's frivolous hopes of having a royal sponsor at her wedding. More promising was Prince Leopold of Hohenzollern-Sigmaringen[36] (referred to burlesquely by Pilar and other non-linguistic Spaniards as *Ole-Ole*). Since this candidacy sparked off the Franco-Prussian war,[37] the opportunity is provided for a detailed examination of some aspects of a major European conflict, again intertwined with the fictional story. Halconero reports on the hostilities from Paris; he

[33] Compare Galdós's chronicle of 13 Jan. 1872 in the *Revista de España* (*3a*, *Artículos políticos* 29): 'Ya no existen otros obstáculos tradicionales que los que creemos nosotros mismos con nuestra condición inquieta y díscola, entorpeciendo todos los caminos, desbaratando hoy *como niños impacientes* lo que hemos hecho ayer, dejándonos arrastrar por los primeros impulsos de una sensibilidad rebelde y desenfrenada, *como los adolescentes mal educados, a quienes ninguna regla enseña ni amaestra ninguna experiencia*' (my italics).

[34] Hay gives a detailed eyewitness account of this session in 'A Field-night in the Cortes' (*1b*, 313–46).

[35] Widower of Maria da Glória and father of the reigning king Luis I.

[36] Compare Alarcón's leaflet *El prusiano no es España* (1870) (*4a*, 1984: 365–74), in which he vehemently opposes the Hohenzollern candidature, even though he considers Prim 'ireemplazable, dada la nulidad a que se han reducido los demás prohombres de la nación' (374).

[37] The war made a deep impression on the young Galdós. The story 'Dos de Mayo de 1808. Dos de Setiembre de 1870', discovered by Hoar (*3c*), links the Spanish rising with Sédan.

meets Santiago Ibero, accompanied by Teresa Villaescusa, as they go off to take part. Galdós's three tentacular and archetypal families—Calpena, Ibero, Halconero-Ansúrez—come together in Bordeaux, in confrontation with the three intriguers whom Halconero, reflecting no doubt his creator, most dislikes: Urríes, his newly acquired wife, and Carolina de Lecuona.

The war offers an example of how even an aggressive, bombastic, and apparently impregnable empire like Napoleon III's could crumble and give way to revolution and a republic. The incompetence of the imperial government, the military reverses, Napoleon's surrender to the Prussian king Wilhelm, the proclamation of the Republic, and the flight of the empress (who had been responsible for many of the empire's mistakes) are all rehearsed. The resounding defeat momentarily dismays such a cultured Francophile as Vicente and the many Spaniards of all conditions accustomed to visit Paris. The effect of the French withdrawal from Rome, thus allowing Garibaldi's assault, is noted, and some discussion on papal rights is reflected by the garrulous Pilar; the ironical suggestion is even made that the Pope should transfer his court to Spain.

In the novels the war is not absent, but it is incidental. Aurora's patriotic French husband Fenelon is killed in action, and his not too disconsolate widow uses a lively metaphor drawn from two famous seiges during the conflict to describe Manuel Moreno-Isla's amorous conquests: 'Yo fui Metz, que cayó demasiado pronto; y ella [Jacinta] es Belfort, que se defiende; pero al fin cae también' (**V**: IV. 1. xii, 444).[38]

At the moment of triumph at the election of Amadeo, Prim stands at the head of a bridge 'por donde la Nación había de pasar de la interinidad a un estado efectivo' (**III**: 997), but the situation is full of dangers. One is reflected in Prim's meditations on the 'dos fuerzas enemigas', federalism and aristocracy, the latter more subtle and effective than the more strident Republicans. Another danger is a personal weakness: his fatalistic overconfidence: 'la bala que a mí me mate no está hecha todavía' (**III**: 550), he declares in foolhardy fashion.[39] As Segismundo says (**III**: 1005), 'Al General le ha perdido la vanagloria de su valor.'

Prim's murder is referred to without analysis from distinct points of view in

[38] Bismarck is referred to incidentally in some of the novels (*El amigo Manso* **IV**: 1207), as are other contemporary foreign politicians like Gladstone, e.g. *La incógnita* (**V**: 747; with Bismarck) and *Torquemada en el Purgatorio* (**V**: 1041–2).

[39] The phrase is traditionally associated with him. See his round assertion in Valle-Inclán's *Baza de espadas*: '¡Mi vida, señores, la respetan las balas! Soy providencialista' (*4a*, 163). For further evidence see Sinclair (*4a*, 82–3). In this respect Galdós's characterization bears some slight resemblance to the famous more hostile portrait by Castelar in the parliamentary debate of 3 Nov. 1870, when Amadeo was elected: '¿Sabéis cuál es el dios del general Prim? El acaso. ¿Sabéis cuál es su religión? El fatalismo. ¿Sabéis cuál es su ideal? Lo presente. ¿Sabéis cuál es su objetivo para el porvenir? Vincular al Poder a su partido. A esto lo sacrifica todo' (*1b*, Castelar 1873: iii. 418; quoted in Cambronero, n.d.: 147, and Fernández Almagro 51).

two novels. In *Fortunata y Jacinta* it is significant, as previously mentioned, that Doña Isabel Cordero dies at the same time as Prim, just as she has achieved the triumph of marrying her daughter Jacinta to Juanito. The Santa Cruz couple start their married life, therefore, with the turbulent period, lurching between order and revolution, ushered in by the assassination and paralleled by Juanito's private life. And in *La desheredada*, Isidora sets out on her road of personal dishonour in the arms of Joaquinito Pez at the exact spot in the Calle del Turco [= *Marqués de Cubas*] where Prim was shot: 'aquella pared donde a balazos estaba escrita la página más deshonrosa de la historia contemporánea' (IV: I. 17, 1060).

España trágica deals with the murder much more consequentially as a contemporary event. The tragedy is brought about in the highly effective final chapters through the general's failure to take heed of the warnings given him. The many slight details drawn from contemporary anecdotes lend a cumulative effect to the event: Sagasta's entry and descent from the coach; the signal given by the lighting of a cigarette;[40] the wounds which were apparently not grave, since the general was able to reach his bedroom unaided; the dramatic announcement at the Masonic dinner; the reassuring forecasts and the false hopes of recovery; Prim's failure to identify the voice of one of the assailants;[41] the momentary solidarity of the revolutionary leaders with Serrano and the selection of the loyal Topete to serve as delegate to greet the newly elected king at Cadiz; and, finally, the obsession with the passing of time by the dying leader.

The *episodio* is much more than a collection of documents or an examination of evidence. Pío Baroja raises an interesting question concerning the way in which Galdós uses his material. He chides him for not revealing the information he has acquired about Prim's assassins: 'tenía muchos datos acerca de los preparativos del crimen, de quiénes habían participado en él, de lo que habían cobrado y de lo que habían hecho con el dinero . . . Yo suponía que Galdós hablaría de esto cuando dedicara un episodio a Prim.'[42] Baroja returns to the question in his *Memorias*, noting that 'me contó una serie de detalles muy curiosos de gente que había intervenido en el asesinato de Prim', but that when he published the *episodio*, 'De cuanto me había contado no decía nada. Y me quedé en el mayor asombro. Yo, en su caso, hubiera contado todo cuanto supiera' (*4a*, viii. 34–5). Montesinos assumes arbitrarily that 'ello se debió, es casi seguro, a que esas gentes vivían aún, o varios de ellos, como vivirían los confidentes, que, es natural, exigirían sigilo. En casos

[40] A part of the legend which grew up about the murder and which Pedrol Rius dismisses as an unproven 'novelesco detalle' (*1b*, 25).

[41] The most convincing evidence of Paúl's complicity is that his very distinctive voice was recognized by Prim himself and by one of his aides. See Pedrol Rius (*1b*, 54–5).

[42] 'Los Carbonarios' (*4a*, v. 1150). See Regalado García (*3d*, 460).

así, no siempre se dice lo que se quiere' (*3c*, iii. 249). Baroja's criticism and Montesinos's explanation correspond in my view to a quite mistaken conception of Galdós's literary motivation, which is not to solve historical mysteries but to present the current attitudes of the time of the action.

The first appearance of the prime suspect, José Paúl y Angulo, is skilfully interwoven into the fictional story in the train journey in which several characters are returning from France. Paúl's intransigent views, violent, psychotic behaviour, corrosive sarcasm, and provocatively offensive demeanour are contrasted with the petty-bourgeois concerns of Portillo (*Sebo*) and his socially ambitious wife and the middle-of-the-road, compromising attitude of Halconero. Segismundo attributes Paúl's excesses to over-drinking, and notes his extremes of temperament: 'Yo he visto en él rasgos de bondad admirables; le he visto también pasar de la dulzura de carácter a la grosería más soez' (**III**: 976). Thus Paúl is unexpectedly considerate when Vicente is accidentally wounded by Ducazcal's men, so that Halconero's final assessment of him is not entirely unfavourable—'le inspiraba menos odio que lástima'—though he continues to find him 'un loco irresponsable, peligrosísimo' (991).

In Paúl's opinion Prim betrayed the revolution, and he unleashed a continuous stream of abuse on his erstwhile leader in the journal *El Combate*[43] which he founded in November 1870, culminating in a virulent special final issue (25 December 1870) in which Paúl is quoted as saying: 'Una mayoría facciosa, prostituída y encenagada hasta la hediondez . . . *maniató traidoramente la Soberanía* a la espuela del dictador don Juan Prim' (**III**: 998). Other examples of Paúl's venom against Prim and the threats which go with it are recorded: 'yo inicié la Revolución de septiembre, yo traje la Libertad, y Prim la vende . . . ¿No es un miserable, no es un bandido? . . . ¿Estoy o no cargado de razón cuando digo: *Hemos de matar a ese hombre*?' (987).

Despite all this, Paúl is not seen unequivocally as the murderer; in fact, as Brian Dendle has indicated, Galdós 'distorts, and at times suppresses, the testimony of historical sources; to fit historical events to the literary ends of the novel, evidence against individuals and dynastic or political groups is presented in a deliberately ambiguous manner' (*3b*, 1969: 67). Even if one might add that 'suppression' may be no more than selection, and demur at the term 'impressionistic', it is clear that not all the evidence against Paúl is marshalled, to my mind in order not to make the case against him too overwhelming. Galdós wishes the perpetrators of the murder, not definitively

[43] The numbers of *El Combate* I have consulted (Newspaper Library, British Library, Colindale, London) contain violent denunciations of Amadeo, Montpensier, Ducazcal, *La Iberia*, Sagasta, and Rivero, but Prim is the main target. His speech in favour of Amadeo's election (4 Nov. 1870) is described as 'Notoria es la torpeza de D. Juan Prim en lances parlamentarios; torpeza comparable tan sólo a la que ha demostrado como hombre de gobierno; pero esta vez se escedió [*sic*] a sí mismo; tan desconcertado, balbuciente, inconexo y extravagante fue su original discurso.' There are also veiled threats, like 'A un César un Bruto' (20 Dec. 1870).

established even today,[44] to remain as undetermined as they were to contemporary observers. For the same reason he emphasizes, as we have seen, what Montpensier had to gain from the elimination of the general, as 'los familiares, los allegados y los amigos de Prim' (*1b*, Pedrol Rius 82) had done at the time. Dendle also notices that Ruiz Zorrilla is excluded from critical attention; but Zorrilla was never a prime suspect.

In accordance with his renewed anticlericalism, Galdós dwells on the suspicious and sinister activities and the gloating satisfaction at Prim's death of the so-called *Ecuménicas*, who now represent the extreme clerical faction. These are three women, all emotionally twisted in various ways, with chequered careers from previous *episodios*. The most dominant is Domiciana Paredes, Lucila's fierce antagonist for Gracián's love. Donata, extracted from Juan Ruiz Hondón's harem by Santiuste, is seduced by Segismundo Fajardo and then has an affair with the renegade priest Andrés Romeral. Rafaela del Milagro is the veteran Progressive José's daughter and Riego's god-daughter. Known in her profligate days as *Perita en dulce*, she meets her old friend Santiago Ibero on the street and they both realize wistfully how far their paths, personally and politically, have diverged.

Dendle's final summing up seems to me on the whole judicious: 'Galdós did present contemporary reaction to the events described; hence the importance of gossip, which gave the immediate popular interpretation, and of *proteísmo*, which, with its temporary enthusiasm for all viewpoints, momentarily enabled the novelist to suspend judgement and to avoid the risk of hindsight' (*3b*, 1969: 67). I, however, take a more positive view of what Dendle calls *proteísmo*, which corresponds to 'indeterminacy' in reader-response criticism, as defined by Iser (*2*, 1974: 283). Indeed, as I have been consistently arguing, such openness is one of the outstanding qualities of the better *episodios*, even if it is not always appreciated by critics who like clear-cut solutions. It is by no means a 'temporary enthusiasm' which enabled the novelist 'to suspend judgement' nor an indecisive attitude such as Pepe García Fajardo or Tito Liviano display,[45] but a deliberate and sustained ability, not

[44] The vast documentation gathered together for the official inquiry yielded no positive result. Gallenga, in 1875, speaks of the 'Huge folio volumes of reports [which] have piled up several fathoms deep; but no result has ever been, or ever will be published' (*1b*, ii. 183); and he visited Prim's still open tomb at Atocha. The latest study of the assassination by Pedrol Rius (*1b*, 1980; 1st edn. 1971) concludes that the assassins were directed by Paúl y Angulo and José María Pastor, a member of Serrano's staff. Money may have been provided by Montpensier's aide, Solís y Campuzano. There is no evidence that either Serrano or Montpensier knew about the plot. See also Regalado García (*3d*, 461–4), who holds that Galdós is obliquely arguing against Paúl y Angulo's involvement and in favour of 'una conjura, o del gobierno y de los diputados . . . o de los altos personajes del palacio'. I do not share his view that Galdós 'quiere dar indicios de los autores, pero tan entre nieblas que nada descubren'.

[45] Dendle is far more critical, wrongly in my view, of *proteísmo* in his 1980 book: 'A trait related—as symptomatic of inner insecurity—to evasiveness is *proteísmo*, the ability to defend all points of view' (*3d*, 1980: 200).

confined to any individual character, to present historical situations sympathetically from a variety of angles and to keep the options open. It is significant that what would presumably be substantial evidence of the identity of the murderers—the rectified list of dissidents given to Halconero,[46] which we may perhaps equate with the suppressed evidence Baroja complained about—is deliberately burnt by Lucila.

Galdós is not writing a judicial case history, nor even a historical novel in which a solution to the mystery might be offered, but an *Episodio 'nacional'* in which the *national* issues of the time, confusing and inconclusive as they might be, can be suitably ventilated. He is not concerned to bring forward fresh evidence: hence the symbolic burning of the document. Moreover, Dendle is right in realizing that part of Galdós's purpose was to establish a sense of the collective responsibility of all Spaniards, seized as they appear to be by a fit of mass insanity (*3b*, 1969: 68). Anticipating the event itself, Segismundo Fajardo declares that, as a result of its epic frame of mind, Spain required a tragic hero 'para dar cumplimiento al trágico designio de la fatalidad histórica . . . Y ésta nos dice con acento de oráculo infalible: ¡Españoles, matad a Prim!' (III: 975). Prim's murder was the supreme example of the inability of Spaniards to follow a coherent and progressive lead, between the extremes, as the various assessments of the dead leader make clear.

No doubt is expressed in the *episodios* about the significance of Prim's disappearance. Earlier in *España trágica*, Ibero Senior had declared that he was 'la clave de la libertad y el porvenir de España . . . Si aquel hombre faltase, volveríamos, tarde o temprano, al reino de las camarillas' (III: 945). Of course Galdós knows, his contemporary readers knew, and we know that Prim would *faltar* and very promptly. The conclusion, at the moment of his death, takes the same view; he was 'la puerta de los famosos *jamases*', the only bulwark against the return of 'seres e institutos condenados a no entrar mientras él viviera' (1008), that is to say, against a clerically dominated Restoration. The epitaph given to him in *Amadeo I* reiterates both the power he exercised and his capacity for occupying the middle ground, limiting the scope of the revolution while keeping back the forces of reaction: 'Así fue llevado al sepulcro el hombre que ejerció durante veinte y siete meses una blanda dictadura, poniendo frenos a la revolución y creando una monarquía democrática como artificio de transición, o *modus vivendi*, hasta que llegara la plenitud de los tiempos' (1011). Galdós, around 1909, is clearly aware of the incompleteness of the 1868 revolution and evidently aspires to an eventual, more radical, Republican solution. Looking back to 1870, however, he sees the impelling need at that time for a period of consolidation to curb the

[46] Such a list of potential assassins undoubtedly existed. Pedrol Rius severely censures the governor of Madrid, Rojo Arias, for not arresting the suspects (*1b*, 31–2) and notes that Prim took no notice of it (19).

powerful clerical and reactionary forces. For such a consolidation Prim was the indispensable key.[47]

[47] Galdós never lost his deep admiration for Prim. One example from 1908, when he was writing *España trágica*, is the message he addressed to Miguel Moya in the campaign to oppose the law against terrorism: 'Ninguno de los aquí presentes dejará de sentir en su alma una secreta voz que reproduzca, sin ninguna variante, un concepto del primer estadista español del siglo XIX, del glorioso, del inmortal Prim: "¡Radicales, a defenderse!"' (*El Liberal*, 29 May 1908; *3b*, Fuentes 64). The quotation is the final phrase of his speech in the *sesión de San José* (19 Mar. 1870).

5

Political Experimentation, 1868–1874

PRIM's death left the incoming monarch Amadeo in a very precarious situation. Opposed on the right by the adherents of the Bourbon dynasty, on the extreme right by the Carlists, and on the left by the Republicans, Amadeo did not even enjoy the unconditional loyalty of the political factions within the centre.[1] Among intense and irreconcilable rivalries, the king attempted to be impartial. On the one hand he denied Serrano's request for the suspension of guarantees, and on the other opposed Ruiz Zorrilla's army reforms; after signing the latter, he abdicated in February 1873.

In his periodicals *El Debate* and *La Revista de España* José Luis Albareda and his associates León y Castillo and Pérez Galdós[2] favoured an English model of divergent but mutually tolerant rival parties, exercising a *turno pacífico*. In this respect Galdós's articles of 1872 in *La Revista de España* (*3a, Artículos políticos*) are symptomatic. He vehemently opposes the extremes of Carlism on the one hand and the International and the Paris Commune on the other; he is unsympathetic towards the Republicans. He deplores the breakup of the Conciliation government of the first year of Amadeo's reign and the personal animosity of the divided leaders. In particular, he angrily denounces Ruiz Zorrilla's destructive policy of *retraimiento* and his opportunistic coalition with the Democrats, the supporters of Alfonso, and the Carlists. Lack of solidarity among Progressive forces, then and always, is his constant complaint.

Because it lasted only a bare two years, the permanent impact of Amadeo's reign was very small. Accordingly, it occupies relatively little place in the novels. In *Fortunata y Jacinta* Doña Guillermina pursues the newly installed royal couple for support for her orphanage as implacably as she does everyone else and is received as a friend by the queen María Victoria,[3] largely shunned by high society; but she sheds no tears over their departure: 'Todo sea por Dios!' (V: I. 7. ii, 81) is her sole comment. For the Santa Cruz circle the king's abdication is merely an example of instability which causes stocks and

[1] The main groups were the Radicals under Ruiz Zorrilla, the more conservative 'Constitutionalists' under Sagasta, and the so-called 'frontiersmen' ('fronterizos') of Serrano. The Democrat Party was divided between Rivero and Martos.

[2] For a detailed account of Galdós's spell as editor of *La Revista de España*, with an index of the articles published under his editorship, see Ballantyne (*3b*).

[3] María Victoria was renowned for her charitable works. See Ballesteros Baretta (*1b*, 181).

shares to fall disastrously: another instance of the fickleness of Spain faithfully reflected in the personal life of the *señorito* Juanito Santa Cruz.

In *La desheredada* the short-lived Savoyan dynasty has a somewhat greater role. Altogether, Isidora's story is well integrated into the historical setting. As befits the immediate successor of the first batch of *Episodios nacionales*, there is considerable density of historical allusion: care is exercised to provide a greater degree of historical continuity, particularly at the beginning of volume ii, than in any other novel. In his stimulating book tracing the evolution of Galdós's fiction towards the supreme achievement of *Fortunata y Jacinta*, Stephen Gilman (*3c*, 102–10), characteristically, brings many sharp insights to *La desheredada*, but it does not seem to me correct, for reasons I have given earlier, to speak of *La desheredada* in terms of an *Episodio nacional* (91). Moreover, the links he establishes rather insistently (91, 114–15, 206, 207 n.) with *episodios* from the second series are based on an erroneous reading: neither Canencia nor Tomás Rufete has any direct connection with the series. The usually pacific scribe who has been an inmate of Leganés for thirty-two years[4] has nothing except his surname in common with the Bartolomé Canencia who plays a substantial role in the first and second series from *La batalla de los Arapiles* onwards; elderly even in 1813, an *afrancesado* intellectual associated with Santorcaz, and a prominent member of the Masonic lodge 'El grande Oriente', he is finally killed by the absolutists in Seville during the French invasion of 1823 (*Los cien mil hijos* I: 1704). Similarly, the loquacious *ayacucho* Liberal and associate of Aviraneta's, captain Rufete, of *Un faccioso más*, is not Tomás Rufete himself but his father.[5] The link is significant, for it establishes a chain of deranged behaviour over three generations as well as between the *episodios* and the first of the contemporary novels; it is not, however, an example of 'pseudoreappearing characters' (*3c*, Gilman 92), but a case of heredity.

I have, moreover, serious reservations about the symbolic role assigned to Isidora by Gilman. I am in full agreement with the persuasive argument he develops that Prim's death is a moral watershed and that recollection of his assassination, with Amadeo's departure its natural belated sequel, is crucial to the structure of the novel and the characterization of Isidora. *La historia grande* and *la historia pequeña* cross and intertwine, mutually illuminating each other, but this does not justify the complete mythical identification Gilman makes between them. When he claims that 'she and the nation *for which her biography*

[4] There is another Canencia in *Tormento*, who as a 'protegido de la casa' is 'muy torpe' in assisting the Bringases in their move to a new flat (IV: 1462); there seems no reason to associate him with our figure either. For other Canencias in a forgotten story see Hoar (*3c*).

[5] Isidora describes him as follows: 'Fue, en no sé qué tiempo, de la Milicia Nacional, hizo barricadas, hablaba mucho, y para él todos los que gobernaban eran ladrones. Cuando yo era niña jugaba con el morrión de mi abuelo' (IV: I. 1. iii, 973). Compare 'el remate y coronamiento de tan singular cabeza había de ser uno de aquellos morriones de base estrecha y anchísima tapa' (*Un faccioso más* II: 238).

was the latter-day myth were the victims of themselves' (*3c*, 105), I would accept the main statement but reject the subordinate clause I have italicized. Isidora does not in my view represent Spain as such; she is not even 'the symbol of her epoch', from 1868 to 1872 or from 1868 to 1878;[6] nor do I find it helpful to equate her various lovers with future governments or her hydrocephalic child *Riquín* with the Restoration.[7] The public and the personal affairs are similar, but by no means identical; what they share, essentially, is that the inflexible, if illusory, hopes of both nation and individual are shattered. Isidora is an original creation: one should not take away from her that marvellously created impression of independence by forcing her into a mould of symbolic representation any more than we should see her as inexorably conditioned by heredity or environment in a Zolaesque manner.[8]

The first substantial reference to a current event concerns the famous *mantillas blancas* episode when the aristocratic ladies who support Alfonso turn out in force to snub the foreign monarchs.[9] Father Luis Coloma's once famous novel *Pequeñeces* (1890) describes the ostracism practised against the Savoyan queen:

eran el brazo derecho de los políticos de la Restauración las señoras de la grandeza . . . Elías, con sus alardes de españolismo y sus algaradas aristocráticas, habían conseguido hacer el vacío en torno de don Amadeo de Saboya y la reina María Victoria, acorralándolos en el palacio de la Plaza de Oriente . . . Las damas acudían a la Fuente Castellana, tendidas en sus carretelas, con clásicas mantillas de blonda y peinetas de teja, y la flor de lis, emblema de la Restauración, brillaba en todos los tocados que se lucían en teatros y saraos. (*4a*, 72)

When the despicable Currita Albornoz agrees to become lady-in-waiting of *la Cisterna*, as her snobbish coterie called her, the action is treated as scandalously reprehensible by both author and her peers.

The well-integrated episode in *La desheredada* shows, through Miquis's

[6] Bly (*3c*, 1983: 14). In my view Bly greatly overestimates the degree of Relimpio's historical perception ('keen feeling for contemporary history', 10; 'superb political insight', 14). Nor do I see any evidence in support of his 'tempting hypothesis' (3) that Rufete represents a fictional counterpart to Amadeo or that Doña Laura is 'tentatively presented as a symbol of the unity and honour of all Spain' (17).

[7] Gilman (*3c*, 123), following Ruiz Salvador (*3c*) and Chad Wright (*3c*, 1971). Wright takes the concept much further by seeing in Isidora's house a representation of the First Republic and in *Riquín* the danger of over-concentration of power in Madrid.

[8] For example, Ruiz Salvador speaks of 'el rígido determinismo histórico, que en esta novela se tiñe de naturalismo' (*3c*, 60).

[9] Galdós had written about it at the time, alluding circumspectly to Ducazcal's counter-parade: 'El grupo moderado . . . refugiado a los tocadores y en los salones . . . halló en la inhumación de ciertos trajes españoles, pertenecientes a cierta época de desvergüenza e ignorancia que es página de rubor en nuestra historia, una fórmula de protesta contra la nueva dinastía. Pero aquella sátira de mal gusto produjo efecto bien distinto del que se proponían sus autores, los cuales no consiguieron sino . . . sugerir al público comparaciones nada favorables por cierto a personas y cosas justamente anatemizadas por la revolución' (*3a*, *Artículos políticos* 23).

disdainful remarks to Isidora, the hollowness of the demonstration, as well as the precariousness of the new ruler. It reveals at the same time both Isidora's propensity for day-dreaming and her unashamed attachment to outward show and rank.[10] This craving for rank applies equally to her attitude towards Amadeo ('el rey es rey') as towards the Alfonsine society ladies, who, she insists, in the face of Miquis's scepticism, must be right: 'Es la gente principal del país, la gente fina, decente, rica; la que tiene, la que puede, la que sabe' (**IV**: I. 4. iv, 995). It seems strange that the mocking parody of the Alfonsine demonstration by Madrid prostitutes[11] organized by the impresario and politician Felipe Ducazcal is referred to, at most, in a highly oblique fashion, but Bly (*3c*, 1983: 4–5) and Dendle (*3c*, 1982) have made the point that to contemporary readers Isidora, by associating herself with the demonstration, was also identifying herself, indirectly, with the prostitutes and anticipating her eventual fate.

Galdós could not fail to deal with the incident in *Amadeo I*, where Ducazcal hires the typically popular garments he needs from none other than Torquemada. In this work, written many years later, he is equally scathing about the society ladies, but in indicating the incongruity of their attitude he shows more concern to express his moral indignation than to let the implications emerge from the circumstances:

> Ridículo, afectado y artero resultaba el españolismo de nuestras clases altas. Las que desde el segundo tercio del siglo habían renegado de todo lo castizo, arrojando al montón de las prenderías las modas españolas y vistiéndose, comiendo y hablando a la francesa, salían ahora con la tecla de adoptar preseas sacadas del Rastro indumentario. Bien hicieron los pícaros de la política en poner frente a ellas el manchado espejo de un Rastro moral. (**III**: 1101)

More effective to my mind is the manner in which he takes pleasure, with his habitual *socarronería*, in listing the ladies and the prostitutes in uncomfortably close proximity and in giving the latter pseudo-noble titles; the first includes members of 'la aristocracia burguesa' like our old acquaintance Eufrasia, Marchioness of Villares de Tajo; among the second are: 'Paca *la Alicantina, marquesa del Cieno*; *la Eloísa*, muy conocida en todos los círculos . . . viciosos', etc. Once more, however, he insists on ramming the point home with a statement from his privileged status as narrator: 'La fatalidad política había confundido lo más aristocrático con lo más villanesco. Y sobre la bullanga femenil oíamos una estupenda carcajada de la moral pública' (**III**: 1101).

Other incidental political references in *La desheredada* reinforce attitudes already revealed. One of José de Relimpio's heterogeneous political beliefs is

[10] Alicia Andreu demonstrates that Isidora has a literary model for her aspirations in the sentimental story *La cruz del olivar* by Faustina Sáez de Melgar (*4a*).

[11] In order to accommodate Isidora's story to the incident, Galdós had to force the timing slightly.

the true if anodyne conviction that Amadeo 'es una persona decente', which adds in a small degree to the censure of those who by denying him their full allegiance brought about his abdication. The outside observer Gallenga (*1b*, ii. 181) confirms this opinion by noting that 'many persons' called Amadeo 'Un Rey demasiado decente para nosotros Españoles' [*sic*].[12]

At the time of the abdication Amadeo's departure is linked with Isidora's distressed state of mind after being rejected by the Marchioness of Aransis. She consciously relates herself to the political event, with its reiterative double refrain—'Ya somos iguales' and 'el rey se va'—but in a subtle and ambivalent way. On the one hand, in her bruised pride at her rejection, she takes bitter pleasure in renunciation, identifying herself quite falsely with the departing monarch: 'ella también despreciaba una corona. También ella era una reina que se iba' (**IV**: I. 17, 1058).[13] At the same time, she rejoices in the overturning of the established order which has scorned her:

> pensaba que aquello de ser todo iguales y marcharse el rey a su casa, indicaba un acontecimiento excepcional de ésos que hacen época en la vida de los pueblos, y se alegró en lo íntimo de su alma, considerando que habría cataclismo, hundimiento de cosas venerables, terremoto social y desplome de antiguos colosos. (**IV**: I. 17, 1057)

In common with many other characters in Galdós—Juan Pablo Rubín, Izquierdo, and Villaamil are good examples—Isidora momentarily adopts a revolutionary or subversive stance, not out of conviction, for her inclinations are decidedly against equality, but out of sheer personal frustration; this characteristic will of course find its strongest expression in Mariano's pathetic and desperate attempt at regicide later in the novel. For the moment, the Republic is ushered in with no sign of enthusiasm,[14] 'como por disposición testamentaria . . . Se le aceptaba como un brebaje de ignorado sabor, del cual no se espera ni salud ni muerte' (**IV**: I. 17, 1059). With its vague aspirations to equality, it raises hopes of destroying the existing order which has rejected the claims of the individual; and a dishonourable national tragedy is paralleled in some degree by a personal one.

The *episodio* devoted to the Savoyan's brief reign, *Amadeo I* (1910), has several startling new features, which are consolidated and extended in subsequent volumes. From Galdós's revealing letters to his last lover, Teodosia Gandarias,[15] it is evident that the change in technique was deliberate, and

[12] A contemporary visitor remarks on the royal couple's 'trying and lonely existence', which made their situation 'more and more untenable and insupportable' (*1b*, Harvey 70).

[13] Later on, during her imprisonment, Isidora compares herself with none other than Marie-Antoinette, imprisoned in the Conciergerie (**IV**: II. 13. i, 1129).

[14] Historically, this appears somewhat exaggerated. See Hennessy: 'ugly crowds, waving republican slogans and agitated by club leaders, milling around the *Cortes* building, emphasized the dangers of delay and the threat of armed rising if the Republic were not immediately proclaimed' (*1b*, 172).

[15] These letters (239 in all) are unpublished, except for 17 excerpts and other fascinating glimpses given by Benito Madariaga in his *Biografía santanderina* (*3b*, 88–95; transcriptions

began with *El caballero encantado*, which he started in the summer of 1909 at Santander and expected to have finished in September in Madrid. He demonstrates a possibly excessive concern for innovation and refers pointedly to the fact that he is now composing his forty-third *episodio* (later Nougués, in a letter written on his behalf, refers to the forty-fifth *episodio*). The result is that he is over-concerned not to bore his audience; perhaps the relatively low sales of the new batch of *episodios*[16] caused this preoccupation. At all events, the enthusiasm he expresses at times knows no bounds; with an uncharacteristic 'franqueza de auto-bombo', he writes

> con un ardor, que me recuerda los años floridos de mi oficio literario. La obra me domina; es un vértigo que me arrastra, una hoguera que me caldea. Ya voy por cerca de la mitad. Va saliendo con chorro afluente como el de un manantial de roca viva, que no desmaya. No sé si me equivocaré; pero creo que me va saliendo muy bien, y con extraordinario interés. (26 August 1909; *3b*, Madariaga 351)

At other moments, it must be added, as the letters collected by Phoebe Porter (*3b*) indicate, he reveals uncertainty and lack of confidence.

It is of great interest to see how he describes his 'filón nuevo':

> Es un método de humorismo encerrado dentro de una forma fantástica, extravagante, algo por el estilo de los libros de caballerías, que desterró Cervantes, y que a mí, en guasa, se me ha ocurrido rematar para poder decir con la envoltura de una ficción lo que de otra manera sería imposible . . . (2 September 1909; *3b*, Madariaga 354)

At the same time he stresses that 'en el fondo hay realidad o realismo y una pintura que yo creo justa de la vida social tal como la estamos viendo y tocando' (26 August 1909; *3b*, Madariaga 351). He adds that

> en esta obra presento algunos cuadros de la vida española en aspectos muy poco conocidos, la vida de los labradores más humildes, la de los pastores, la de los que trabajan en las canterías en obras de carretera y en otras duras faenas. Son cuadros de verdadera esclavitud que en la vida hay en estos tiempos, aunque no lo parezca. (2 September 1909; *3b*, Madariaga 354)

For her part, Teodosia collaborated by sending him some maps of Soria made by a pupil of hers, which Galdós utilized in Chapter XVI (VI: 1075; *3b*, Madariaga 352).

342–63) and the 7 letters reproduced by Sebastián de la Nuez (*3b*, 1989). I have had the benefit of consulting a substantial unpublished article by Phoebe A. Porter (*3b*) on this correspondence, and she has kindly shared with me photocopies of some of these letters. Teodosia was a woman of mature years and considerable intelligence. A letter of 21 July 1907 refers to a pregnancy (*3b*, de la Nuez 1989: 210–11), and Madariaga mentions in passing that she bore him a child who died young (95), but gives no further details. Biscayan by origin and an elementary schoolteacher, Teodosia was interested in languages and actively collaborated with Galdós. When his failing eyesight prevented his writing to her, Pablo and Lydia Nougués did so on his behalf.

[16] Botrel (*3b*, 1984: 134) gives sale figures of 104 volumes a month for the third series and 60 a month for the fourth during the period 1905–8, compared with 402 and 331 respectively for the first and second series.

The allegory of *El caballero encantado* is sustained and continuous, with a complete abandonment of concrete historical referentiality: names are symbolical—Mudarra, Sisones, Bálsamo, Becerro—dates and specific events are avoided. Its social criticism is, as Galdós claimed, powerful but generalized. It is a fully fledged Utopian fantasy based closely on *Don Quijote*,[17] in which the exploitative landowner and *cunero* politician Tarsis is punished for his egoism and abuse of his privileged station by an external supernatural force—*La Madre*—and returned reformed to the world. There he is reunited with his lover, the schoolteacher Cintia,[18] and their son Héspero, who represents the future hope of Spain. Whatever one may think of the work (and in my opinion it suffers severely from its divorce from actual circumstance and from too specific recipes for change), the structure does not falter. All accusations of senility[19] seem inappropriate. Rather, Galdós's restless urge towards experimentation—possibly a last fling—has operated again.

At his best moments (for there were also times of discouragement), Galdós clearly took just as positive a view of the change in technique in *Amadeo I*. In a letter to Teodosia he describes his intention:

Como necesito variar los asuntos, los personajes y hasta el método descriptivo para que la obra total no se haga pesada (el tomo actual es el 43 de la serie) en *Amadeo I*, me propongo hacer una obra parecida a las del género picaresco que es la más interesante tradición de la novela española. En este tomo predomina pues el elemento cómico. Ya he *pasado el Rubicón*, es decir, ya he hecho más de cien cuartillas. Una vez dominado el asunto, lo demás irá rápidamente hasta el final. (21 August 1910; *3b*, Madariaga 355–6)

A few days later he indicates once more the affiliation with the picaresque: 'En Amadeo I verás una obra extraña, del género que llaman picaresca, que es el género más castizo de la novela española, como el Lazarillo de Tormes, Guzmán de Alfarache y Rinconete y Cortadillo del maestro de maestros' (7 September 1910; *3b*, Madariaga 92). Madariaga adds another phrase from one of the letters: '"He tenido que buscar formas nuevas de narración para evitar la monotonía," le dice' (*3b*, 92). Galdós's insistence on the imperative of innovation may be connected with his doubts about when to abandon writing *Episodios nacionales* discussed in Chapter 4. His enthusiasm for the new technique adopted in *El caballero encantado* evidently induced him to apply it to the *episodios*, and to press on, with the apparent intention of completing a full ten-volume series.[20]

[17] See Rodríguez-Puértolas's introduction to his edition (*3a*, 29–31; also *3c*).

[18] Cintia may, as Madariaga suggests (*3b*, 91), be influenced by Teodosia, who shares the same profession.

[19] For example, Eoff (*3c*, 1961: 16, 155) and Hinterhäuser (*3d*, 144).

[20] In her study of the later *episodios*, Urey (*3d*, 1989: 148–9) underplays the differences between the first two volumes and the rest and treats the six we have as self-sufficient: 'The six volumes can only be read now as they are written' (153).

A further factor is his increasingly failing eyesight, an infirmity which Berkowitz (*3b*, 1948: 409) implies Galdós was stubbornly unwilling to recognize. Apart from the physical difficulty of writing, poor sight obviously inhibits ease of access to documents and reference books and makes checking and revising what has already been written very burdensome. It thus affects structure, especially for such an inveterate corrector and reviser as Galdós. Dictation makes these matters even more onerous.[21] *Amadeo I* is the first narrative in which he was obliged, though only towards the end, to have recourse to dictation; he struggled against it by reverting several times to writing himself (*3b*, Berkowitz 1948: 410; *3d*, Urey 1989: 252). In my view the incoherence admitted by the narrator Tito of much of *Amadeo I* and subsequent narratives stems from this circumstance as much as from adherence to a new less formal technique.

At all events, the abrupt modification of technique has been subject to much controversy. A contemporary review of *La razón de la sinrazón* by 'Andrenio' (Eduardo Gómez de Baquero) is typical in its lack of enthusiasm:

> ¿Cómo ... ha venido a parar, [*asks 'Andrenio'*] a este arte, un tanto vago y difuso, aunque elevado y noble, de las alegorías, de las personificaciones de ideas, de los simbolos? ... Diríase que la fantasía de nuestro gran novelista y dramaturgo, cansada de imitar las figuras concretas de la vida, las formas a fenómenos individuales, se ha ido elevado a la región de las ideas y de los arquetipos sociales, y que el espíritu de Galdós, eminentemente observador, por lo común apacible y ecuánime, ha pasado de la contemplación de lo exterior a la idea o representación interna de los individuos y se complace en divagar por este círculo sereno, donde los accidentes particulares no son más que sombras proyectadas por la luz interior de cada sujeto y de cada asunto. (*3b*, quoted in Percival 1985: 63–4)

The influential scholar Montesinos is likewise severely critical of Galdós's 'persistente manía alegorizante' (*3c*, iii. 327–37), and he is by no means alone.[22] More recently, however, the late *episodios* and novels have been defended no less vehemently. Let us therefore examine the new structure in more detail.

Instead of Vicente Halconero, who effectively reflected the shifts of political opinion, a new narrator appears, the result of a curious dialogue between two figures, *el celtíbero* and *el guanche* or *isleño*, about who shall tell the story. The latter has been left with a commitment he no longer welcomes, as a con-

[21] '... esto del dictado debió de ser una tortura horrible ... La memoria de don Benito, ello es constante, flaqueaba tanto como la vista, y las relecturas y las vueltas sobre lo hecho debieron de ser muchas y fastidiosísimas; para un hombre como Galdós, acostumbrado a escribir vertiginosamente, lentitudes y titubeos fueron sin duda exasperantes' (*3c*, Montesinos iii. 246).

[22] Regalado García (*3d*, 501) finds that Tito 'raya en la bufonería, y pasa rápidamente de lo sublime a lo grotesco'. Fletcher (*4a*, 48–50) is dismissive of the whole fifth series as 'a kind of *reductio ad absurdum*' of previous techniques, but she takes no account at all of the first two *episodios*. Dendle (*3d*, 1980: 153–81) likewise does not distinguish between the first two and the rest.

sequence of 'una promesa indiscreta' (III: 1027). He is evidently based on Galdós himself and possesses a series of deliberately tell-tale characteristics: he is a Canary Islander, an associate of Albareda, a summer resident in Santander; even Galdós's exact residence in Madrid (Olivo, 9) is given as his. More important still, Galdós's self-portrait as a writer is recognizable, if slightly ironic: 'él despuntaba por la literatura; no sé si en aquellas calendas [1871] había dado al público algún libro; años adelante lanzó más de uno [*!*], de materia y finalidad patrióticas, contando guerras, disturbios y casos públicos y particulares que vienen a ser como toques o bosquejos fugaces del carácter nacional' (1010). Some pages later he is further described (again an ironic self-portrait): 'Mi amigo no llevó mal sus años maduros, y su rostro alegre y su decir reposado me declaraban mayor contento de la vida que el que yo tenía. Hablamos de trabajos y publicaciones; díjele yo que había leído las suyas, y él, replicándome que algo le quedaba por hacer, saltó con esta idea [*of his friend's undertaking the work*]' (1026–7), which *el celtíbero* accepts.

The *isleño* arranges to hand over the task to *el celtíbero*, thirty-seven years—spelt out as 13,305 days!—from the date of the end of Amadeo's reign, in return for a specific favour received at that period. Also attracted by literature, the *celtíbero* confesses that, by temperament, he lacks the perseverance of his friend; both had at that time cultivated journalism ('matábamos el tiempo y engañábamos las ilusiones haciendo periodismo, excelente aprendizaje para mayores empresas'). Following this he had travelled (to the colonies of Cuba and the Philippines—the traditional way to make money), and had worked and written without success.

The new narrator is normally called Tito, but his full name is Proteo Liviano. Thus he is 'protean', in a very different sense from that previously discussed, being capable of readily adapting form and attitude to differing circumstances. He is also a historian, a burlesque imitation of Livy; the name also includes a punning reference to his lasciviousness. 'Celtíbero' like the Ansúrez family, he is very unlike them in every other way. He is puny in stature, volatile, prone to exaggeration, subject to illnesses, beset by fantasies, and incurably amorous. What precisely is his function? His own claim is to record 'true history', emphasizing the mixture of private and public occurrences and the primacy of the former: 'De asuntos privados, confundidos con los públicos, hablaré, para que resulte la verdadera historia, la cual nos aburriría si a ratos no la decalzáramos del coturno para ponerle las zapatillas' (III: 1023); the aim of avoiding boring material—equated with official history—recalls the letters to Teodosia. Dendle sees his role as at least partly representational—'symbolizing the make-shift illusory nature of the ramshackle monarchy of Amadeo' (*3d*, 1980: 168)—or, alternatively, as 'a fitting representative of a society that has abandoned all hope of reform' (178). Urey argues for continuity with the previous *episodios*, seeing Tito as 'degenerated Halconero, just as Halconero is a degenerated reflection of past protagonists', but in a literary, not a historical, context: 'Tito functions not to present a lifelike personality, but to serve as one more absurd note of discord among the absurdities and discontinuities of the series' (*3d*, 1989: 163–4).

For me, it is clear that Galdós has deliberately set out to abandon conventional realism. To be sure, Tito takes in a wide range of regions and classes: he is 'un ciempiés o cien ramos' and 'un queso de múltiples y variadas leches' (III: 1014). Nor is he completely free from family ties; he has a Carlist father in Durango. None the less, he has little socio-historical plausibility compared with his predecessors, and he has no natural base in society from which to operate. He therefore cannot be fully representational, as Dendle contends. Yet historical referentiality, if inconsistent, is still strong: Tito's affairs, though far-fetched and presumably intended to have a comic effect, are frequently used to reflect historical occurrences.[23] The most evident case is María de la Cabeza Ventosa, characterized by her 'abolengo liberal, rancio y clarísimo' (1032), who evidently demonstrates the continuity of the past: she is Benigno Cordero's granddaughter, the niece at some remove of Calvo Asensio and José Abascal, and related to Mariana Pineda; she was overwhelmed with grief at Prim's death and is now a supporter of Ruiz Zorrilla. Through another of Tito's lovers, Nieves, we gain access to the royal palace, while through Obdulia we are informed about the ladies of the aristocracy, who prophesy, rather inopportunely, 'esa cosa que llaman la Restauración, que es como decir Alfonsito, el niño de doña Isabel', and demonstrate their hostility to María Victoria (1016). And when Tito has to produce an excuse for his absence from Cabeza with Graziella (who is herself the 'numen de la nueva Italia', 1106), he writes a lengthy report supposedly prepared for Ruiz Zorrilla. At the same time, despite his displacement of *el guanche*, he is close in some respects to Galdós's own biography (the *desdoblamiento* is thus double), as when he loses his sight, when he is offered a seat in Congress, and possibly, too, in the innumerable sexual adventures in which he engages. He provides his creator with frequent opportunities for self-indulgent reminiscences.

Tito, in my view, then, is not a clear, fully fledged allegorical figure, but retains much historical and ideological baggage. He is not quite an authorial voice, nor a reflector, nor a full participant, but has something of all three: 'a grotesquely "hybrid" alter ego', Gilman calls him (*3d*, 1986: 47). In view of Galdós's letters to Teodosia, we can hardly doubt that Tito owes much to the picaresque tradition, as a peripatetic figure who witnesses a range of

[23] As Tito indicates explicitly—'Las aventuras que me sirven de trama para la urdidumbre histórica' (III: 1055)—and when he speaks of 'sensaciones personales que no carecen de miga histórica' (1081). Varela (*3d*, 39) makes a rather rigid classification, as follows: 'Obdulia (monarchy), Felipa (lower class), María de la Cabeza Ventosa (capitalism and the mercantile middle class), Graziella (liberalism), Pepa Hermosillo (aristocracy), Lucrecia (lower class), Delfina Gil (Alfonsinists [*sic*]), Facunda (Carlism) and Josefa Izco de Larrea (Carlism).'

characters and occurrences. An outsider made insider, he seems even more like a grotesque Asmodeo of *El diablo cojuelo*, commenting sarcastically on events. Galdós seems to have felt that the ironic self-portrait of *el guanche* and the eccentric *alter ego* of *el celtíbero* brought in elements of picaresque humour which would obviate the monotony associated with a string of so many similar *episodios*. For my part, however, I find it difficult to see in them more than an unnecessary and inapposite desire for self-justification, in an enterprise to which he was perhaps no longer fully committed and which had only an age of pernicious stagnation (as he saw it) to chronicle.

Clío's allegorical role is much more evident; in accordance with the strong pull of *intrahistoria*, the Muse of History is made humble and domestic, though without suffering the sardonic degradation to a syphilitic streetwalker imposed on her by Schopenhauer. She is introduced as an elderly dealer in antiques, *la tía Clío*, who sometimes sleeps in the porter's entrance of the Academy of History[24] (III: 1043), from which she and Tito receive a small stipend. It is another attempt, it appears, at humour. She is called Mariclío and later *Madre Mariana* (with an obvious pun on the sixteenth-century historian), and her appearance varies according to the historical circumstances, so that at times she will revert to the ancient world or take on more august roles, wearing classical footwear ('cotornos') instead of slippers (1057); at others she almost disappears. She has at her command supernatural devices, like the pen whose text fades away after two days if the event recorded is ephemeral. She has frequent recourse to magic. The most spectacular instance is the grotto of Graziella, to which Tito is enticed and where he is then obliged to stay, but which has disappeared when he looks for it later. Tito is magically transported from one place to another and, transformed into a kitten, witnesses the final departure of the royal couple for Italy.

Clío comments frequently, with authority, on events, since she represents truth ('mi boca, que no miente. Mi único guiso es la verdad', III: 1058): for the first time in the series, we now have an undisputed final word on matters of history, albeit from an incongruous source. Although her consistency increases in later *episodios*, the supernatural interventions jar continually with the subsisting framework of real events; and Clío is reduced either to inaction or to commenting on happenings she is powerless to modify. As a consequence the effectiveness of the interplay between the fictional story and the historical events is impaired. With this increasing artificiality, the relation

[24] The text adds that 'por la mañana le cepilla la ropa al gran don Marcelino, por quien siente ardoroso cariño'. Engaging as this may be, Menéndez Pelayo was only 16 years old at the time (Montesinos [*3c*, iii. 291] credits him with 20 years). Distinctly preferable to my mind, in any case, is the sort of sly humour by which Galdós portrays the porter of the Academy of History as a silent self-made expert on historical events in the discussions between Izquierdo, Ido, and others in *Fortunata y Jacinta*: (V: IV. 5. i, 488): 'echaba al concurso miradas desdeñosas, no queriendo aventurar una opinión que habría sido lo mismo que arrojar margaritas a cerdos.'

between *la historia grande* and *la historia chica* has changed decisively; the two elements tend to be juxtaposed, as they had been in *La Fontana de Oro*, rather than integrated,[25] so that the metonymic structure is weakened; and the latter has become subordinate to *la historia alegórica*.

The new technique evidently involves the use of the fantastic, always a strong undercurrent in Galdós.[26] In earlier novels (*El amigo Manso*, *Realidad*, *Misericordia*) fantastic elements had served as an effective support to a basically realistic structure, corresponding to the first or at most the second of the distinctions established by Todorov between the strange (finally resolved according to reality), the fantastic (ambiguous), and the marvellous (completely divorced from reality) (*2*, 1970). In the case of these *episodios*, the use of the fantastic veers towards the marvellous, but without a decisive break with reality. It appears arbitrary, indeterminate, neither subordinate to real events nor integrated with them, nor independent of them. It seems incompatible with the still dominant historicist intention and gives the impression of being little more than a convenient device to provide an alternative solution for a permanent difficulty confronting the *episodios*: how to secure the narrator-protagonist's presence at key events. Miguel Enguídanos (*3d*) makes a spirited defence of a technique which he sees as a quest for truth reflecting the personal anguish of the novelist. While he is right in seeing in the words 'concomitancia(s) y simpatía(s)' (*Cánovas* III: 1340) a key indication of Galdós's attitude, such agonized pursuit of truth offers no proof in itself of success in the literary enterprise he has undertaken. Ricardo Gullón's notion that 'A un espacio cargado de falsedad y mentira corresponde un tiempo caótico que autoriza, si no impone, el desorden de la narración' (*3d*, 1970: 31) seems to me to confuse content with form. Antonio Varela takes a similarly favourable, and for me unconvincing, view of the last *episodios*, as a 'successful literary solution to Galdós's own inner conflicts in the last stages of his career and life' (*3d*, 38).

Nor is the use of time innovatory, for the manipulations of the time-scale are questions of distance; chronology is not basically disrupted. Thus the tremendous leap of thirty-seven years which occurs from the time of the events portrayed to the present time of writing, 1910,[27] is basically diachronic, not synchronic; it is intended to emphasize the close relation between the two periods, not the fluidity of the concept of time. Events right up to 1909 are

[25] Rodríguez speaks acutely of the 'open bifurcation of novelistic activities: one direction represented by the dissipated woman chaser; the other by the official historical observer' (*3d*, 181).

[26] See the pioneer study by Clavería (*3b*, 1953). Alan Smith has in preparation an extensive study of Galdós's fantastic short stories throughout his literary career, which shows how enduring and consistent this tendency was. It may be noticed that Valera turned to fantasy, within a historical context, in his last novel *Morsamor* (1899).

[27] Urey gives the date 1908 (*3d*, 1989: 219), a miscalculation based on counting thirty-seven years from Amadeo's inauguration (1871) rather than from his abdication (1873).

explicitly recalled, so that a preoccupation with the historical present is evident:

Corrió el tiempo arrastrando sucesos públicos y privados; se fue don Amadeo; salió por escotillón la República; feneció ésta, dejando paso a la Restauración . . . Reinó Alfonso XII; pasó a mejor vida; tuvimos Regencia larga; se fueron de paseo las Colonias y entraron a comer manadas de frailes y monjas . . . El niño Alfonso XIII fue hombre; reinó, casó . . . Vino lo que vino: agitación de partidos, inquietud social, prurito de libertad, alerta de republicanos, guerra con moros, semanas de fuego y sangre . . . (III: 1026)

In fact, we are given here a sort of foretaste, indicating Galdós's current priorities as well as the salient facts, of the whole fifth series as it was then planned, up to and including the *Semana trágica* of July 1909. We may note his concern with the loss of the colonies, with the invasion of religious orders, with social unrest. These concerns coincide quite well with the rather less specific titles proposed for the unwritten *episodios*: *Sagasta*, *Las colonias perdidas*, *La reina regente*, *Alfonso XIII* (*3b*, Berkowitz 1948: 345). None of this supports an overriding concern with temporality as such or with fictional technique. What is apparent is a certain fatalism ('vino lo que vino') and a deep pessimism resulting from the long and (for Galdós) essentially unproductive period of time between the era depicted and the moment of writing.

Gullón's view is that novelistic considerations are now put before historical ones. He offers an explanation of the autobiographical intervention on these lines: 'Incluyéndose en la novela, como hizo tantas veces, el autor se noveliza e iguala con los personajes, presentándolos como seres de la misma especie (semejantes, digo otra vez, y no criaturas), con libertad de acción y decisión' (*3d*, 1970: 25). The narration becomes a treatise on how to write a historical novel: '*Amadeo I* podría haberse titulado, anticipando a Unamuno, *Cómo se escribe una historia* . . . Desde el instante en que acepta la encomienda de su amigo "el guanche" y comienza a historiar los sucesos pretéritos, el foco narrativo se dirige más al acto de contar que a lo contado, y ese acto se presentará como algo singular y hasta fantástico' (26). Gilman and Urey likewise both downplay the role of history, and see in it a deliberate cultivation of a self-reflexive criterion designed to break down the conventional author–reader relationship, to produce what Gilman calls a 'liberation from the constriction of identity' (*3d*, 1986: 49). Urey interprets the last sentence of the chapter in which Tito accepts the commission, 'mirando al cielo y a los tranvías que de un lado a otro pasaban' (III: 1027), as illustrating 'the indeterminacy between self and other, real and imaginary, past and present, history and fiction. All documents, memories, histories, and fantasies can only be distinguished arbitrarily, since they rely on the same textual processes' (*3d*, 1989: 219). None of this seems to me convincing; Galdós's narrator has never before thrust himself into his novels in such an explicit, if still oblique, way; and he takes no further part in the action. Nor is there any sign of an Unamuno-like intent to establish an existentialist parallel between the character and life in the style of *Como se hace una novela*. In my view these arguments read twentieth-century criteria into a text still rooted (without much conviction any longer, however) in nineteenth-century values.

From the point of view of Galdós's attitude to history, the essential point is whether the author is still basically concerned with the conveying and interpretation of objective fact and with the search for meaning in history. Urey puts forward an interesting argument: 'If Spain's history', she argues, 'is only a repetition or distortion of itself, if there is no effective change, no functional difference between past and present, then there is no progress, no movement, no lesson that is to be learnt' (*3d*, 1989: 194). The description is convincing as far as it refers to the specific history of modern Spain—as seen by Galdós—but not as far as history itself is concerned. If he finds no change or little sense in Spanish history, this is not because all history is by its nature unchanging or meaningless, but because it appears so in the specific context of Spain. Both in the period chronicled and at the time of writing, Spain is for Galdós a country which has lost all its energy and sense of direction. For Galdós as for Valle-Inclán, it is Spain which is anomalous: 'una deformación grotesca de la civilización europea' (*4a*, *Luces de bohemia* 168), but the comparison with Valle-Inclán's innovatory and dynamic creation of the *esperpento* indicates in my opinion how inadequate Galdós's new technique is. Galdós despairs about what has gone on and is going on in Spain, not so much about history itself. That is why despite everything he clings so strongly to detailed historical data, even though he treats them with a measure of contempt. This is not to deny that he is suffering a crisis about history, which caused him to abandon historical realism in his novels after *Misericordia*. The images of mimetic movement (Stendhal) or of organic growth (the tree) or of constant flow (the stream) no longer apply. A more appropriate image might be drawn from those offered in the Introduction—waves, distorting mirrors—or the very Spanish one of the Arabic water-wheel, the *noria*, which constantly, monotonously, draws up water which has no free-flowing outlet.[28]

It is the great accumulation of historical facts found in the *episodio* which most undermines Urey's argument. Many such incidents are largely unassimilated or dealt with rapidly, without any reference to Tito's story and largely separate from him; they are seen as essentially similar and fundamentally unimportant because they do not lead to the radical change Galdós yearns for. What Galdós does with historical events is not to abolish them, but

[28] Compare Antonio Machado's 'La noria': 'Yo no sé qué noble, | divino poeta, | unió a la amargura | de la eterna rueda | la dulce armonía | del agua que sueña [. . .] Mas sé que fue un noble, | divino poeta, | corazón maduro | de sombra y de ciencia' (XLVI; first published in *Soledades* [1903]: *4a*, 155).

to make them less equivocal, more black and white, more programmatic. The concern with the present means that there is a greater urgency and stridency in the political assertions. Historical distance is no longer maintained, nor is hindsight eschewed: there is an effect of telescoping a period of more than a generation. If Galdós's colloquial language remains as lively as ever, it is given a new, not always justified, vehemence as a result of his pessimistic assessments.

The figures who formed part of Galdós's immediate circle of the early 1870s are recalled in very considerable circumstantial detail; Gilman notes acutely, 'In strange disguise and with vivid details, the aged Galdós can now tell all he remembered about the crucial years of his young manhood' (*3d*, 1986: 48). On occasion, the opportunity is seized to rectify some standard or previously held opinions. This is particularly the case with José Luis Albareda and his pro-Amadeo group. Thus the conciliation of interests and the establishment of a bipartite system on the English model so staunchly advocated by Albareda, and supported by Galdós at the time, is now rejected by Tito as 'infecunda' (III: 1023).[29] Furthermore, the dubious source of income—from Cuban slavery interests—for such enterprises is proclaimed. Story-line, history, and autobiography are linked by the fact that the favour *el isleño* did for *el celtíbero* was to introduce Tito to Albareda's *El Debate*. Affectionate personal respect for Albareda himself remains strong:

> El hombre que sostenía con fatigas y el apoyo de sus amigos 'La Revista de España', fue un grande y desinteresado propulsor de la cultura de este país. Fue el más aristócrata de los periodistas y el más elegante de los políticos. Las campañas que él inspiraba llevaron siempre el sello de distinción exquisita . . . se mantuvo fiel a los ideales de la soberanía de la nación . . . Y viviendo entre millonarios siempre fue pobre . . . (III: 1025)

Once more the suitability of material comes into question: such an evidently personal tribute would fit better in memoirs or a biography than in a historical novel.

Another similar figure is Ramón Rodríguez Correa, 'un espíritu liberal metido en la armadura de un eclecticismo elegante y conservador', who, like Albareda, argues against Tito in favour of slow progress: 'no cambiemos de siglo antes de tiempo' (III: 1049). Yet another is Pepe Ferreras, the well-known journalist and friend of Galdós's,[30] who is quoted at length at the time of the assassination attempt on the king. Ferreras attributes blame widely for the present crisis: to Ruiz Zorrilla's government for its lack of direction and to

[29] For perceptive opinions on Galdós's political views in the 1870s see Goldman (*3b*, 1969) and Estébanez Calderón (*3b*, 1982). In the interests of 'conciliation', Galdós was then strongly opposed to Ruiz Zorrilla, an attitude which was reversed by the time he wrote *Amadeo I*.

[30] In a moving obituary of May 1904 (VI: 1227–8) Galdós speaks of *el Maestro* Ferreras as 'el mayor altruísta de los tiempos presentes' and as an example of exceptional loyalty and devotion to Sagasta.

the Republicans and Socialists for inertia (here Galdós is evidently thinking also of 1910). The worst thing in politics, Ferreras declares, is confusion: 'En política todo puede admitirse, menos el barullo, el caos y la falta de orientación' (1102). All this, Ferreras argues, aids the supporters of the Bourbons; if there is no change of attitude, he prophesies that the Restoration will come (a no-lose prolepsis).

One political figure, Nicolás Estévanez, a fellow Canary Islander, who was a friend of Galdós's and a minister under the Republic, begins to play the disproportionate part which is to characterize the later *episodios*. Galdós drew extensively—to the point of glossing or even quoting—on long passages from Estévanez's memoirs, turning the latter's prose narrative about his journey to and stay in Cuba into direct speech; he particularly emphasizes his acute distress at the shooting of students for a harmless protest, the painful relinquishing of his army commission, and his return home (*3c*, Montesinos iii. 268). In the case of Estévanez, Galdós abandons his usual critical distance, for he is praised unconditionally as a hero worthy of the name: 'la precisión, elocuencia y donaire' of his writing and 'las tremendas causas que lo motivaron, y el admirable tesón que vigorizaba su alma generosa' (**III**: 1048). Appreciated by Beramendi as a consistent revolutionary, he wins a popular seat in the election by merit, not by influence or expenditure. His statements are treated as little less than an oracle,[31] as when he claims that the 1868 revolution is now dead, proclaims the need for a barbarian like Pizarro to rule the country, or voices an immediate pessimism and a more remote and problematical optimism, the only sort of forecast which could afford the Galdós writing in 1910 about 1871 a modicum of consolation:

> Debajo del pesimismo de mi gran amigo latía, como es ley en todo ser superior, un fuerte optimismo. No desconfiaba de la idea, sino de los hombres en el telar político . . . la oposición republicana . . . se contagiaba de esa legalidad indigesta que siempre resulta infecunda, y cándidamente, hacía el juego de sus naturales enemigos. (**III**: 1055)

As for King Amadeo himself, in the first pages of the narrative, before we know the identity of the narrator, we see his entry into Madrid and his visits to Prim's body lying in state at Atocha and to his widow. The question is raised as to whether Amadeo, like Prim, is a Mason. In a curious parenthentical note which seems intrusive in novelistic discourse the suspicion that Amadeo belonged to an Italian lodge is rectified (the result, perhaps, of an enquiry carried out by Nougués) and the dismissal of the rector of Atocha, where a

[31] The modern historian Alistair Hennessy by no means shares this opinion, speaking of 'men like the ex-officer deserter Estévanez who were temperamentally incapable of accepting any society which gave no opportunities for conspiracy and the *élan* of the barricades and who, like him, were prepared to try and make a revolution even though they knew it was bound to fail' (*1b*, 152).

Masonic service was held, is noted. Thanks to Mariclío's magic powers, we are taken to Amadeo's dinner-table, which is magically and picaresquely observed by Tito from a vantage-point under the table, where he can view the guests from below. Amadeo's personal characteristics are duly recorded: his poor Spanish, his liking for Virginia tobacco and Italian food like *lesso* and *guindillas en aguardiente*, as well as the queen's firm grasp of politics in her doubts about the Amorevieta treaty (III: 1046). Tito follows Amadeo on his clandestine excursion outside the palace to see his lover Adela Larra. A worthy daughter of Figaro, Adela gives him sound advice concerning Serrano's manœuvring about the treaty. The king's famous refusal to suspend constitutional guarantees (*Yo contrario*) is anticipated. Later, other erotic incidents are recounted: his love for an Englishwoman married to *The Times* correspondent at a rapidly evoked Santander, his abandonment of Adela and the purchase of her letters, together with various other anecdotes, true or false, about one who, like his father, fully deserved the title of *il re gentiluomo*. In general, his decision to abdicate after having signed the measure to reform the artillery corps he has opposed is respected; he has been consistent in his honest determination to abide by the constitution. Even greater respect is accorded the queen, María Victoria, for her interest in consolidating the dynasty, her desire for ecclesiastical reform (she wanted to make the extremely liberal Don Hilario a bishop), and for her personal qualities, such as nursing her own children. According to Ferreras, she is relieved to leave Spain. The overall mediocrity of politicians ('esta cuadrilla de politicajos por oficio y rutinas abogaciles, hombres de menguada ambición, mil veces más dañinos que los ambiciosos de alto vuelo': 1041) is directly castigated by Mariclío. Ruiz Zorrilla, however, is respected for his reformist energy ('trajo a la política oxígeno abundante y frescura de reformas': 1028), although the quarrels between his faction and Sagasta's over-trivial differences continue to be deplored by Mariclío herself ('Sagastinos y zorrillistas le traen mareado [i.e. Amadeo] con sus necias enemistades por un quítame allá esas pajas': 1058).

Somewhat wider concerns are introduced during Tito's visit to his father at Durango. Clearly influenced by Teo (Porter *3b*), Galdós reveals some respect for the firm character of the Basques[32] and meets the ex-guerrilla priest Don José Miguel Choribiqueta, who is nostalgic for the heroic Carlist past and vehemently opposed to Maroto for having signed the Peace of Vergara. Against this background the compromise Peace of Amorevieta, negotiated by Serrano in what Mariclío terms 'su inocente optimismo' (III: 1080), has no chance of success. Tito struggles with the Basque language and acquires a local fiancée, just as Galdós, in his letters to Teo (Porter *3b*), uses some terms

[32] 'La verdad, yo comprendía y admiraba las sólidas virtudes de la raza, su contumaz apego a la tradición, cualidad meritoria cuando sirve de punto de partida para el progreso, como acontece en Inglaterra; me agradaba la lealtad de los hombres, la lozanía de las mujeres' (III: 1064). The reservation about tradition and progress (and perhaps the praise of the women) is significant.

of endearment in Basque. His satirical lecture on the desirability of a Catholic Republic under the direct rule of the Pope introduces an amusing new burlesque dimension to the subject. It is the best comic piece in the volume. Its spectacular success is, not surprisingly, succeeded by serious doubts about his intentions, and Tito is removed from the scene by a fictitious invitation from the Vatican to discuss his scheme in Rome.

Figures from earlier novels and *episodios* appear frequently; as Gilman puts it: 'a whole roster of lodgers in Galdós' (and our) mental boarding house put in cameo appearances: Torquemada, Ido, Nicanora, Estupiñá, Bringas, Pedro Polo, and even don Basilio de la Caña' (*3d*, 1986: 49). José Ido del Sagrario has reached the relative prosperity of maintaining a boarding-house (at much the same period, in *Fortunata y Jacinta* he is depicted as living in the direst poverty in very different circumstances).[33] It is difficult to justify his presence functionally; Galdós simply seems reluctant to abandon a character so clearly delineated physically and uses him in rather an indiscriminate fashion. Bringas's role as a compliant servant of his current master—previously Isabella II, now Amadeo—his reappointment by Ruiz Zorrilla, and the authority he still commands over Rosalía do not square either with his early ideology or his physical decline; and the pompous Basilio de la Caña turns up unchanged. These not very apposite reappearances seem to indicate that Galdós has laid hand on convenient ready-made goods instead of providing custom-tailoring.[34]

Are we to conclude that the work is all bitter humour, all pessimism? Not quite: Galdós, writing about the period over a generation before, and conscious that no real improvement has occurred up to his own day, can only project into the future. Mariclío sees soldiers and maidservants courting in a park, and then children playing at soldiers, using incorrectly pronounced names of heroes of the past—'Plim', 'Napolión', 'Pecinado': once more we have petrified historical references reminiscent of *La desheredada*. Both groups represent none the less a tentative hope for the future: the first of creating a new honourable generation ('Traed acá nuevos hombres de quienes yo pueda referir acciones altas y nobles'); and the second of establishing a responsible citizenry; even if they now throw stones at her, she commands them to be 'guerreros chiquitos para que de grandes seáis buenos ciudadanos'. Her conclusion is: 'Tito mío, estas diabluras de los rapaces y el embeleso de las parejas de enamorados me consuelan de la mísera vida que arrastro en esta tu decaída tierra' (III: 1060). As with Estévanez, a possible optimism launched well into the future (i.e. beyond 1909) lies beneath the pessimism of the present.

[33] In *Fortunata y Jacinta* he lived in a slum in Mira el Río Alta; at that same time Maxi had an apartment in the Calle del Amor de Dios, where Ido kept his boarding-house in *Amadeo I*.

[34] Rodríguez notes that Galdós 'selected only typed characters, figures requiring virtually no further development' (*3d*, 188).

Comments on the abdication are not dissimilar to those made in *La desheredada*. In *Amadeo I* there is slightly more agitation, but it does not pass the bounds of moderation: 'Ello fue como un plácido regocijo lugareño festejando *la traída de aguas* o la elección de un alcalde muy querido en la localidad' (III: 1111). What Tito and Obdulia do notice is the spontaneous illumination of the balconies, which they term 'vergonzosa y repugnante'; only Republican homes remained dark. The houses of the Alfonsine and Carlist aristocracy naturally 'brillaban con espléndido alumbrado, signo de lisonjeras esperanzas'; what scandalizes them is 'la cínica refulgencia' of the homes of the erstwhile supporters of 'el viejo progresismo' and of Don Amadeo. This metonymic feature reflects the beginning of the *episodio* and gives it a circular ending. The lack of adornments on aristocratic dwellings as Amadeo entered Madrid is matched by the bareness of the balconies owned by supporters of the Federal Republic. The strength of the Bourbon cause and the breakup, so much deplored by Galdós, of the left–centre coalition so carefully constructed by Prim are evident at both the beginning and the end of Amadeo's reign.

On Amadeo's abdication Spain was faced with the same alternatives as before: a Republic or *la solución alfonsina*. Nicolás Rivero, as president of the Congress, immediately called together the two chambers, Senate and Congress, in a National Assembly, at which, with dubious legality, the Republic was proclaimed, its form—Federal or Unitarian—to be decided subsequently by the Constituent Cortes. In the process Rivero was replaced by his rival Cristino Martos. Four presidents succeeded each other in the course of the year before Pavía's *coup d'état* in January 1874 ushered in a regency by Serrano, which was in turn overthrown by the Restoration.

A typical summary of these options is supplied at the beginning of the second part of *La desheredada*. As he marks the passing of a long period without providing any information about Isidora, the narrator offers a fleeting review of the four leading developments since Amadeo's abdication: 'La República [*1*], el Cantonalismo [*2*], el golpe de Estado del 3 de enero [*3*], la Restauración [*4*], tantas formas políticas sucediéndose con rapidez, como las páginas de un manual de Historia recorridas por el fastidio' (IV: II. 1, 1063). Moreover, in order to provide the bare bones of the events between March 1873 and 1875, he draws on the garrulous memory of José de Relimpio, whose mental diary' or *Efemérides*, offers a marvellous pot-pourri of fictional story, telegraphic political news, and comments from the characters; the short-lived and chaotic experiment in federalism is succinctly recalled in note form, alongside Isidora's no less turbulent personal life and events in his own family. For convenience I shall divide up the historical events and comments into groups and take them one by one, starting with the purely parliamentary incidents:

1873. *1 de Marzo.*— ... Disturbios en Barcelona; cunde la indisciplina militar ... *Abril.*—Desarme de la Milicia por la Milicia. Dos cobardías se encuentran frente a frente, y del choque resulta una página histórica. No corre la sangre ... Célebre discurso de Pi ... *Junio.*—Reúnense las Cortes Constituyentes ... Nuevo Gabinete ... *Julio.*— ... Ministerio Salmerón ... *Septiembre.*— ... Castelar, ministro ... *Diciembre.*—Castelar reorganiza el Ejército. La patria da un suspiro de esperanza ... La marea revolucionaria principia a bajar. Se ve que son más duros de lo que se creía los cimientos de la unidad nacional ... 1874. *Enero.*—El día 3 Pavía destruye la República ... (IV: II. 1, 1065)

What is referred to here is the struggles between the Federalists and the Radicals, joined with the more conservative, unitarian Republicans. In a dramatic confrontation on 23 April 1873 Pi y Margall, now acting president of the Republic, succeeded, by a show of force, in forestalling a rising by the rival radical militia led by Martos, and dissolved the hostile Assembly. He refused, however, to use his strength to set up a Federal Republic without elections, waiting instead for the election results which made him president on 11 June. It is notable that Galdós, deploring as always dissension among Progressive forces, treats the conflict as a sham: 'Dos cobardías ... frente a frente'. Four years after the event, in an article published in *La Prensa* (15 July 1885), on the occasion of a parliamentary debate between Pi y Margall and his friend León y Castillo, he is highly critical of Pi:

Fue el verdadero padre de la desdichada propaganda federal, y si los republicanos que tomaron parte en aquella brillante Asamblea hubieran persistido en sostener la forma unitaria, otra hubiera sido la suerte de la revolución de Setiembre y los destinos del país habrían ido por muy distintos caminos. (*3a*, *Cartas desconocidas* 183–9)

In *La Primera República* we shall see a further evolution. As is laconically recorded in the *Efemérides*, Pi y Margall was replaced in July by Salmerón, and he in turn in September by Castelar, with a tougher military policy against the Cantonalists.

Next let us see how the Cantonalist insurrections are rapidly characterized: '*Junio.*— ... Asesinato del coronel Llagostera ... *Julio.*—Alcoy, Sevilla, Montilla. Sangre, fuego, crímenes, desbordamiento general del furor político ... *Septiembre.*—Cartagena, excursión de las fragatas ...' These events—the murder of Martínez Llagostera at Sagunto, followed by the bloody revolt at Alcoy and elsewhere and then by the rising of the fleet at Cartagena—are intertwined with family circumstances (Doña Laura's death, *Riquín*'s birth, Mariano's return home, Emilia Relimpio's marriage, her sister Leonor's disappearance) and insubstantial comments, such as the following patriotic effusion by Relimpio: '¡Oh! Don José les perdonaría a los cantonales en su calaverada si aprovecharan el empuje de las fregatas para irse a Gibraltar y conquistar aquel pedazo de nuestro territorio, retenido por la pérfida Inglaterra' (IV: II. 1, 1066).

In *Fortunata y Jacinta* José Izquierdo falsely claimed to have played a key part in these uprisings, but, like his fulminations against Republican figures (Castelar, Pi, the Cantonalist Roque Barcia, etc.) and his threats of incendiarism, his claims are patently absurd. This diatribe serves to show the widespread disillusion, especially among the less sophisticated classes, with the effects of the revolution. Evidently, no clear political approach emerges, or is intended to emerge, from these references, other than a sense of absurdity and waste. Indeed, in the novels and especially in *Fortunata y Jacinta*, the word 'republic', like the term 'canton', often becomes a simple metaphor for revolution, disorder, chaos, to be applied to the private lives of the fictional characters (see *3c*, Ribbans 1970: 102–5). Juan Bou's similar denunciations in *La desheredada* encompass all contemporary politicians (even venturing into France to cite Thiers), but from a more coherent political position, reflecting pro-anarchist doctrines prevalent in his native Barcelona: 'Mira tú qué tipos. ¿Prim? Un tunante. ¿O'Donnell? Un pillo. Tiranos todos y verdugos. Olózaga, Castelar, Sagasta, Cánovas. Parlanchines todos. ¿Y ese Thiers, de Francia? Otro que tal' (IV: II. 4. ii, 1983). It is hardly more constructive.

At the other extreme, the main events of the Carlist war are covered in the *Efemérides* equally rapidly but thoroughly, and with equal admixture of fictitious occurrences:

> *Junio.*— ... La guerra toma proporciones alarmantes, y en Navarra se ven y se tocan las desastrosas consecuencias de la desgraciada acción de Eraul ... La guerra, la política, ofrecen un espectáculo de confusión lamentable ... *Julio.*— ... La guerra civil crece. Cada día le nace una nueva cabeza y un rabo nuevo a esta idea execrable ... *Septiembre.*— ... Horrores del cura Santa Cruz ... *Diciembre.*— ... Háblase ya de la sima de Igusquiza, y se cuenta horrores del feroz Samaniego. 1874. *Enero.*— ... La guerra civil avanza ... *Marzo.*—San Pedro Abanto. Inmenso interés despiertan en toda España el estado de la guerra y el sitio de Bilbao ... Sangrientos combates del 25, 26 y 27, que ocupan la atención pública ... *Mayo.*—Bilbao es libre. Alegría, repiques, farolitos. Crece a los ojos del país la gran figura militar del marqués del Duero ... *Junio.*—Muerte del general Concha. Pánico y luto. Retirada. La patria, que creía próxima su salvación, gime. *Diciembre.*—La guerra sigue. La Restauración toca a las puertas de la patria con el aldabón de Sagunto ... (IV: II. 1, 1066–7)

In this lengthy passage the main events of the war in the north are rehearsed: Carlist successes like Eraul and San Pedro Abanto, the savagery of the guerrilla leaders Santa Cruz[35] and Samaniego (the latter threw his prisoners into the Igusquiza chasm outside Estella), together with the failure of the siege of Bilbao and Concha's unexpected death. The final pacification will come only after the Restoration.

[35] This ferocious guerrilla-priest also attracted the attention of Valle-Inclán (*El resplandor de la hoguera* and *Gerifaltes de antaño*), Unamuno (*4a*, *Paz en la guerra* ii. 161), and Baroja (*Zalacaín el aventurero*). See Extramiana (*4a*, 241–67, 337–44).

Typically interspersed with these multiple disturbances are the fortunes of the Pez and Relimpio families: Melchor de Relimpio is appointed governor of a third-class province, but quickly loses it again and plots fraudulent deals, by which new fortunes are made ('Sobre las ruinas de las fortunas que desaparecen, elévanse las colosales riquezas de los contratistas': IV: II. 1, 1066); all the Pez clan are dismissed from their posts, but this is followed by a speedy recovery and involvement in these same speculations; Joaquín Pez is insolvent once more and seeks help from Isidora. Also integrated into the *Efemérides* are satirical political comments from the characters: Relimpio blames the war on the amount of the redundancy pay ('la cesantía de treinta mil reales': 1066) paid to dismissed ministers; Don Manuel Pez, compared to Isaiah, makes pompous pronouncements on peace and freedom; *la Sanguijuelera* lends support to Alfonso; Augusto Miquis suggests that the best way of dealing with the Carlists is to give them power: 'El carlismo guerrero se sostiene. El carlismo establecido no podrá durar un mes' (1067).

Fleeting references are made to the Carlists in *Fortunata y Jacinta*: we see Juan Pablo Rubín as a gun-runner for the faction in the summer of 1873; the peninsula is described as 'una inmensa pira a la cual cada español había llevado su tea y el Gobierno soplaba' (V: II. 1. ii, 160); later, in the café conversations of Juan Pablo's circle, the siege of Bilbao and Concha's death are mentioned in passing (V: III. 1. iii, 298), as frozen references of the third type delineated in Chapter 1.

When we return to the *episodios*, we find that *La Primera República* continues and accentuates the characteristics of *Amadeo I*. Tito introduces himself exuberantly as the same as ever: 'chiquitín, travieso, enamorado, con tendencias a exagerar estas cualidades o defectos, si es que lo son' (III: 1115). In both the historical coverage and the fictional plot there are more reminiscences of Estévanez concerning his revolutionary career, taken largely from his *Memorias*.[36] Tito displays a 'veneración fanática' (1125) for Pi y Margall personally and succeeds without difficulty—or plausibility—in working for him as an aide; he is portrayed as 'un hombre afectuoso, reposado, esclavo del método' and of frugal habits. In collaboration with Estévanez, Pi is shown as energetic in reacting to the disturbances in Barcelona and to Martos's attempts in February and April 1873 to take over the Republic. Although Pi is criticized by Tito, in his role as interpreter of history, for his excessive

[36] As Galdós indicates in a letter to Estévanez: 'He reproducido, extractándola de sus *Memorias*, la campaña revolucionaria de usted en Despeñaperros, a fines del 72, y después los actos de usted como gobernador y ministro.' He adds, slightly apprehensively, 'espero que no le desagradarán las páginas que le dedico; después verá usted, en el curso del libro, que me he permitido presentarle en sucesos más novelescos que históricos; pero como en ellos no he alterado el carácter de usted, ni hay en ello nada que para usted no sea lisonjero, confío en que no lo tomará como impertinencia o abuso de confianza' (*3b*, Armas Ayala 1989: 292). Estévanez, for his part, seemed intrigued by his new role: 'eso de que usted me haga personaje novelesco (era lo que me faltaba) excita mi curiosidad y redobla mi impaciencia por recibir el tomo' (293).

legalistic zeal in not combating from the beginning 'el serpentón monárquico-radical' (1131), Galdós's sympathy for an incorruptible figure of the left has, by 1911, largely overcome his aversion towards Federal Republicanism.

He also waxes sarcastic about the falsity of the political struggle, comparing it, through Tito, to a contemporary *zarzuela*:

> No tuvo nada de epopeya; no fue tragedia ni drama; creí encontrar la clasificación exacta diputándola como entretenida zarzuela, con música netamente madrileña del popular Barbieri. No hubo choques sangrientos ni encarnizadas peleas, ni atronó los aires el horrísono estruendo de los cañones. El *acto* de Congreso fue un paso de comedia lírico-parlamentaria, con un concertante final en que desafinaron todos los *virtuosos*. Los *actos* de la calle fueron un continuo ir y venir de nutridas comparsas, que disparaban víctores y exclamaciones de sorpresa o de júbilo . . . Concluía la obra con un gran coro de generosidades ridículas y alilíes de victoria, sin luto por ninguna de las dos partes. (III: 1137)

Through Estévanez we have a portrait of Figueras, the first president of the Republic, who, profundly affected by the death of his wife, fled to France without explanation while still in office (1164). Antonio Orense, son of the demagogic Marquis of Albaida, is particularly praised for his vehement denunciation of subversion against the Republic.

A new angle on the incessant seeking for government sinecures is provided. Tito's influence with prominent Republicans is continuously sought by such old stagers as *Sebo*, Basilio de la Caña, and others, and he is profusely thanked for his good offices; it later turns out that this was the work of Floriana's magic circle.

His search for Floriana, apparently a daughter of the late priest Don Hilario but in reality a purely mythical creature, leads to the most substantial flight of fantasy in the series, one which is reminiscent of *El caballero encantado*. He is taken for a long sojourn in a timeless imaginary world, controlled by Mariclío and her sister Muses and peopled by the educational Establishment, under Floriana's leadership, of *Doña Gramática*, *Doña Aritmética*, *Doña Geografía*, and *Doña Caligrafía*.[37] In the midst of this episode, he meets a survivor from the battle of Trafalgar, significantly named Elcano, just as he had met one (Galán) in real life (*3b*, Berkowitz 1948: 98); it is a curious effort to establish historical continuity from the very beginning of the series. He also catches a glimpse of the Cantonalist risings at Alcoy and Orihuela.

The stirrings of the more important Cartagena canton are described to Tito, intertwined with the Floriana fantasy, by Fructuoso Manrique, one of Tito's informants and guides in Cartagena. He explains to Tito the organizational powers of Manolo Cárceles and the enthusiastic reception given to

[37] The impulse which led Galdós to dedicate *La desheredada* to schoolteachers and his concern with pedagogy in *El doctor Centeno* is now stronger than ever. It should be noted that Teodosia Gandaria was a schoolteacher (*maestra*); Madariaga identifies her with Floriana and with Atenaida in *La razón de la sinrazón* (*3b*, 93, 95).

Antoñete Gálvez. Contreras is created president of the provisional government of the Spanish Federation established in Cartagena and then quickly replaced by Roque Barcia as Contreras takes charge of military operations. The conflict with the German fleet is described. All the Cantonalist operations are seen as Quixotic, ineffectual, and absurd.

Federalism is treated as an attractive but remote idea, like the centuries-old 'Anfictionado de Tesalia' described by Mariclío:

> La idea federal es hermosa; es mi mayor encanto, la ilusión de mi vida en esta y en todas las tierras que visito, pero dudo, ¡ay!, que pueda implantarla de una manera positiva y duradera un pueblo que ayer como quien dice ha roto el cascarón del absolutismo . . . yo, que soy vieja eterna espero ver algún día . . . algún día triunfante y dichoso el 'Anfictionado español'. (III: 1204)

There follows a more purely allegorical development: the new Spaniard, in contrast with Tito the 'hombre muñeco', who suffers from *el morbo europeo* and is 'romántico de estofa ínfima y barata' (III: 1206), has to be practical as well as theoretical. As a consequence Tito witnesses godlike men working in a forge; as 'forjadores de los caracteres hispanos del porvenir' (1210), they are leaving behind an unacceptable present. Floriana's school contains poor children, who are taught with love, not coercion; and the two processes—new men, enlightened education—are linked: 'Floriana es mi novia,' declares the superman-craftsman. As in *El caballero encantado*, Galdós is playing with a Utopian ideal of a new perfectible race: 'Las divinidades que gobiernan el mundo han dispuesto que el Fuego plasmador se una en coyunda estrecha con la Feminidad graciosa y fecunda, para engendrar la felicidad de los pueblos futuros' (1212). Galdós's strengths do not lie, in my opinion, in these allegorical extravagances.

De Cartagena a Sagunto, begun in the summer of 1919, after the cataract operation performed on his left eye in May, was dictated to Pablo Nougués, at the slow pace of two or three *cuartillas* a day (4 August 1911; letter to Teodosia Gandarias, courtesy of Phoebe Porter). By 1 September he appears satisfied with what has been done: "Avanzo mucho en la obra *De Cartago a Sagunto*, y creo no equivocarme decir que va saliendo muy bonita." The task will be continued, with Teo's help, on his return to Madrid.

The *episodio* revolves round three major historical events: the Carthaginian canton once more, Pavía's *coup d'état*, and the Carlist war; despite its title, the narrative does not reach Sagunto, the *pronunciamiento* that brings in the Restoration. The canton is dealt with in much detail, as a ridiculous toy government with its own coinage, ministries, and foreign policy. It possesses its special idealism and heroism, represented by Cárceles and Coláu[38] (III:

[38] An interesting comparison can be made with Ramón Sender's *Mr. Witt en el Cantón* (1935). Jover, in his extremely thorough edition (*4a*, 78–9), notes that Sender, preoccupied by the

1224); the Trafalgar veteran Elcano dies in the fighting, and even the worthless *Pepe el Empalmao* becomes a hero who also perishes in action. Yet the canton is absurd—a 'jactanciosa hinchazón de nuestras fantasías' (1217)—its fleet sailing into battle is a 'Lepanto en zapatillas' (1223), its president Roque Barcia is consumed in confused biblical rhetoric, Contreras[39] is a purely Quixotic figure. Its elegy is pronounced by Manrique: '¡Adiós Cantón! ¡Adiós República ingenua y romántica, que a la Historia diste más amenidad que altos y fecundos ejemplos!' (1262).

With the fall of the canton, Tito catches glimpses of the features previously idealized in *La Primera República*, now fallen into disrepair: a real forge and school, very dilapidated; elderly teachers who have some resemblance to *Doña Aritmética*, *Doña Geografía*, *Doña Gramática*; and from time to time he fancies he sees Mariana. It is an allegorical portrayal of lost opportunities. There is one good characterization, another of Galdós's endearing prostitutes, *Leona la Brava*, who seeks to improve her racy speech by acquiring the clichés of society so that she can enter the *demi-monde* of Madrid (III: 1217). Such an aspiration fits in well with the coming world of the Restoration.

The one major event in which novel and *episodio* still coincide is the incident which spelt the doom of the First Republic: the *coup d'état* carried out by the Captain-General of Madrid, General Manuel Pavía, on 3 January 1874.[40] After a suspension of Congress lasting two and a half months, Emilio Castelar, the fourth president of the Republic, has to face a parliamentary debate which is virtually certain to bring down his government. When the coup is referred to factually in Relimpio's *Efemérides*, attention is drawn to the lack of bloodshed: '1874. *Enero.*—El día 3 Pavía destruye la República sin disparar un tiro. Desaloja el salón del Congreso y pone en la calle cañones que no hacen fuego' (IV: II. 1, 1066).

The use to which the incident is put in *Fortunata y Jacinta* shows great care and imagination. As an example of emplotment, it stands comparison with the famous *comises agricoles* in *Madame Bovary*, so praised by Percy Lubbock (*2*, 70–1). The narrator at first proclaims lyrically the importance of the action, which has become the subject of discussion and comment of all Spaniards.

revolutionary situation in 1935, eliminates any incident which tarnishes the picture of 'la noble utopía de una revolución sin violencia'. He makes the *pueblo* the real protagonist of the canton, relegating all the leaders except Antonete Gálvez to a lower plane. Coláu, for his part, is closely integrated into the novelistic action by his relationship with Milagros. Carrasquer (*4a*, 87–8), basing himself on inessential stylistic criteria, makes a rapid comparison, extremely unfavourable to Galdós, between the two novels.

[39] On Contreras, Gálvez, and Barcia see Hennessy's negative opinion: 'men of negligible political ability' (*1b*, 224).

[40] Pavía's coup was very like the abortive attempt of 23 Feb. 1981 to stifle the nascent Spanish democracy; Tejero's rising might even have been inspired by Pavía's example.

Expectancy is enhanced by the delayed appearance of Jacinto Villalonga,[41] a deputy and hence an eyewitness of the coup. Just as important, he has been Juanito's friend since their college days and his companion when they sought amorous adventures among working-class girls. When the two friends finally come together a few days later, the crucial political event is not the most important thing on their minds: their priorities lie in a different mode of illegality. Thus, as soon as they are alone, they speak ecstatically of Fortunata's reappearance in Madrid and switch to the political incident only when the suspicious Jacinta re-enters the room. The interaction between the two disparate narratives is considerable and skilfully executed. Caught by the entrance of Jacinta and Bárbara in the middle of a word, *embebe-cido*, Villalonga is forced to apply it rapidly to a political speech by Salmerón.[42] He is even so taken up with his story that he continues for a moment after the ladies have gone. Especially notable is the pretext, very typical of a spoilt *señorito*, which Juanito uses of the loose button on his waistcoat to make his wife leave the room; he suggests sending if necessary for Colonel Iglesias,[43] the colonel of the Civil Guard in charge of the invading troops, thereby neatly demonstrating the degree of imperious authority the bourgeois male exercises in both domestic and national situations. In the course of the scene a large amount of historical detail is transmitted almost casually: we are informed of the vote against Castelar and the election of a successor, Eduardo Palenca; the previous president, Figueras, is mentioned in passing as appearing curiously detached; the speeches of Salmerón and Castelar are referred to, including, during the lunch[44] which follows, the latter's famous denunciation of the Federalists: '¿Y la Constitución federal? La quemasteis en Cartagena' (**V**: I.

[41] It is curious to note that one of the deputies was named Villalonga: Antonio Villalonga Pérez; he voted against Castelar. See Comín Colomer (*1b*, 564).

[42] Nicolás Salmerón, 1838–1908, was a *krausista* and a professor of philosophy. The third president of the Republic, he resigned because of his opposition to the death penalty. See Trend (*1b*, 133–52) for a sympathetic portrait. Llorca (*1b*, 1966: 185–220) is vehement in attributing to him the blame for the fall of Castelar's government.

[43] In a letter to Galdós dated 22 Feb. 1912 (*3a*, Casa-Museo de Galdós), Nicolás Estévanez notes a contradiction concerning 'Colonel Iglesias' in *De Cartago a Sagunto*. On p. 98 [the Pavía coup: **III**: 1248] he is said to be 'alto', on p. 260 [the defence of Cuenca: **III**: 1304; see also 1306–7] there is a Don José de la Iglesia, who is described as 'chiquitín'. Estévanez explains: 'Sin duda ignora V. que esos dos personajes son una misma persona. El coronel del Congreso y el brigadier de Cuenca son el mismo don José de la Iglesia; no hay tal Iglesias. Era alto. Yo conocí mucho al personaje por haber sido uno de mis maestros en Toledo.' Ballbé (*1b*, 220) also calls him 'La Iglesia'. In *Fortunata y Jacinta* he is twice (**V**: I. 11. ii, 154, 155) called 'Iglesias'. Armas Ayala (*3b*, 1989: 293) refers to the letter, but transcribes 'José de la Iglesia' incorrectly as 'José Iglesias'. *La Época*, 26/7756 (4 Jan. 1874), quoting *La Correspondencia*, also speaks of 'Coronel Iglesias'. Perhaps, if (as seems likely) Estévanez is right, the newspaper reference was the direct or indirect source of Galdós's mistake.

[44] Even in the call to lunch there is a historical reference: '¡Estómagos a defenderse!' parodies Prim's famous cry '¡Radicales, a defenderse!' before a crucial parliamentary vote in the *sesión de San José* (19 Mar. 1870).

11. ii, 155).[45] One not unimportant deputy, Francisco Díaz Quintero,[46] a contributor to *La Discusión* and friend of Rivero's, is noted as resisting the occupation, but there is no political interpretation.

Villalonga's part is clearly that of an implicated informer; with Zalamero, another of Galdós's fictitious *señoritos*, he runs to and fro passing information to Serrano, Topete, and others—not directly to Pavía—about what was happening in the Chamber and is as aware as they are of the movement of troops. When Jacinta, fastening on a chance remark of her husband's, revels in the idea of her hated namesake hiding under a bench, it is clear that as one of the conspirators ('si yo estaba en el ajo', as he says) he was in no danger. The conclusion recalls the description in *La desheredada*:

> Os diré el último detalle para que os asombréis. Los cañones que puso Pavía en las boca-calles estaban descargados. Y ya veis lo que pasó dentro. Dos tiros al aire, y lo mismo que se desbandan los pájaros posados en un árbol cuando dais debajo de él dos palmadas, así se desbandó la Asamblea de la República. (V: I. 11. ii, 154)

The clear but unexpressed insinuation is that there was a considerable measure of bluff in the intervention and a consequent failure of nerve by the Republican politicians.

The coup is described in detail in *De Cartago a Sagunto*. More purely documentary information—names, reactions, determinations—is given: for example, the informers who relayed information to Pavía are named as Galdós's fellow Canary Islanders Fernando León y Castillo and Antonio Matos y Merelles.[47] The eyewitness Galdós once more provides is not a fictitious but very plausible deputy, but rather our not very realistic narrator Tito. He is carried by Mariclío's magic powers into the Congress, where he is plied with information by a number of deputies and reporters, real or imaginary. Among them, almost incestuously, is Pablo Nougués, Galdós's secretary and amanuensis at the time of writing, and a supporter of a unitarian Republic.

The religious question, in accordance with Galdós's preoccupations in 1911, is seen as just as important an issue as maintaining a hard line against the Cantonalists; the *modus vivendi* with the Pope negotiated by Castelar, praised by Moreno Rodríguez, is bitterly opposed by Salmerón (III: 1242).

[45] In *De Cartago a Sagunto* (III: 1245) it is quoted, incorrectly (from memory?), as '—"¿Y el proyecto de Constitución?" . . . —"Lo enterrasteis en Cartagena".'

[46] Díaz Quintero is referred to somewhat differently in the *episodio*: 'Por la puerta que da a la misma calle se escabulleron cantando bajito los que más habían alborotado en los pasillos, queriendo desarmar a la tropa: eran Olías, Casalduero, Díaz Quintero, el marqués de la Florida, y otros' (III: 1249).

[47] Ballbé (*1b*, 220–1) quotes León y Castillo's *Mis tiempos* (i. 140) as indicating that he supplied details of the debate to Víctor Balaguer for him to pass on to Pavía. Matos y Merelles could easily have collaborated too.

The 'eterna discordia' (1243) between Unitarians and Federalists is once more deplored and Castelar's scathing remark on the latter is again quoted. Intransigence among the Federalist leaders, in both the technical[48] and normal sense, is singled out for special censure; an anonymous assertion—not dissimilar to that attributed to Figueras by Prim in Chapter 4—that a monarchy is preferable to a non-federal Republic is quoted, as is Salmerón's 'implacable sentencia: "Sálvense los principios y perezca la República"' (1243). Tito, on Galdós's behalf, does not conceal his view that this was a betrayal of the cause: 'El espantable fallo del Presidente de las Cortes arrancó lágrimas a los leales republicanos que más de una vez jugaron su vida en las conspiraciones y en las barricadas' (1243). We are told that Figueras, indecisive in the novel, personally abstained, but instructed his followers to vote against the government: an act of irresponsibility which is implicitly condemned. Castelar retains his dignity and demonstrates his courage,[49] but the mass of deputies are portrayed as pretentious, selfishly individualistic, and in the final resort cowardly. Tito is moved to act as a *deus ex machina* by voicing his indignation: 'Majaderos fuisteis; sed ahora ciudadanos y dejaos matar en vuestros asientos' (1246); and when Salmerón stated that the intervention has wiped out all differences between them, he comments sarcastically that it is too late: '—*¡Tarde piache!*' The poor soldiers, bashful and distressed by the role they have to play, are treated with conspicuous consideration, in contrast to the deputies whose disunity and obstinacy are responsible for their humiliation. Long passages of dialogue are unnecessarily transcribed, presumably from the official record.[50] One small incident typifies the scene. A deputy we have met before in the company of Halconero, Emigdio Santamaría, snatches a rifle from a sergeant, but is persuaded by another real politician, Antonio Fernández Castañeda, to hand it back rather than indulge in meaningless heroics.

Finally, Pavía, having, in Galdós's presentation,[51] broken his solemn

[48] 'The Intransigents were those who rejected any sort of understanding with monarchical parties' (*1b*, Hennessy 151). The more accommodating Republicans like Castelar were called 'benévolos'.

[49] The question of whether Castelar had foreknowledge of the plot is not clear. It is obvious that he could not have been unaware of the risk of a coup if he lost the vote of confidence, but he refused to postpone the session. Llorca (*1b*, 1966: 191) has no doubt that Castelar knew of Pavía's intentions (though these obviously contradict his promise of allegiance), but still hoped to persuade Salmerón to support him. Castelar always claimed an absolute regard for legality and was outraged by the use of force. See his address to the nation that same day (*1b*, Llorca 1966: 210–11) and his letter of 4 July 1874 to Adolfo Calzado (*1b*, *Correspondencia* 4–5): 'En lo que ahora sucede, no tengo parte ni responsabilidad. El 2 de Enero se hizo contra mí y a pesar mío. ¿Por qué he de cargar yo con las consecuencias? Además, nuestra España padece de un mal gravísimo, de grande menospecio por la legalidad.'

[50] The text, with minor deviations, corresponds to the transcript from the *Diario de las sesiones* given by Llorca (*1b*, 1966: 194–209).

[51] Llorca (*1b*, 1966: 221) refers to this promise, and suggests that Pavía expected Castelar to acquiesce in being restored to power by military action. He is known to have seen Castelar on 24

promise to Castelar not to resort to arms, is ironically depicted as being surrounded by a general staff as numerous and brilliant as Napoleon's at Austerlitz; the whole tragic incident is reduced to a 'carácter y matiz enteramente cómicos' (III: 1249).

The more radically committed Galdós of 1911 is now prepared to be categorical rather than 'protean' or indeterminate in his pronouncements. Thus Mariclío calls the incident a 'fácil y criminal hazaña', from which Republicans should learn. Galdós is disgusted on two counts. One is the lack of unity among the Republicans, the other is their failure of nerve. The country is far too prone, as he sees it, to easy accommodations and compromises which leave the situation essentially the same:

> El grave mal de nuestra Patria es que aquí la paz y la guerra son igualmente deslavazadas y sosainas. Nos peleamos por un ideal, y vencedores y vencidos nos curamos las heridas del amor propio con emplastos de arreglitos y anodinas recetas para concertar nuevas amistades y seguir viviendo en octaviana mansedumbre. En aquel día tonto, el Parlamento y el Pueblo fueron dos malos cómicos que no sabían su papel, y el Ejército suplantó, con solo cuatro tiros al aire, la voluntad de la Patria dormida. (III: 1249)

As Galdós's considered view in 1911, this passage has some interest and validity. Whatever its value as a political assessment, however, it seems to me too direct, too unequivocal, and too unmediated a statement to emerge from the narrator of a historical novel.

One group, we learn through Pepe Ferreras, is very pleased at the change of government: the shopkeepers, for whom it means more business. As Ferreras sardonically puts it, forty-nine new dress-suits will be needed for the forty-nine new provincial governors.

A final word comes from Nicolás Estévanez. Because of a family bereavement he attended the Assembly just to vote—against Castelar,[52] though Galdós does not say so—and is afterwards quoted as saying literally: '—*Lo de ayer ha sido una increíble vergüenza. Todos nos hemos portado como unos indecentes.*' He despairs, moreover, of any Republican counteraction: 'Esto está perdido. Cantemos a nuestra pobre República el debido responso' (III: 1252).

Exceptionally, Pavía's action involved no political ambitions and was not carried out on behalf of any party: it was *la flor de un día*. The same day Pavía

Dec. 1873 and again on 1 Jan. 1874 and advised him to postpone the session. Serrano was certainly conspiring to bring about a coup, so that the scene in *Fortunata y Jacinta* rings true. See also Hennessy (*1b*, 237–41) and Headrick (*1b*, 203–5).

[52] See Comín Colomer (*1b*, 562–7) for the voting-figures. Of those mentioned by Galdós, Díaz Quintero, Casalduero, and Santamaría were among those who voted against Castelar; Olías, Antonio Orense, and Fernández Castañeda voted in favour of the government.

handed over to a government presided over by Serrano, who became regent for the second time. His attempt to build up a unitarian Republic with himself at its head[53] foundered on the increasing waves of pro-Alfonso sentiment sedulously cultivated by Cánovas, who approved of the coup but urged an immediate declaration of Alfonso as king.

The treatment of the Pavía coup in this late *episodio*, lively as it is in certain respects,[54] suffers from severe limitations. In comparison with the way historical events like the lynching of Chico or the execution of the sergeants of San Gil are handled in the fourth series, the historical coverage has become too detailed, the narrative structure too artificial, and the political moralizing too obtrusive. The hedgehog remains as irreducible as ever, but its defensive or negative qualities, its spikes, are too much in evidence. The richness of concrete experience, *la historia pequeña*, with its multifarious activities, represented by the sly fox, has all but disappeared. If a disrespectful modification of the metaphor may be permitted, the fox has been replaced by a chattering and intrusive parrot. By contrast, the *Fortunata y Jacinta* scene is much more effective, for here historical vision and detailed incident interact imaginatively: fox and hedgehog in this instance collaborate.

The third subject of the volume is the Carlist war.[55] Tito is sent as a special delegate to suborn Carlist leaders. He travels north with the volatile *Chilivistra*, and is arrested and imprisoned by Carlist forces for fifty-two days. This enables him to meet and describe sympathetically the general Antonio Dorregaray, who after serving in Isabelline Spain has joined the Carlists. A marked change of emphasis concerning the conflict can be observed. Whereas previously Galdós has placed the blame squarely on the Carlists and has seen some hope in the Liberal cause, now his aversion towards the Restoration and the regency which followed is such that he treats the two sides as equally culpable.[56] Mariclío describes the war once more in terms of farce: a

[53] 'El propósito del duque de la Torre es consolidar la República unitaria con su presidencia vitalicia' (*1b*, Fernández Almagro 216).

[54] Montesinos (*3c*, iii. 275), for instance, praises it highly.

[55] In Unamuno's *Paz en la guerra* (1897) the Carlist war is seen through Basque eyes, from the double perspective of the Carlist army and besieged Bilbao. In a letter to Clarín, Unamuno claimed that *Paz en la guerra* had influenced *Luchana* and other *episodios* (*4a*, Menéndez Pelayo 99; *OC* ii. 8); this appears highly doubtful. Cardona (*3d*, 1968: 141) reports that Galdós left the pages from 74 to 343 uncut in the edition Unamuno dedicated to him. The war is also the subject of Valle-Inclán's *Sonata de invierno* (1905), which takes considerable liberties with chronology, and of his trilogy *Los cruzados de la causa* (1908), *El resplandor de la hoguera* (1909), and *Gerifaltes de antaño* (1909). Baroja's *Zalacaín el aventurero* (1909) also deals with the war. See Extramiana (*4a*). For the course of the war see Holt (*1b*).

[56] Compare his sarcastic comment in the prologue to Ricardo Fuente's *Vulgarizaciones históricas* (1910): 'Al Pueblo español debieran llamarle el Cristo de la Paciencia, pues ha tolerado la institución monárquica partida por gala en dos ramas, que, en perfecto compadrazgo con el dichoso *Altar*, se han desvivido por hacernos felices, imponiéndonos la teocracia, el centralismo administrativo, la ineducación, y amarrándonos cuidadosamente, para más fácil dominio, a la cola de las naciones europeas' (*3a*, *Prólogos* 100).

'Carnaval sangriento' (III: 1277), 'juego y travesuras de chicos malcriados' (1285). When Tito describes the jubilation he witnessed at the lifting of the siege of Bilbao (2 May 1874), in a spectacular but for me inopportune prolepsis he suddenly leaps into the present—the time of writing—to claim that the victory was a farce in which the only gainer was his implacable enemy, the Church:

Al recordar hoy los sublimes momentos de aquel día, ayes de gozo, alaridos de esperanza, me parece que oigo burlona carcajada del Destino. Sí, sí, porque la Restauración primero, la Regencia después, se dieron prisa a importar el jesuitismo y a fomentarlo hasta que se hiciera dueño de la heroica Villa. Con él vino la irrupción frailuna y monjil, gobernó el Papa, y las leyes, teñidas de barniz democrático, fueron y son una farsa irrisoria. (III: 1285)

As they prepare for the coming battle, Tito asks himself how it is that Dorregaray and his former friends and comrades from the African war are fighting against each other. His conclusion is that they are both equally devoted to despotic rule:

Debajo del emblema de la Soberanía Nacional en los unos y del absolutismo en el otro, latía, sin duda, este común pensamiento: establecer aquí un despotismo hipócrita y mansurrón que sometiera la familia hispana al gobierno del patriciado absorbente y caciquil . . . Concluirán por hacer paces, reconociéndose grados y honores, como en los días de Vergara, y la pobre y asendereada España continuaría su desabrida Historia dedicándose a cambiar de pescuezo a pescuezo, en los diferentes perros, los mismos dorados collares. (III: 1288)

A close description of the battle of Abárzuza is given, with an excess of detailed figures. The key event is the heroic death of Manuel (Gutiérrez) de la Concha, Marquis of the Duero (III: 1292), one of the most ambitious and decisive among the political generals; a well-known *alfonsino*, he was expected to proclaim Alfonso as king after the battle. Galdós retains some admiration for heroism and leadership: he calls Concha 'el primer soldado español de aquellos maldecidos tiempos' (1292), but what is more striking is that the Carlists now command as much respect or more as the Liberals: their convictions are firmer and Galdós is now less interested in conciliation than in decisiveness.

The denunciation of the war—'*infantil y sangrienta*' (III: 1298)—is confirmed by Mariclío herself, who also explains the symbolic role of *Chilivistra*: her frenzied changes of mood are like those of Spain itself:

—Has de persuadirte, hijo mío, de que en el carácter borrascoso y tornadizo de tu *Chilivistra* tienes un perfecto símbolo de la vida española en el aspecto político, y estoy por decir que en el militar. Tan pronto es cariñosa y tierna como altiva y marimandona. El amor la dulcifica hoy, y mañana la endurece el orgullo. Inventa con lozana invención fábulas absurdas y acaba por creerlas. Se finge deshonesta sin fundamento real de sus mentirosos pecados. En ella habrás observado que al fuego del sentimentalismo sustituye rápidamente el hielo de los negocios menudos, todo ello sin criterio fijo, sin noción alguna de la realidad. (**III**: 1298)

Politicians, Clío continues, govern Spain in exactly the same way, and *Chilivistra* will give him a constant 'lección viva' of the 'defectos capitales de tu patria' (*3d*, Gilman 1986: 50).

I do not find this at all convincing. While Juanito Santa Cruz or Isidora have individual qualities which reflect those of the society in which they live, and Teresa Villaescusa represents an evolution which corresponds to the stimulus of the political society she inhabits, *Chilivistra* appears to have been created with grotesque characteristics simply in order to parody the distortions of Spain as Galdós sees them. 'The perfect symbolism', imposed rather than organically evolving, does not come off.

La Madre then dispatches Tito south to see the other half of the Carlist rising. He is accompanied by *Chilivistra* and a Father Carapucheta, who turns out to be José Ido searching for his daughter Rosita (a most awkward and improbable use of a fictional character). They end up in Cuenca, occupied by the Carlists, after La Iglesia's heroic resistance. Here the barbarity of the fanatical traditionalists reaches a new height. It has two aspects: the cruelty of the Carlist troops or *zuavos*, who terrorize the city, and the vicious barbarity of María de las Nieves, the wife of the mediocre Alfonso of Bourbon (**III**: 1308). She is called 'Atila con faldas'; even the bishop is treated with contempt.[57] Rosita is finally discovered under the urbane protection of the priest Pagasaunturdua. Galdós sardonically makes Ido content with the relative well-being his daughter has achieved as a priest's concubine. When Cuenca is captured not a single supporter of the faction is left (1315).

To conclude, in referring in his novels to the important revolutionary period between the end of Isabella II's reign and 1874, Galdós's technique of pinpointing specific historical incidents and incorporating them into his fictions is sure and skilful. It does not get in the way of the action but adds a solidity of fact to the structure. For their part, the *episodios*, at their best, find an adequate formula to illuminate a historical action which is more detailed, coherent, and continuous than in the novels, through fictional characters acting in a substantially autonomous manner who are none the less subordinate to historical necessities. In this way Galdós takes care to open up avenues of choice or demonstrate inevitable consequences as the historical process unfolds. This applies essentially to the first two *episodios* of the fifth series as it has to the whole of the fourth series, though the tendency to

[57] Holt calls it an 'orgy of looting and murder, which Alfonso and his wife were believed to have actively encouraged' (*1b*, 260). The Carlist historian Oyarzun (*1b*, 464–5) speaks of 'algún exceso' not authorized by the leaders and exaggerated by the Liberals. His explanation of their cool relations with the bishop—that the bishop had sheltered loyalist soldiers in the episcopal palace—is in itself evidence that atrocities were being committed.

allegorize through certain semi-archetypal literary figures like Don Wifredo-Don Quixote and Urríes-Don Juan is on the increase without becoming overwhelming.

From *Amadeo I* onwards, unfortunately, when Galdós's considered presentation of the key events of his youth in a fictional context would have been especially valuable, these characteristics no longer fully hold. He draws, sometimes in excess, on detailed recollection and provides much factual information, while he succumbs to a new technique in which the narrative structure itself is undermined. The constant use of allegory, applied in a largely clumsy and obvious way, the inopportune characterization of the stunted narrator, with his sordid affairs, and the major and minor magical elements introduced all detract from the authenticity of background without in my view replacing it with any new method of substantial value. Thus abstract figures largely displace the earlier fictional narrators and reflectors and thereby remove those elements (*la pequeña historia*) which, within a recognizable social structure, lent variety and depth to the historical focus. The new procedure seems to me on the one hand to take too far the dissatisfaction Galdós has come to feel with realist narrative,[58] and on the other to indulge to excess in the despair he experienced at the political morass Spain found herself in. The Utopian aspirations do not seem to me to offer any great compensation and the new technique is not thoroughgoing enough to create an effective metafictional structure. The balance previously achieved between the presentation of a wealth of empirical fact and an overall historical perspective, varied in proportion though it was between novels and *episodios*, has now been lost. The hedgehog remains, in an excessively prickly form, but the fox has got away, leaving a garrulous parrot to repeat itself endlessly.

[58] This dissatisfaction gives rise to the distinctive attitude to historical truth described in Chapter 7.

6

The Restoration and the Established Order

THE Restoration was a watershed and the culmination of the period of revolutionary fervour which fascinated Galdós. As Peter Bly has observed, there is an important continuum in nineteenth-century Spanish history; the Restoration is closely linked with the pre-revolutionary years without any substantive ideological change from one period to the next.[1] In one essential respect, however, it was unique in nineteenth-century history: it brought political stability for the first time. The price paid for this stability is a question of dispute. It is clear that its architect, Antonio Cánovas del Castillo, setting out from pessimistic and pragmatic premisses, took great pains to reconcile, or to give the appearance of reconciling, the previously irreconcilable. Thus the Sandhurst manifesto of November 1874, issued by Prince Alfonso at Cánovas's behest, seeks to commit the new monarchy to both Catholicism and liberalism. And, despite the unconstitutional means by which, to Cánovas's annoyance, the Restoration was effected—the military intervention at Sagunto—Cánovas was determined to keep the army out of politics and to restrict the royal prerogative severely. This was achieved by a series of expedients which sought to secure a wide spectrum of support from the dominant classes of society. The system depended on an oligarchy founded, as Llorens has pointed out (*3b*, 118), on largely Andalusian-based landownership; it took little heed of the new industrialized complex of Catalonia and Biscay. Initially, severe curbs on the freedom of the press and of academic institutions were imposed as sops to the traditionalists. Thanks to a systematic use of *caciquismo* and blatant manipulation of the electoral process, a viable parliamentary structure was secured by the deliberately ambiguous constitution of 1876. By this means two 'dynastic' parties, roughly modelled on the English pattern and committed to a peaceful alternation in power, were consolidated. Operating under the leadership of Cánovas and Sagasta respectively, the parties speedily became known as Conservatives and Liberals; the working classes were excluded from political representation. The difficult religious question was subject to a compromise by which Catholic

[1] Compare Fontana, who affirms that, 'Más que una auténtica restauración, que hubiera significado una vuelta a la etapa anterior a la revolución, el golpe de estado de 1874 fue una corrección de la trayectoria seguida desde 1868. Cánovas completaba y perfeccionaba la obra iniciada por los Prim, Serrano, Sagasta y compañía' (*1b*, 141).

supremacy was upheld but other religious practices were largely tolerated.[2]

By the end of the century intellectual opinion is almost universally hostile to the Restoration settlement, which was seen as hypocritical, anaemic, and complacent. Moreover, the separation between the political process and public opinion has become notorious. When Ortega y Gasset launched his programmatic call for a new start in *Vieja y nueva política* (23 March 1914), the Restoration was his immediate target:

Qué es la Restauración, señores? ... La restauración significa la detención de la vida nacional.... La vida nacional se repliega sobre sí misma, se hace hueco de sí misma ... La Restauración, señores, fue un panorama de fantasma, y Cánovas el gran empresario de la fantasmogoría ... la corrupción organizada ... una omnímoda, horrible, densísima incompetencia. (*1b*, i. 97–9)

The same attitude is expressed in *Meditaciones sobre el Quijote* (1914). Galdós's approach, as we shall see, is, by 1912 at least, not dissimilar.

The same two novels cover the time of the Restoration as had covered the revolutionary period. In *La desheredada* it plays a substantial part, though not as significant a one as the reign of Amadeo. In *Fortunata y Jacinta* it has greater importance, both as setting and as point of reference. In addition, the Cánovas era is echoed as a past event in other novels.

Let us begin with *La desheredada*. We have seen that part II, Chapter 1, entitled 'Efemérides', opens with a rapid synthesis of historical events from the advent of the Republic until the Restoration settlement. In the mental diary by Relimpio which follows, the entry relative to *Diciembre* [1874] is:

La guerra sigue. La Restauración toca a las puertas de la patria con el aldabón de Sagunto. Asombro. La Restauración viene sin batalla, como había venido la República. La Providencia y el Acaso juegan el ajedrez sobre España, que siempre ha sido un tablero con cuarteles de sangre y plata. (IV: II, 1, 1067)

The idea of the Restoration knocking on the door strikes a slight note of comedy, as Bly says (*3c*, 1983: 17), but it also implies an occurrence whose time has come: hence the lack of commotion. The skilfully turned metaphor of chess leaves, as we have seen, ample scope for chance ('el Acaso') as well as for design or fate ('la Providencia'). Galdós does not share Thomas Hardy's powerful but negative view of the aimlessness of history, nor Zola's rigid determinism. For him, history, though predisposed by the laws of necessity, is none the less affected by chance events.

Under the entry for *1875* we find the archetypal bureaucratic Pez family momentarily in disarray as a result of the new dispensation:

Los Peces grandes y chicos se ven desterrados de las claras aguas de sus plazas y oficinas. Bien quisieran ellos aclamar también al rey nuevo; pero la disciplina del

[2] For an excellent account of the historical background consult Jover Zamora (*1b*, 1983). Despite its age, Gerald Brenan's account is the one which best captures the spirit of the time (*1b*, 1–16).

partido les impone, ¡ay!, una consecuencia altamente nociva a sus intereses. Tienen que poner un freno a sus agallas. Además, la lucha por la existencia, ley de las leyes, ha llevado a los Pájaros al Gobierno, y éstos no encuentran en la Administración bastantes ramas en que posarse. Algunos Peces de menor tamaño y del género *voracissimus* quedan en oficinas oscuras. Son Peces alados, transición entre las dos clases, pues la triunfante tuvo en situaciones anteriores sus avecillas con escamas. (IV: II. 1, 1067–8)

Here we have, expressed in extremely vivid style, a recurrent topic of Galdós's assessment of the Restoration: hungry state employees avidly demanding posts from their leaders. The only difference from later versions of the theme is that, at this stage, some are temporarily discomfited, though evidently not for long. Galdós ingeniously uses an image drawn from the new science of evolution to demonstrate the genetic ability of the Pez species to adapt itself to the temporary dominance of the Pájaros.[3] What emerges essentially is the impelling urge of the government to provide for its followers (the Pájaros if not the Peces), the success of some undeserving individuals—such as Alejandro Sánchez Botín, of regular occurrence in the novels—in the commercial and political ambience of the time, and the efforts of others, notably Melchor de Relimpio, to exploit the system.

Later in the same passage a second key theme surfaces, when we learn that the Pez's eclipse inclines them towards the Carlist camp: 'Todos los Peces, confirmando la antigua idea de que en España *el despecho es una idea política*, se alegran de las ventajas de los carlistas' (IV: II. 1, 1068; my italics). The idea that political extremism is produced by personal frustration—'el despecho es una idea política'—is very deep-rooted in Galdós. Juan Pablo Rubín is a clear example of this tendency, and so is the young Ángel Guerra.[4] Characteristic rather of the pre-Restoration period, it is attenuated by the broad liberality of the new Establishment. In *La desheredada* the two prime examples are of course Mariano and Isidora; their deep-rooted frustrations are not quelled but exacerbated by the Restoration settlement. Degraded and epileptic, Mariano seems arbitrarily damned from the start by his nickname *Pecado*. He is driven to make an attempt on the king's life[5] by blind resentment and envy of the wealthy within whose ranks he has been led to believe he rightly belongs (IV:

[3] As Bly points out (*3c*, 1983: 15 n.), the Republican government of Feb.–June 1873 was known as 'el ministerio de los pájaros' because of the names (non-Castilian) of some of its members: Pi, Tutau, Chao, Sorní.

[4] '¿Y qué menos podía hacer el desgraciado Rubín que descargar contra el orden social y los poderes históricos la horrible angustia que llenaba su alma?' (V: IV. 5. iv, 495). For his part, Guerra imagines himself declaring to his despotic mother: 'Soy revolucionario por el odio que tomé al medio en que me criaste y a las infinitas trabas que poner querías a mi pensamiento' (V: I. 3. xi, 1251).

[5] Mariano's attack is reminiscent of two real attempts on Alfonso XII's life, by Juan Oliva and Francisco Otero; both are described in *Cánovas* (III: 1384, 1391). See Ruiz Salvador (*3c*, 55), Bly (*3c*, 1983: 20–1), and Gordon (*3c*, 1972: 70–3), who attaches greater importance to the Otero case.

II, 16, 1147–8);[6] at the same time he is not wholly conditioned, since a more personal impulse also influences him: a certain 'Herostratism' ('quiere ser célebre, aunque para ello tenga que hacer una barbaridad', II, 4. ii, 1083), which anticipates Unamuno.[7] If he thus participates in Isidora's illusions, his sister in turn has much in common with him, as she turns against 'la gente grande' without losing her contempt for 'la grosería y suciedad de las personas bajas' (II. 13. iii, 1134). She is entirely lacking in that sense of the middle ground which is the Restoration's strength: she declares that 'no me gustan términos medios' (II. 10, 1110). As Miquis, representing the voice of reason, says: 'Su hermano y ella han corrido a la perdición: él ha llegado, ella llegará. Distintos medios ha empleado cada uno: él ha ido con trote de bestia, ella con vuelo de pájaro' (II. 17, 1154). For her part, *La Sanguijuelera* is well contented with a new traditionalist government, full of show and splendour, where 'los señores son siempre señores y los burros siempre burros' (II. 1, 1067), though her enthusiasm for the new sovereign does her little good when she seeks clemency for *Pecado*.

In *Fortunata y Jacinta* the references are more frequent and more nuanced. It is remarkable how the event which inaugurated the Restoration—the *pronunciamiento* at Sagunto by Martínez Campos—is given prominence, indirectly and in anticipation, in the year 1871, during the Santa Cruzes' honeymoon.[8] The young couple pass the station in a train:

> ¡Sagunto!
>
> ¡Ay, qué nombre! Cuando se le ve escrito con las letras nuevas y acaso torcidas de una estación, parece broma. No es de todo los días ver envueltas en el humo de las locomotoras las inscripciones más retumbantes de la historia humana. Juanito, que aprovechaba las ocasiones de ser sabio sentimental, se pasmó más de lo conveniente de la aparición de aquel letrero.

Jacinta is not over-interested: '¡Ah! Sagunto; ya . . . un nombre. De fijo que hubo aquí alguna marimorena. Pero habrá llovido mucho desde entonces' (V: I. 5. iv, 56–7). In effect, much has happened since 218 BC, when Hannibal sacked Saguntum during the Second Punic War, to which Galdós refers in this very oblique manner.[9] Employing a well-calculated prolepsis, he is concerned of course with the modern 'marimorena' which is to occur three and a

[6] See Labanyi (*3c*). In stressing Galdós's belief in meritocracy, Labanyi points out the role of both envy and hereditary factors in Mariano's personality. Her argument for a political interpretation of the word 'desheredada' is strengthened by its use in *Cánovas*.

[7] For Unamuno's 'Herostratism' see Ribbans (*4a*) and Ulmer (*4a*).

[8] 'Un extraño viaje de novios', as Raphaël's title suggests (*3c*).

[9] It is worth noting that in the manuscript draft at Harvard (*3a*) the text is more explicit; it refers to 'los romanos [que] andaban por aquí', to their 'carros de marfil', and to Suetonius. In *Cánovas* too Mariclío recalls the classical event—'Días ha encontrábame yo en las ruínas del teatro romano de Murviedro, rememorando la espantosa ocasión de la caída de la heroica Sagunto en poder del furioso Aníbal' (III: 1326)—and links it with Martínez Campos's *pronunciamiento*.

half years later.[10] The coup is also mentioned as a contemporary event in Juan Pablo Rubín's *tertulia*, when Juan Pablo, typically, vehemently denounces the action only to have to eat his words shortly after (V: III. 1. vi, 308–9).

Moreover, the very word *Restauración*, with its counterpart *revolución*,[11] mirrors one of the main themes of the novel, the order which follows a period of disruption and innovation. In the private life of Juanito Santa Cruz, and to a lesser extent that of Maxi and Fortunata, the political terminology becomes both a domestic metaphor (V: III. 2: 'La restauración vencedora'; III. 3: 'La revolución, vencida'; III. 4: 'Otra restauración') and a social parallel (*3c*, Ribbans 1970).

What do the characters expect from the Restoration? Amid the general enthusiasm of his class, the disillusioned Anglophile Moreno-Isla, who has no confidence in his compatriots, is pessimistic, though well disposed towards the new monarch. He is depicted as seeing him off on his departure from England:[12] 'Yo le dije: "Vuestra majestad va a gobernar el país de la ingratitud; pero vuestra majestad vencerá a la hidra." Esto le dije por cortesía; pero yo no creo que pueda barajar a esta gente. El querrá hacerlo bien; pero falta que le dejen' (V: III. 2. iv, 319).

The idea of the Restoration is welcomed wholeheartedly by the Santa Cruz family circle. The fickle Juanito, after short-lived enthusiasms for Montpensier and the Republic, 'Por graduaciones lentas . . . llegó a defender con calor la idea alfonsina', evidently the most appropriate for his class and his temperament. Only Don Baldomero, old-fashioned Progressive in the mould of his namesake Espartero as he was, at first contradicts his son, bringing forward the famous *jamases* of Prim. By the time Alfonso enters Madrid, Don Baldomero too 'estaba con la Restauración como chiquillo con zapatos nuevos' (V: III. 2. i, 309). Doña Bárbara is equally ecstatic ('reventaba de gozo'). Aparisi, like Pez with regard to the revolution in *La desheredada*, bows very readily before one of the characteristic complacent clichés of the time: the *fait accompli* or 'los hechos consumados'. For his part, Juanito is now no longer so

[10] On this question Francisco Caudet, editor of the excellent Cátedra edition, seems unduly cautious: 'Como el viaje de novios transcurre en 1871, no sería pertinente ver una alusión al pronunciamiento saguntino de Martínez Campos (29 de diciembre de 1874) aunque Galdós, que escribe en 1885, puede jugar con la resonancia que entonces tendría el renombre de la ciudad' (*3a*, i. 221). Ortiz Armengol comments: 'Pocas veces ha empleado Galdós tanta sutil malicia' (*3c*, 1987: 216).

[11] Jover Zamora points out that 'en el lenguaje coloquial de las clases medias tradicionales la palabra "república" pasa a ser sinónimo de desorden' (*1b*, 1983: 281). See also Estébanez Calderón (*3b*, 1985). 'República' is frequently found in this sense in Galdós. The Bringas couple make no distinction between political revolution and moral laxity: witness their indignation, put into political terms, at the elopement of Agustín Caballero and Amparo at the end of *Tormento*: 'La revolución no tarda; vendrá el despojo de los ricos, el ateísmo, el amor libre' (IV: 1568).

[12] Ortiz Armengol points out that Don Alfonso left for Paris on 30 Dec. and that it was historically impossible for Moreno and other Spaniards to see him off from Dover as king on that date, so soon after the Sagunto *pronunciamiento* of the 29th (ed. *Fort. y Jac.*: *3a*, ii. 1029–30 n. 351). It is a good case of poetic vs. historical truth.

convinced of the desirability of the Restoration; he does not venture to contradict the overwhelmingly strong support for Don Alfonso expressed by his father and friends, but he does voice doubts about its permanence (his own return to legality will certainly be temporary), adding sententiously that he does not approve of the illegal coup which brought it about: 'lo que a mí no me gusta es que esto se haga por otra vía que la de la ley.' Coming from him this is particularly ironic, and produces an inner expression of disgust from Jacinta: '¡Pillo, tunante! . . . ¿Qué sabes tú lo que es ley? ¡Farsante, demagogo, anarquista! Cómo se hace el purito . . . Quien no te conoce . . .' As for Doña Guillermina, she demonstrates by her use of the first-person plural both her own involvement with the Establishment and her individual priorities: 'lo hemos traído con esa condición: que favorezca la beneficencia y la religión' (III. 2. i, 310).

The exception is Jacinta. Earlier, like all the upper-class ladies of the time, live and fictitious, described by Galdós, she had been entranced with the young prince, in her case simply because he was only a child: 'Jacinta dejaba muy atrás a las más entusiastas por Don Alfonso. "¡Es un niño! . . ." Y no daba más razón' (V: I. 8. ii, 86). Now, having just received incontrovertible proof of her husband's infidelity, she has her own personal reasons for being sceptical about apparent returns to law and order. For her, private life takes absolute precedence over political upheavals; a similar reaction had occurred at the time of the departure of another king, two years earlier, in February 1873: '¿qué le importaba a ella que hubiese República o Monarquía, ni que don Amadeo se fuese o se quedase? Más le importaba la conducta de aquel ingrato que a su lado dormía tan tranquilo' (V: I. 8. i, 83). On the latter occasion she turns, significantly though silently, on Don Baldomero, who exercises a benign but all-embracing despotism over her and the Santa Cruz household: 'Pero a este buen señor, ¿qué le va ni viene con el Rey?' (V: III. 2. i, 309). She has gone with Doña Bárbara to witness the procession from Eulalia Muñoz's house on the Plaza Mayor, and Amalia Trujillo takes her aside:

> Hallábanse las dos solas en el balcón de la alcoba de Eulalia, y ya sonaban los clarines anunciando la proximidad del Rey, cuando Amalia, ¡plum!, le soltó el pistolazo.
>
> —Tu marido *entretiene* a una mujer . . . (V: III. 2. i, 310)

The image of the pistol-shot[13] in a domestic situation has several implications. It hints once more at the intertwining of private and public life, for this domestic violence echoes the many political assassination attempts, successful or unsuccessful, made on Isabella II, Prim, and Amadeo; Don Alfonso was to be the target of various attacks, including the fictitious one by Mariano

[13] It might also be associated with Stendhal's famous remark quoted in the Introduction.

Rufete. More remotely, it anticipates Fortunata's frenzied plea to Maxi to buy a revolver and kill Aurora and Juanito (V: IV. 6. ix, 525–6). In short, the news has an effect on Jacinta as violent and as devastating as a physical assault.

Quedóse yerta. . . . Desde aquel aciago instante, ya no se enteró de lo que en la calle ocurría. El Rey pasó, y Jacinta le vio confusa y vagamente, entre la agitación de la multitud y el *tururú* de tantas cornetas y músicas. Vio que se agitaban pañuelos, y bien pudo suceder que ella agitara el suyo sin saber lo que hacía . . . Todo el resto del día estuvo como una sonámbula.[14]

The *Episodio nacional* devoted to the Restoration, *Cánovas* (1912), was written some thirty-seven years after the event and twenty-five after the writing of *Fortunata y Jacinta*. The emphasis at the king's arrival is naturally more political, though in one respect there is a parallel with the novel. Don Alfonso, it is noted, made a generally favourable impression on the men, but among the ladies produced once more an altogether more enthusiastic response: 'Entró el rey a caballo. Vestía traje militar de campaña y, ros en mano, saludaba a la multitud. Su semblante juvenil, su sonrisa graciosa y su aire modesto le captaron la simpatía del público. En general, a los hombres les pareció bien; a las mujeres agradó mucho' (III: 1331).

What concerns Galdós, though, is to establish historical comparisons. Not only does Alfonso's procession recall past parades following military coups; it also brings back memories of a very different procession: Prim's funeral cortège; and Tito does not fail to indicate the irony of this recollection, given Prim's vehement opposition to the restoration which is now taking place:

Una procesión de carácter bien distinto, tétrica y desesperante, y que marchaba en sentido inverso, dejó en mi alma impresión hondísima: la salida del cortejo fúnebre de Prim para el santuario de Atocha. Señaló una coincidencia que me resultó irónica: en el mismo sitio donde vi la entrada de don Alfonso de Borbón había visto pasar el entierro del grande hombre de la revolución de septiembre que dijo aquello de 'jamás, jamás, jamás'. (III: 1331)

The hopes of the 'glorious' revolution of 1868 which deposed the Bourbons and in which Prim played the principal part are implicitly seen to be finally dashed.

To the café habitués surrounding Juan Pablo Rubín, Feijoo, and the so-called *curas de tropa*,[15] the Restoration brings a golden opportunity to profit from government patronage. Towards the end of 1874 the apparently

[14] This scene is very like the one in *Los apostólicos* when Sola learns, as María Cristina is entering Madrid, that Salvador Monsalud is getting married (II: 121).

[15] Priests with no regular parish. Ortiz Armengol explains that they are 'curas "sueltos" como los llama el autor, sacerdotes en situaciones irregulares de todo tipo' (*3c*, 1987: 385). Ricard (*3b*, 1966) perceptively links these unemployed priests, a serious social problem of the time, with Cayetano Galeote, the assassin of the Bishop of Madrid-Alcalá Martínez Izquierdo in 1886. See also Boo (*3b*).

impending uprising had been the subject of spirited discussion. Basilio Andrés de la Caña throws out mysterious hints of conspiracies involving Cánovas and Romero Robledo, and contends, against Rubín and others, that Alfonso will be on the throne within a month. The most spectacular example of benefit is the totally undeserved success of Juan Pablo himself. Immediately after Sagunto, Villalonga, who in Feijoo's words 'es de la situación' and 'uña y carne de Romero Robledo' (V: III. 1. vi, 308), secures a minor post for him, at the colonel's instigation, whereupon Juan Pablo, after some token resistance, renounces his resolute anti-Alfonso position. Subsequently, Villalonga, now a minister in the Cánovas government,[16] offers him a senior position far above his deserts, as *gobernador de una provincia de tercera*.[17] Juan Pablo quickly sheds all previous allegiances and loyalties, including personal ones to his lover Refugio, subject, like all women driven to illicit relationships, to such sudden and undeserved reversals of fortune. His uncouth brother Nicolás, thanks once more to Feijoo, also benefits from the situation by becoming a canon.[18] The hollow and pretentious Basilio Andrés de la Caña attains his coveted *credencial*, the ex-military priest *Pater* receives a post manipulating elections. Since all the reinstalled employees order new clothes, the tailors too participate in the new prosperity, just as they do in *De Cartago a Sagunto*. Even Izquierdo, thanks to Doña Guillermina, is a beneficiary; he becomes an artist's model for heroic characters like Hernán Cortés and James the Conqueror: a case of outward show and inner hollowness. Only poor Villaamil fails, despite Feijoo's recommendation, to benefit from the new regime, for reasons we shall see later. Needless to say, no question of merit enters into any of these appointments. The overwhelming increase in jobs made available by the new regime for old and new supporters—*resellados*, as the latter were called—is now firmly established.

The manner in which compromise and abandonment of principle have become second nature is shown in a famous passage from *Fortunata y Jacinta* describing café life in Madrid:

Allí brillaba espléndidamente esa fraternidad española en cuyo seno se dan mano de amigo el carlista y el republicano, el progresista de cabeza dura y el moderado implacable. Antiguamente, los partidos separados en público, estábanlo también en las

[16] Villalonga continues his political career, as opportunistic as ever, in *Lo prohibido* and *La incógnita*. Juan Pablo Rubín and Basilio de la Caña are still enjoying official positions in *Miau* and *Ángel Guerra* (V: III. 1, 1228), and the latter reappears in *La Primera República* (III: 1124). In *Miau* Juan Pablo's life history is modified; he once served Villaamil as his clerk and had the effrontery to offer him *medio duro* in his financial distress (V: 554).

[17] It is interesting to note that Galdós, discussing the acute problems new administrations face in choosing provincial governors, indicates that 'todos los candidatos quieren las provincias de primera clase, y no hay ninguno, por insignificante que sea, que no haga ascos a las de segunda o tercera clase' (*3a, OI* iii. *Política española* 102). Juan Pablo is in no position to object, but Galdós makes sure to emphasize that the province assigned to him, like Melchor Relimpio's in *La desheredada* (IV: II. 1, 1066), was *de tercera*.

relaciones privadas; pero el progreso de las costumbres trajo primero cierta suavidad en las relaciones personales, y, por fin, la suavidad se trocó en blandura. Algunos creen que hemos pasado de un extremado mal a otro, sin detenernos en el medio conveniente, y ven en esta fraternidad una relajación de los caracteres. Esto de que todo el mundo sea amigo particular de todo el mundo es síntoma de que las ideas van siendo tan sólo un pretexto para conquistar o defender el pan. Existe una confabulación tácita (no tan escondida que no se encuentre a poco que se rasque en los políticos), por la cual se establece el turno en el dominio. En esto consiste que no hay aspiración, por extraviada que sea, que no se tenga por probable; en esto consiste la inseguridad, única cosa que es constante entre nosotros; la ayuda masónica que se presta a todos los partidos, desde el clerical al anarquista, lo mismo dándose una credencial vergonzante en tiempo de paces, que otorgándose perdones e indultos en las guerras y revoluciones. Hay algo de seguros mutuos contra el castigo, razón por la cual se miran los hechos de fuerza como la cosa más natural del mundo. La moral política es como una capa de tantos remiendos, que no se sabe ya cuál es el paño primitivo. (V: III. 1. i, 15)

It would be difficult to find a more accurate evocation of what the Restoration represented:[19] all its key characteristics are here. In a world of constant civil strife and insecurity, principles and ideas have given way to an intricate network of personal relationships which provide protection and insurance. The means of achieving these conditions are specifically mentioned: the *turno pacífico*. The narrator is careful not to express his own views ('Algunos creen . . .'), but it is clear that political morality is—literally—a thing of shreds and patches. As a practical consequence, since the state exercises no discrimination, let alone moral judgement, in bestowing favours, it leaves the initiative to individuals to exploit it as best they can or otherwise pursue their own private advantages.

The typical use of a crude vocabulary of synecdoches to denote basic sustenance ('pescar', 'mamar', 'garbanzo', 'el pasto', 'conquistar o defender el pan', etc.) emphasizes the down-to-earth nature of these concerns. Idealism and reformist zeal, as expressed by this highly complacent narrator, participator in this bourgeois society and friend of its members (*3c*, Ribbans 1977*a*: 37–42), are conspicuous for their absence, but so is anger or indignation.

The way the political process which culminated in the restoration functions accounts precisely for the attitude and actions of one of the most fascinating characters of the novel, Don Evaristo González Feijoo.[20] He views politics with complete scepticism, as a 'comedia'.[21] As he declares to Juan Pablo

[18] Both Fortunata and Doña Lupe comment on this injustice. See Ribbans (*3c*, 1977*a*: 116).

[19] Gallenga (*1b*, ii. 427–45) gives an interesting bird's-eye view of Restoration politics without, however, understanding what led to its success.

[20] Feijoo is discussed from four different viewpoints in *Conflicting Realities* (*3c*, Goldman 1984); I have a divergent reading of his significance (*3c*, 1987).

[21] Compare also the corrosive political cynicism expressed by Don Carlos de Cisneros in *La incógnita*.

Rubín: 'Yo . . . soy progresista desengañado, y usted tradicionalista arrepentido. Tenemos algo en común: el creer que todo esto es una comedia y que sólo se trata de saber a quién le toca mamar y a quién no' (V: III. 1. i, 15). Politics has become a way of distributing largesse and nothing more. The logical consequence is that, believing only in tolerance and personal relations, Feijoo uses what opportunities arise, as we have seen, to benefit his friends or his interests.

Living in society, for Feijoo, means respecting outward forms and appearances, acting prudently to conceal any infringement of the rules, and keeping one's emotions carefully under control. What separates him from the hypocritical and self-seeking caste of society as a whole is his awareness of his motivation and his benevolence towards others. This goes a long way to explain, in my view, the apparent paradox that Feijoo can at the same time reflect so faithfully Restoration behaviour and yet have a clearly independent, tolerant, and humane approach to life. The way in which the novel deals with political and social issues—oblique, incomplete, ambiguous—is necessarily very different from the procedure revealed in an *episodio* or an article on current affairs, but the underlying attitude is the same. In his outward conformity and inner subversion, therefore, Feijoo is not unlike Beramendi. His attitude is the natural outcome of the whole administrative structure of Spain and of political accommodations made over a period of decades.

In *El amigo Manso*, set in the early Restoration period, similar forces are at work. In Máximo's Olympian philosophical pronouncements, the accommodations of the time are clearly indicated, but are given a different slant; the development of bourgeois society, in the figure of his *indiano* brother, is described in a generalized, non-specific fashion: 'José reproducía en su desenvolvimiento personal la serie de fenómenos generales que caracterizan a estas oligarquías eclécticas, producto de un estado de crisis intelectual y política que eslabona el mundo destruído con el que se está elaborando . . .' (IV: 1203). Manso goes on to describe at great length, in similarly abstract terms, the transformation of society currently taking place, but he fails to see what is going on under his own nose. José María, for his part, eager to become a marquis, pushes the urge towards the reconciliation of opposites to an absurd degree ('hasta el delirio', according to Máximo); the model cited is the archetypal example of conciliation without resolution of conflict, the *abrazo de Vergara* ending the first Carlist war: 'No existe rivalidad histórica y fatal que él no se proponga resolver con un abrazo de Vergara. Eso es: abrácense como hermanos el separatismo y la nacionalidad, la insurrección y el ejército, la monarquía y la república, la Iglesia y el libre examen, la aristocracia y la igualdad' (IV: 1209).

Doña Javiera avidly seizes the opportunity provided by the political environment, concerned to bring together the effective sources of power—wealth or talent[22]—in a new coalition, to enable her son Manolo Peña to leave behind their past in trade. Despite some difficulties—he is called the 'hijo de la carnicera' and finds himself fighting a duel—Peña does make his way. As Blanco Aguinaga elucidates, 'Manolo Peña . . . inicia, pues, su entrada en la "oligarquía ecléctica" de la Restauración en casa de quien, desde un mostrador, llegaría a marqués . . .' (*3c*, 31). If Doña Javiera is frustrated in her desire to make a brilliant marriage for Manuel, he and Irene, an eminently practical woman—no studious Nordic type, as Manso believed—will none the less prosper in the new climate, in which there is no place for the abstract philosopher, of whom indeed scarcely any recollection remains after his disappearance.

Other novels contain isolated references to the Restoration era. Even a novel set as early as 1863 like *El doctor Centeno*—as we have noted, the earliest of the whole series—has an ironic reference, indicated by Bly (*3c*, 1983: 35–6), to the year of composition, twenty years later, in full Restoration period. As a prank, students from Doña Virginia's boarding-house send the crazy educational reformer Jesús Delgado, accustomed to sending letters to himself, a letter of their own, but signed with his name. In it they pretend that twenty years have passed (it is dated 8 November 1883) and make far-fetched prophecies, in a high-sounding tone, for the next twenty years:

> Ya no existen aquellos rutinarios moldes que se oponían a la *Educación Completa*. Todo ha variado, egregio hierofante: la sociedad ha vencido su letal modorra, y despabiladísima aguarda las ideas del legislador de la enseñanza.
>
> En este lapso, ¿no sabe usted que, al fin, ha sido derrocado el trono secular, y con él han desparecido las prácticas añosas y las ideas rancias? Cual generosa espada cubierta de orín, que en un momento es limpiada y recobra su hermosura, temple y brillo, así la nación ha limpiado su mugre. Nuevas instituciones tenemos ya, ¡oh!, y nuevos caracteres y principios. (IV: 1388)

The irony is palpable. No doubt the reader of 1883 could not restrain a wry smile at this travesty of what the Restoration might have been: 'nuevas instituciones', 'nuevos caracteres y principios', indeed! The contrast with the views expressed in *Cánovas* could not be more marked. Moreover, as Hazel Gold acutely indicates,

> By portraying these years as interchangeable, the novel in a sense equates late Isabeline society with Restoration society, thereby denying that the revolutionary and republican interlude of 1868–1874—still an anticipated event for the characters in the novel, but

[22] 'El dinero y el ingenio, sustituídos a menudo por sus similares, agio y travesura, han roto aquí las barreras todas, estableciendo la confusión de clases en grado más alto y con aplicaciones más positivas que en los países europeos, donde la democracia, excluída de las costumbres, tiene representación en las leyes' (IV: 1182–3).

a reality for the narrator and the novel's extrinsic readers—had exercised any lasting historical impact. (*3c*, 1989: 235)[23]

Lo prohibido, set in the early 1880s, heyday of the Restoration, also has evident links with its values. José María Bueno de Guzmán considers himself conditioned by his times: 'producto de mi edad y de mi raza ... hallándome en fatal armonía con el medio en que vivo' (IV: 1776). Whether this is a justification, the result of genetic and environmental determinism, or the excuse of an irresponsible individual is precisely what the novel encourages us to examine (see, for example, Terry, *3c*). From the political point of view, José María is a *diputado cunero* who does not take his duties seriously, in contrast with the earnest Anglophile Pepe Carrillo. Both his closest friends, Severiano Rodríguez and Villalonga, are *diputados* too, and they carry out a sort of private *turno pacífico* among themselves: 'su rivalidad política era sólo aparente, una fácil comedia para esclavizar y tener por suya la provincia' (IV: 1692). There is, however, much more insistence on economic than on political activity, and fortunes are made (Cristóbal Medina) and lost (Torres, José María) on the Stock Exchange: another characteristic aspect of the Restoration, typified by 'the gold fever'[24] of reckless investment.

The same naturally applies to the Torquemada series. The moneylender's prosperity, comprehensibly, is intimately linked with the Restoration.[25] Setting out from a low rung on the social ladder as a halberdier, Torquemada had a period of relative penury, from 1851 to 1868, which is still reflected in his miserly ways and squalid habits in *Fortunata y Jacinta*, though by then he owns two houses; hence his nearly clean shirt, his greasy and frayed number two cape, his ill-fitting trousers and squeaky boots (V: II. 3. i, 194–5; *3c*, Folley). His prosperity dates from 1870, when the full effects of the change in landholdings (*desamortización*), as well as changes in political fortunes, begin to be felt. With the new circumstances comes a different kind of moneylending, distinct from the old miserly type which Galdós calls mystical or metaphysical. The classical usury of Balzac's Gobseck[26] (see *3c*, Fernández-Cifuentes

[23] Similarly, in *Tormento* the society of sixteen years ago (i.e. 1867, time of the action of the novel) and that of the present are explicitly linked: 'En una sociedad como aquélla, o como ésta, pues la variación en dieciséis años no ha sido muy grande' (IV: 1465).

[24] 'En 1876 se produce una crisis intercíclica ... que, en España ... inaugura el período más brillante del siglo XIX: el decenio 1876–1886, sobre cuya dorada cresta se cimentó la Restauración' (*1b*, Vicens Vives 1972: 244). It was not without its risks, however. *La febre d'or* (1890–2) is the striking title of an important novel by Narcís Oller, which traces the disastrous financial projects in Barcelona of the self-made man Gil Foix. Gilabert (*4a*) makes a rather simplistic comparison between Oller and his Castilian contemporaries.

[25] He is, as Peter Bly says, 'a very historically-moulded character' (*3c*, 1983: 152), though I do not agree that 'generalized historical echoes which are used as backdrops' replace the more specific historical references of earlier novels.

[26] Until the mid-19th cent. such moneylending was typically in Jewish hands in other European countries. Torquemada conforms only implicitly to the Jewish stereotype of the usurer, and even less to that of a *converso*; both aspects seem to me exaggerated by Schyfter (*3b*, 55–77).

71–2) has given way to one which is in harmony with the acknowledged and self-satisfied materialist society:

don Francisco ... no pudo eximirse de la influencia de esta segunda mitad del siglo XIX, que casi ha hecho una religión de las materialidades decorosas de la existencia. Aquellos avaros de antiguo cuño ... eran los místicos o metafísicos de la usura ... *Viviendo* el Peor *en una época que arranca de la desamortización*, sufrió, sin comprenderlo, la metamorfosis que ha desnaturalizado la usura metafísica, convirtiéndolo en positivista. (*Torquemada en la hoguera* V: 908; my italics)

During this long transitional period moneylending has thus become a very profitable undertaking.[27] Parvenues society women who vied with each in pretensions to luxury and elegance (Rosalía) and the crop of *señoritos* (Joaquín Pez, Melchor de Relimpio, etc.) who emerged from the mobility of both aristocracy and bourgeoisie made easy pickings. Each change of administration produces a pressing demand for cash, as one group of government employees is replaced by another. Thus we read at the beginning of *Torquemada en la hoguera*:

El año de la Restauración, ya había duplicado Torquemada la pella con que le cogió la *gloriosa*, y el radical cambio político proporcionóle bonitos préstamos y anticipos. Situación nueva, nóminas frescas, pagas saneadas, negocio limpio. Los gobernadores flamantes que tenían que hacerse ropa, los funcionarios diversos que salían de la oscuridad, famélicos, le hicieron un buen Agosto. Toda la época de los conservadores fue regularcita; como que éstos le daban juego con las esplendideces propias de la dominación, y los liberales también con sus ansias y necesidades no satisfechas. Al entrar en el Gobierno, en 1881, los que tanto tiempo estuvieron sin catarlo [i.e. *Sagasta's Liberal Party*], otra vez Torquemada en alza: préstamos de lo fino, adelantos de lo gordo, y vamos viviendo. (V: 906–7)[28]

Torquemada's special mixture of success and failure in the developing Restoration world from *Torquemada en la Cruz* onwards shows Galdós's art at its best. The moneylender's successes, while subject to ironic comment, are firmly rooted in the time.[29] His incorporation into the new aristocracy is no less real for being reluctant; and it is engineered as a dying gesture by Doña Lupe, 'la de los pavos', who, envious as she was of Doña Guillermina's status, was not free herself from social aspirations. From then onwards he acquires successively an aristocratic (if impoverished) wife, together with a bossy sister-

[27] '... usura, sencillísimo y elemental arbitrio en todo país pobre, donde se disputan la vida dos fuerzas negativas: la holganza y la vanidad' (II: 1438). The comment is made about Gregorio Fajardo's business activity.

[28] There are other prosperous moneylenders in the Restoration period: the Ramos couple, parents of 'Posturitas', in *Miau*; Medina and Torres, in *Lo prohibido*; a more discreet usurer, the Marquis of Feramor, in *Halma*; and in the *episodios* Gregorio Fajardo and, more aggressively, his widow Doña Segismunda.

[29] Nicholas Round, building on and rectifying Ricard (*3c*, 1963: 61–85), works out a careful chronology of Torquemada's life, noting as a 'source of strength' 'the interaction of this implicit biographical scheme with history' (*3c*, 1971*b*: 85).

in-law, suitable (if absurd) instruction and counsel from Donoso, a title of nobility, a country estate, a position as senator, and a reputation as a financier and even as an orator; Harold Hall has pointed out (*3c*) that his change of status is matched by the modification of his language. As Blanco Aguinaga puts it, 'el ascenso de nuestro personaje equivale, en forma extrema y ejemplar, al ascenso de la burguesía española a lo largo del XIX' (*3c*, 100). But in a skilful balancing-act Galdós parallels his material success story with his private disappointments: the early failure to save Valentín's life, Felisa's death, the abnormality of the second Valentín, his strained relations with Rafael and Cruz, his nostalgic dissatisfaction with his break with the past, the insoluble problem of belief and death, signalled by the stern presence of Gamborena. When Cruz persuades him to leave a third of his estate to the Church, she argues that it is a just restitution for 'La llamada desamortización que debiera llamarse despojo, [*que*] arrancó su propiedad a la Iglesia' (V: 1182). As Bly indicates (*3c*, 1983: 151–2), the wheel has turned full circle: 'viviendo *el Peor* en una época que arranca de la Desamortización' (V: 908), on his death he is to return the basis of his prosperity to the revitalized Church Militant of the later nineteenth century. And the famous ambiguity about *conversión* (religious repentance or conversion of the national debt) has strong political resonances too. Sagasta's finance minister Juan Camacho had carried out a decisive 'conversion' of the debt in 1881–2.[30] Galdós's best fictitious characters are neither outside the course of history of their time nor entirely conditioned by it.[31]

By contrast, we have the crowd of redundant bureaucrats, or *cesantes*, who are cast out by the Establishment. One of them is Isidora's father, Tomás Rufete. As a result of deranged ambition, which his daughter had no difficulty in recognizing ('No tenía más que un defecto, y es que nunca se contentaba con su suerte, sino que aspiraba a más, a más', IV: I. 1. iii, 972), as well as a heartless system, Rufete ends up in the mental asylum of Leganés, playing the political administrator, endlessly drafting decrees and ordinances and calculating illusory debts.

The paradigm of the *cesante* without prospect of re-employment is of course Ramón Villaamil. In *Fortunata y Jacinta* he makes a couple of fleeting appearances as a pathetic mummy-like figure who needs only two months' service to retire on pension; Feijoo's interventions with Villalonga on his behalf prove

[30] In an article of 16 Sept 1886 (*3a*, *Cartas desconocidas* 192) Galdós speaks of the notion that 'tres Juanes habían salvado en diferentes ocasiones el Erario español'; they are Mendizábal, Bravo Murillo, and Camacho. Simón Babel claims credit, in *Ángel Guerra*, for giving Camacho sound financial advice: 'Yo le dije a don Juan Francisco de Camacho cuando se hizo cargo del Ministerio por tercera vez: "Don Juan Francisco, a recaudar, a recaudar a todo trance, y triplicaremos las rentas"' (V: II. 3. vi, 1352).

[31] When Blanco Aguinaga, in his fine analysis of the function of history in the Torquemada novels, declares that 'la realidad socio-histórica *determina* las estructuras significativas de la ficción' (*3c*, 109; my italics), he is, typically, covering only half of the story.

fruitless. In *Miau*, set in 1878, in the full bloom of the Restoration, Villaamil has the role of protagonist. Formed during the Conservative hegemony of Isabella's reign, he enjoyed the support of the tough, reactionary, and efficient Bravo Murillo. Although, as Nicholas Round remarks, he is 'very much the sort of man with whom the Restoration might be expected to deal favourably' (*3c*, 1986: 25),[32] amid the incessant demands for posts[33] and the consequent pushing and shoving Villaamil is too old, too soft, too unprotected, too much of the old dispensation, and too unlucky to be accommodated. Instead he suffers the special indignity of witnessing, through the use of *enchufes* and secret patronage—secular and female in this case—the outrageously unmerited promotion of his son-in-law Víctor Cadalso. As he himself says, 'Con esta Restauración maldita, epílogo de una condenada revolución, ha salido tanta gente nueva' (V: 223), and, speaking specifically to Pura of the Pez tribe: 'Figúrate una gente que ha mamado en todas las ubres y que ha sabido empalmar la Gloriosa con Alfonsito.' Since *La desheredada* the extraordinary ability of the Peces to adapt themselves to any situation has been unquestionably established. Villaamil's story carries, then, 'a perceptible social resonance' (*3c*, Round 1986: 30). At the same time, as befits a novel not exclusively devoted to a social problem, Villaamil has his own individual story; he is the victim of his own illusions as well as of a monstrous system.[34]

In a later novel like *Misericordia* the name Cánovas represents for the beggar community a sort of absolute authority: '¡Ni que *fuás* Cánovas!' (V: 1881). In *Fortunata y Jacinta*, by contrast, he is the exemplar of worldly knowledge, as when Mauricia exclaims, 'Si sabe más que Cánovas esa tía [Doña Lupe]' (V: II. 6. vi, 246). Similarly evoked in *Misericordia* is the king who was able to defy even the Pope: '¡Ni que *fuás* Vítor Manuel, el que puso preso al Papa!' (V: 1883).[35] These are clearly examples of the third type of historical reference discussed in Chapter 1.

Let us now consider the characteristics of the last *episodio* in more detail. The new narrative procedure adopted from *Amadeo I* onwards has been described

[32] Round perceptively analyses the varying attitudes and discrepancies demonstrated in the portrayal of Villaamil and comments in detail on the ministers to whom he owed his career. His first patron (in *Fortunata y Jacinta*), Claudio Moyano, 'superannuated conservative statesman', as Round calls him (*3c*, 1986: 20), was unlikely to do him any good by the 1870s.

[33] Cf. Pantoja's remark, 'Esa gente, que sirvió a la Gloriosa primero y después a la Restauración, está con el agua al cuello porque tiene que atender a los de ahora, sin desemparar a los de antes, que andan ladrando de hambre' (V: 238). For Villaamil and *cesantía* see Lambert (*3c*). The time of the central action is neatly indicated by the marriage of the queen doña Mercedes (V: 565) and her presence at the opera (V: 682); she reigned only five months, between Feb. and June 1878.

[34] For a more detailed analysis of this view see Ribbans (*3c*, 1977*b*). Weber (*3c*) takes a highly critical view of Villaamil, while Parker (*3c*), Scanlon and Jones (*3c*), and Ramsden (*3c*) in differing degrees endeavour to exonerate him from responsibility. See Rodgers (*3c*, 1978) for a sensitive account of the issues.

[35] Culturally, too, the end of an epoch is marked by the funeral of the poet Zorrilla (V: 1883).

in the previous chapter, and the same criticisms of abstract allegorization and implausibility hold for *Cánovas*. What I do discern in this last work is a powerful sense of depression and defeat, even despair, on two counts: first, regarding Spanish history as such, and second, concerning his ability or willingness to finish his project. The promising Republican and Socialist alliance (the *Conjunción republicano-socialista*) Galdós led had little enduring success (it was dissolved in June 1913), and by 1912 he had withdrawn from active intervention (*3b*, Fuentes 39). Likewise, his earlier enthusiasm for the new narrative structure he adopted in *Amadeo I* seems to have largely melted away. Certainly, his letter to Teodosia Gandarias on completing the work in August 1912 strikes an ironically defensive note; the comparison with the Creation may seem to imply that there are no more to come, and he takes refuge in an expression of the personal happiness resulting from his love for Teo:

> Se acabó Cánovas. Como Dios en el último día de su labor constructiva pudo decir: acabó su obra y vio que era bueno. Yo sólo diré que es mediana y pasable. Y digo también que no me cambio por nadie, porque tengo a mi Teo, y los demás mortales no la tienen. (14 August 1912; *3b*, Madariaga 94)

He does not, however, immediately abandon the task, at the same time as he shows some reluctance to resume it. On 25 August he writes to Teodosia:

> Del tomo *Sagasta* no pienso escribir ni una línea todavía. Abraza desde la subida de Sagasta al poder el 81, hasta la muerte de Alfonso XII, en Nov. de 1885, y el nacimiento de este Alfonso XIII en Mayo del 86. Asistí a la ceremonia de la presentación de este Rey, y me parece que estoy viendo a D. Práxedes con la bandeja en la mano, presentando al recién nacido, como un conejo desollado.
>
> El nuevo tomo sólo empezaré bajo tus auspicios, y en él tendré que introducir nuevos elementos sociales para dar novedad a la narración. Es *obra de romanos* buscar y encontrar novedad en un tomo que es el 47 de la serie.[36]

He is certainly finding the search for originality in a well used form very hard going. This may explain why in the long run he did not continue with the *episodios*. Although increasing blindness undoubtedly exacted a toll, exhaustion due to old age is not the answer; he did go on writing for the theatre and *La razón de la sinrazón* (1915) is still to come. It would seem rather that the particular task of writing *episodios* has become increasingly too exacting for him to tackle, though he continued to speak of *Sagasta* and its proposed structure for over a year. He did not finally abandon the project until after January 1914.[37]

[36] Despite the number indicated (47th of the series), Madariaga (*3b*, 93) misdates this letter as 1911 and therefore mistakenly associates it with *De Cartago a Sagunto*.

[37] In a letter to Gerardo Peñarrubia of 1 Sept. 1912 referred to by Shoemaker (*3b*, 1984: 157) he declared: 'Ya estoy trabajando en los preparativos de *Sagasta*, y vamos viviendo, vamos marchando hacia la redención de la patria.' In February 1913, in the prologue to the Nelson

Before we proceed further, it is instructive to look at the political and social commentaries Galdós wrote after the Restoration. In these articles, as in those he wrote before 1875, Galdós opposed all political extremism, and his first priority was governmental stability. Yet, as Peter Goldman, who has done so much to elucidate Galdós's ideology, concludes, this did not mean an unquestioning acceptance of the Cánovas settlement:

Differentiating Galdós' vision of a two-party system from the *turno pacífico* later implemented by Cánovas during the Restoration is the strict maintenance of ideological integrity and its corresponding socio-political purpose by each party; without the 'juego y equilibrado contrapeso' which such political independence affords, no constitutional regime is possible. (*3b*, 1969: 83)[38]

In the articles written around 1885—just before *Fortunata y Jacinta*—Galdós praises to a surprising extent what he sees as its achievements, particularly the maintenance of the constitutional rights of the revolution and the freedom of the press. He treats Cánovas with great deference, condemning only his condoning of Romero Robledo's electoral manipulations[39] and the associated problem of *caciquismo*. He deplores any resort to violence. He does not believe any revolution imminent or desirable. While Sagasta and Castelar enjoy his esteem, he sees as the besetting sin of the Liberals their endemic tendency towards disunity; on this count, he is at that time highly critical of Ruiz Zorrilla.

Some years later, in 1903, in 'Soñemos, alma, soñemos', we have a very forceful critical comment on the role the state machine has played as universal provider:

Las clases más ilustradas reclamaban y obtenían el socorro del sueldo. Había dos noblezas: la de los pergaminos y la de los expedientes, y los puestos más altos de la burocracia se asimilaban a la grandeza de España. Un socialismo bastardo ponía en manos del Estado la distribución de la sopa y garbanzos del pobre, de los manjares trufados del rico. (**VI**: 1258–9)

Surprisingly, he reports, in the familiar down-to-earth language he habitually employs, 'un gran progreso' over the previous fifty years in eliminating this excessive dependence on the state: 'es incalculable el número de los que han aprendido a subsistir sin acercar sus labios a las que un tiempo fueron lozanas ubres y hoy cuelgan fláccidas. Los españoles han crecido; comen, ya no maman' (**VI**: 1258–9).

edition of *Misericordia*, he refers to the 'quinta serie de *Episodios*, ésta no terminada todavía' (*3a*, *Prólogos* 108–10). And at a meeting with the king as late as 7 Jan. 1914 he is reported as saying, Preparo . . . un nuevo tomo de los *Episodios nacionales*: el tomo 47. Se titula *Sagasta*, y termina justamente en la fecha de su nacimiento de V.M., en Mayo de 1886' (*3b*, Dendle 1984: 105).

[38] See also Goldman's other perceptive studies (*3b*, 1971; 1975).

[39] See e.g. 'Procedimientos electorales' (30 Apr. 1885; *3a*, *OI* iii. 33–8) and 'Política menuda' (19 Dec 1885; ibid. 103–10). Robledo split from Cánovas in that year, when power was allowed to pass to Sagasta (*1b*, Carr 363).

These same problems are equally apparent in *Cánovas*, but there is a considerable change of attitude. Felipe Ducazcal, previously involved in the famous counter-demonstration of prostitutes against the Alfonsine *mantillas blancas* procession organized to discredit the Savoyan dynasty,[40] now returns as an early adherent of the Alfonsine faction: 'me llevó al Círculo Popular Alfonsino, hervidero de pretendientes al sinfín de plazas que brindaba la Restauración a los españoles necesitados' (III: 1329). It is another example of the capacity of the renewed dynasty, from its earliest days, to absorb, or *resellar*, its former opponents by creating jobs for them; several other figures, some real, some fictitious, are listed as enjoying the benefits of the Restoration settlement and serve to show the all-embracing influence of the new regime, which can seduce even those who are essentially antagonistic to its ideals.

The most notable case is Segismundo García Fajardo. During Amadeo's reign, after his Federalist phase, he represented the people against the intruder (III: 975). At this time, like other comparable figures—Leoncio Ansúrez, Santiago Ibero—he declares his determination to stay independent at all costs: 'me quedo en la pobreza y la insensatez' (957); though he is saved from the full effects of destitution by Lucila. Now, returning from Paris in absolute squalor and supported by Tito, he, 'el rebelde más tenaz y el revolucionario más gracioso que ha existido bajo el limpio cielo de los Madriles' (1351), ends up by reconciling himself externally to the opportunities vouchsafed by the well-nigh irresistible Restoration.[41] For this reason Fajardo brings in references to his uncle Beramendi, the archetype of the outward conformist and inner rebel, who in turn influences his nephew in the same direction. Such a characteristic figure of earlier *episodios* as Telesforo del Portillo (*Sebo*), the former police agent, intriguer, moneylender, and employee of Beramendi's, is one more person catered for by the Restoration: he is 'colocado ya en un buen puesto del Gobierno civil'.

Vicente Halconero also reappears. Fajardo, who has something of a prophetic function, had declared already in *España trágica* that his friend would come to accept the Restoration: 'Sí, Vicente, joven sensato: quiéraslo o no, tú serás alfonsino, trabajarás por la Restauración... Puede que sea marqués y ministro de un Borbón futuro' (III: 957). And so it turns out: already in the same *episodio* we see Halconero seeking public office. In *Cánovas* he is a fully fledged supporter of the new Establishment; he is working for the Liberals and invites Tito to accept a seat as *diputado* in

[40] By a curious error, the two demonstrations are confused at the time of the bull-fight in honour of the royal marriage: the 'tanda de mujeres, ataviadas estrepitosamente con pañolones de Manila... las que el año 72 hicieron en la Castellana, a las órdenes de Ducazcal, la famosa manifestación *contra* la dinastía de Saboya' (III: 1376; my italics).

[41] In dealing with Segismundo as a 'perfect embodiment of that society' and, together with Halconero and Beramendi, as 'the new nobility created by money' (*3d*, 34), Varela overlooks the fact that they are fully aware of succumbing to the pressures of society and in varying degrees consciously resist its values.

Sagasta's party—as Galdós himself had done—in the next parliament. Tito's vehement refusal may contain an implicit self-criticism from the later Galdós against his more acquiescent earlier self: 'me repugnaba el *cunerismo* y nunca pasó por mi mente pertenecer a esos rebaños parlamentarios' (III: 1382). Unlike Tito, Vicente is confident that some progress is being made and is not apprehensive about what has now become Galdós's *bête noire*: the influx from France of members of the religious orders. His views are not allowed to stand by themselves, though, for they are capped and disparaged by the narrator: '¡Pobre Vicentito, tan joven, tan simpático, y ya contagiado del negro y pestilente virus!'

In the last few pages two consecrated figures are shown fleetingly as fervent supporters of the Church. One is the crafty Eufrasia, Marchioness of Villares de Tajo, now well over fifty and pictured as making her peace with the rapacious clerics 'para mandarla al otro mundo bien limpia de pecados, y aliviada del peso de sus cuantiosos intereses' (III: 1407). The other is Halconero's mother, Lucila Ansúrez, previously the evident symbol ('la Celtibérica') of the primitive roots of Spain; devoted as always to protecting her son's interests, she too has accepted the new clerical predominance.

Tito refuses to take any part or participate in any way in this free-for-all. The extreme degree of patronage exercised, however, is shown very close to him in the offer made to his lover Casiana, who is just beginning to learn to read under Ido's guidance, of a sinecure as 'inspectora de escuelas', with no obligation to turn up except to collect her salary. Rather than let this incident speak for itself, Tito comments: 'Guardé el nombramiento, en el que vi un signo de los tiempos. Todo era ficciones, favoritismos y un saqueo desvergonzado del presupuesto' (III: 1341).

This familiar criticism is repeated directly or by example time and time again throughout the volume. Part of these comments come from García Fajardo and part from Tito, but there is little difference between them: it is clear that the readers are being invited to accept an unequivocal viewpoint,[42] far from the open-ended structure of the previous series. First, Segismundo delivers a violent diatribe against the dynastic wars which had produced half a century of bloody conflict. Then he expresses the opinion that nothing has changed and that all the Bourbons are the same; the only difference is that now, between generals and civil servants, there are more mouths to feed (the point made so vividly in the novels) and there is more show, more pretension. '¡Oh desmedrada España!' he exclaims, 'Cada día pesas menos, y si abultas más, atribúyelo a tu vana hinchazón' (III: 1359). These two factors—crowds of dependents and vain ostentation—typify the Restoration; unfortunately for the fictional structure, we are told so rather than being shown it happening.

[42] Montesinos comments: 'Lo malo de estas jeremiadas, por demás genéricas, es que, no sólo no contribuyen a potenciar la novela, sino que más bien la desustancian, mostrando demasiado a las claras el rencor del novelista' (*3c*, iii. 299).

The topic of hollow ostentation is also repeated *ad nauseam*. Hordes of inflated bureaucrats, parvenus flaunting new titles of nobility from Amadeo or Alfonso, and a new caste of superficial and Gallicized 'señoritos' abound. The birth of María de las Mercedes on 11 September 1880, in particular, brings out all the worst features of Alfonsine society. An ignoble dispute between the patriarchs of Spain and the Indies as to who is to baptize the infanta; the flood of honours which demonstrate 'la degradante frivolidad a que habían llegado las clases superiores del Estado'; the gastronomic and rhetorical excesses to which the event gave rise: all these are related by a narrator committed to a hostile point of view. None of it is as effective as such ironic and apparently tolerant descriptions in *Fortunata y Jacinta* of the fatuous society—Aparisi, the Marquis of Casa-Muñoz, Federico Ruiz—surrounding the Santa Cruz family, the vapid café discussions of Juan Pablo Rubín's circle, or the empty pretensions of the *Samaniegas*. It is, significantly, a similar world to that of the heyday of the Liberal Union portrayed in *O'Donnell*.

Tito, in the company of Casiana, reacts strongly if eccentrically against these pretensions; 'creyendo que la llamada "gente cursi" es el verdadero estado llano de los tiempos modernos', he enjoys 'extremando las formas de la ordinariez presumida ("crema" de la cursilería)' in reaction to 'las elegancias faranduleras y la hinchada presunción traídas a la sociedad española por el cambiazo de Sagunto'. 'Somos cursis', he exclaims, 'por patriotismo' (III: 1346).

Spanish society as a whole is, in his view, a tentacular growth from the values of the Restoration:

> Lo que hemos visto es el familión político triunfante, en el cual todo es nuevo, desde el Rey, cabeza del Estado, hasta las extremidades o tentáculos en que figuran los últimos ministriles; es un hermoso y lucido animal, que devora cuanto puede y da de comer a lo que llamamos pueblo, nación o materia gobernable. (III: 1332)

The concept is clearly related to the intricate social and political ties revealed in the commercial world of the Santa Cruz and Arnáiz families, but the portrayal in the novel, seen from inside, is far more vivid and at the same time more balanced.

Spain is, Tito goes on, a consumer society disputing over its inadequate resources instead of striving to produce more. Its worst feature, according to Tito-Galdós in 1912, in stark contrast with his more encouraging view in 'Soñemos, alma, soñemos' nine years before, is the mass of unproductive individuals who depend parasitically on the state:

> los míseros de levita y chistera, legión incontable que se extiende desde los bajos confines del pueblo hasta los altos linderos de la aristocracia; caterva sin fin, inquieta, menesterosa, que vive del meneo de plumas en oficinas y covachuelas, o de modestas granjerías que apenas dan para un cocido. Esta es la plaga, ésta es la carcoma del país,

necesitada y pedigüeña, a la cual, ¡oh ilustre compañera mía!, tenemos el honor de pertenecer. (III: 1333)

Once again, this is precisely the world—that of the Pez family, Bringas, Melchor de Relimpio, Basilio de la Caña, Villaamil, etc.—that is evoked in a much more lively way within a clearly defined context in the contemporary novels. Tito then talks of

la inmensa tribu de desheredados, y la misión benéfica que desempeñan, en algunos casos y a hurtadillas, las dos mujeres guapas con quienes hemos hablado hace un ratito. . . . España era un país algo comunista. Por los canales contributivos venía todo el caudal a la olla grande, de donde salía para repartirse en mezquinas raciones entre el señorío paupérrimo de la flaca España . . . olla grande . . . a lo que en lenguaje político llamamos Presupuesto. (III: 1333)

The actual analysis of aspects of the same phenomenon—the 'socialismo atenuado e inofensivo'[43] he comments on in *Fortunata y Jacinta*—is much more telling through being less direct and less strident. A certain ironic acquiescence in the realities of life, without overt denunciation or protest, is replaced by a bitter sense of disgust and frustration. As for the sardonic view that women of the *demi-monde* have more influence on the country than its rulers, Galdós had already expressed the same sentiment just before speaking of these same 'damas de las camelias' like Leona la Brava, 'aquellas mujeres guapas que nos hablaron antes ¿también mandan?' asks Casiana. '—¡Que si mandan! Más que el rey. Más que nadie' (III: 1332) is the reply.

Occasionally, it is true, we catch a glimpse of the pristine qualities of open ventilation of issues. Tito does raise the important question, at the beginning of the narrative, of the desirability of the Restoration—'La renovación social y política que se anunciaba, ¿era un paso hacia el bienestar nacional o un peligroso brinco en las tinieblas?' (III: 1322)—but only Sagasta and Halconero, from the real and fictional world respectively, offer, fleetingly, a rather feeble defence of the first view. Again, Casiana makes an interesting general point when, meditating on the issues of violence and conciliation, she voices a feeling closely related to the sentiments of Don Baldomero in *Fortunata y Jacinta* that there is in politics an automatic alternation of periods of storm and tranquillity.[44] Such an unsophisticated snap judgement which takes no account of cause and effect is appropriate to the immediacy of a novel and not incompatible with a historical narrative. It seems, however, to belong to an earlier period, for it hardly matches the overwhelming atmosphere of egoism and inertia Galdós now conveys as the effect of the Restoration.

[43] The *Fortunata y Jacinta* passage is very close to parts of 'Soñemos, alma, soñemos' of 1903 (VI: 1258–60). Augusto Miquis makes similar remarks, in a deliberately provocative way, in *La desheredada* (IV: II. 1, 1064).

[44] 'Lo único que sabemos es que nuestro país padece alternativas o fiebres intermitentes de revolución y de paz' (V: III. 2. i, 310).

A second concern is the lack of an effective degree of interaction between the fictional plot and the historical events. As Montesinos says, 'Galdós desaprovecha ahora las ocasiones de novela como nunca lo hiciera antes' (*3c*, iii. 305). A few events certainly do occur within the action and form a more or less natural part of it. Among them is such a key event as the Sandhurst manifesto. Its contents are given, in a typically indirect way, as it is happening: 'escrito en un colegio no sé si de Inglaterra o Alemania' (**III**: 1321), during the spirited conversations between the now prosperous *demi-mondaine* Leona la Brava and Tito at a performance of the opera after her return from Paris. The essential phrases ('si al igual de sus antecesores será siempre buen católico, como hijo del siglo ha de ser verdaderamente liberal': 1321) are duly recorded, together with an indication of the traditional incompatibility between the two concepts, highlighted in the *Syllabus*: '¡Pero si el Papa ha dicho que el liberalismo es pecado!' (1321). It should be noted that the controversy was by no means stilled by the Restoration. Pius IX's intransigence was forcefully reasserted by Father Félix Sardá y Salvany in his famous book *¿El liberalismo es pecado?* (1884).[45] Moreover, the polemic took on a new relevance in the last years of the regency, when Father José Fernández Montaña, the queen mother's confessor and spiritual adviser to the young king, asserted once more, in opposition to Canalejas, that 'el liberalismo es pecado y no se compadece la profesión de católico con la de liberal' (*El siglo futuro*, 14 December 1900).[46] He was obliged to resign his posts. Galdós, then about to start the fourth series, must have been fully aware of this dispute. It was immediately followed by the Urbao scandal, about an apparently reluctant novice, which may have triggered off *Electra*.[47]

Another issue is the economic boom. Employed by the 'bancos populares' established by Baldomera Larra,[48] Segismundo García Fajardo lavishes heavy sarcastic praise on the perpetrator of the most notorious fraud of the period; he extols her good human relations with her staff and her clients and the moderation of her fraudulent gains of seven million *reales* compared with the enormous profits of the large-scale speculators ennobled by the regime; the wild speculation or 'baldomerismo', the poor man's equivalent of 'la fiebre de oro', is, like its counterpart moneylending, symptomatic of the times.

[45] See also Father Antonio Vicent's *Socialismo y anarquía: La Encíclica de nuestro santísimo padre León XIII 'De conditione opifacum'* (Valencia, 1893), which, in arguing for the establishment of working-class Catholic associations, lays the blame for socialist and anarchist ideology on liberalism.

[46] The intransigent attitude lasted, of course, throughout Primo's dictatorship, the Second Republic, the civil war, and the savage repression which followed. For an excellent literary treatment of a clerical reaction in the special context of the civil war, see Sender's *Réquiem por un campesino español* (*4a*).

[47] For the background to *Electra* and its reception see Fox (*3b*).

[48] Larra never fully acknowledged Baldomera, born before Nov. 1834, as his daughter. See Varela (*4a*, 299). She was evidently named after Espartero.

A further uncommon example of integration is Doña Segismunda García's gift of rare books, via Tito, to Cánovas. Tito is thus enabled to visit the head of the government again and to discuss with him the issue of the unimpeded entry into Spain of the religious orders expelled from France after the Dreyfus affair (*3b*, Llorens 119). Also raised is the more general question of whether Cánovas's historical erudition is an advantage or a hindrance to him as a politician or whether it is as sterile as, say, Casaubon's in *Middlemarch*. Unfortunately, the question is posed, not by a representative figure, but by Mariclío herself: 'Toda esta ciencia arcaica y este fárrago que tuvieron su porqué y sazón en siglos remotos, ¿le sirven al buen don Antonio para consumar y sutilizar sus artes de estadista y gobernador de los reinos hispanos . . . ?' Her reply inclines to the negative: 'Voy creyendo que esto no es más que un bello delirio de coleccionista' (III: 1399). There are not many of these largely integrated episodes. The fact remains that most of the time the reader is told rather than shown, with little encouragement to determine for himself the significance of events.

Galdós brings back various characters from earlier *episodios* and to a lesser extent from the contemporary novels, and this does lend some continuity and plausibility. Fajardo and Halconero we have already discussed. The omnipresent José Ido continues his new role from previous volumes, while another link with the world of the contemporary novels is provided by Matías Luengo, described as Plácido Estupiñá's nephew.[49] As a seller of religious objects, like the Cabrera couple in *Miau*, Luengo is destined in Tito's view to become rich in the ultra-clerical world he sees looming ahead.

A distressing characteristic of the *episodio* is the frequency with which Tito or Mariclío has recourse to definitive *ex cathedra* pronouncements. Thus, Mariclío expresses her opinion directly on Martínez Campos and the Sagunto *pronunciamiento*: 'Ya sabes que aprecio a este general . . . Me ha dolido verle metido en este enredo. Si la Restauración era un hecho inevitable, impuesto por el fatalismo histórico, los españoles debían traerla por los caminos políticos antes que por los atajos militares. Cánovas opinaba como yo' (III: 1326). Opinions of this sort come better from a character within the action than from a pseudo-authorial voice like the Muse of History. We remember that when Juanito Santa Cruz expressed a similar opinion we not only had the opportunity of considering the illegal political act in itself but the discrepancy between the words and actions of a typical member of Restoration society. Equally unmediated and direct are the comments on the subject of Cánovas's difficulties with 'los señores del lastre reaccionario'. Ido, in making the same point in lively speech, is no more than a mouthpiece for a conventional progressive attitude:

[49] The model for Estupiñá, according to Galdós's *Memorias* (VI: 1438), was a stall-holder in the Plaza Mayor named José Luengo.

está tragando mucha quina . . . apretando entre dos muelas cordales, pues de una parte pesan sobre él los malditos "moderados", los Chestes, Moyanos, y Orovios que le piden neísmo, intolerancia y tentetieso, y de otra parte le acosan los alfonsinos que vienen de lo de Alcolea y quieren franquicias, unas miajas de soberanía nacional y vista gorda para el libre pensamiento. (III: 1339)

Criticism of Orovio's measures against free speech is likewise explicit, though enlivened in this case by a picturesque reference to the minister's innovations of fashion: 'Por este acto de brutal intolerancia y por sus pintorescos chalecos, transmitió su nombre hasta los alrededores de la posteridad el marqués de Orovio' (III: 1344). Cabrera's desertion of the Carlist cause serves as the pretext for praising England for qualities so notably absent from Spain: 'el plácido ambiente de un país liberal y protestante, de un país en que imperaba la justicia y el orden, en que los ciudadanos vivían dichosos ejercitando sus derechos y sometidos al suave rigor de las leyes' (1345). The pretender Carlos VII's invitation to the exiled queen and her mother to visit the occupied north is accompanied by a categorical and explicit criticism: 'Ridícula es la carta . . .' (1344). Similarly, Cánovas is directly censured, with a heavy authorial hand, for handing over power to Martínez Campos in order to avoid the hazardous task of implementing the fragile Peace of Zanjón which had been signed in Cuba:

¡Lástima grande que un hombre como Cánovas desestimara el alto ideal que Martínez Campos defendía! Error funesto que don Antonio, por falta de valor para imponerse a los patrioteros, entregase el Poder a un hombre que, si en lo militar era eminente, en lo político carecía de trastienda y travesura para luchar con las pasiones humanas. (III: 1384–5)

There follows a rather facile use of hindsight after the 1898 débâcle: '¡Fatalidad inexorable! . . . De este modo, entre un sabio que no quiere y un valiente que no puede, decretaron para un tiempo no lejano la pérdida de las Antillas' (1384–5).

At times the narrative structure seems to weaken perceptibly, as it had done in *De Cartago a Sagunto*, when the narration fails to reach as far as had been intended.[50] All the *episodios* have a certain tendency to condense masses of historical information in straight mimetic narrative, but this is carried to a new extreme in *Cánovas*. Many incidents are conveyed in chronicle form: the suppression of all the newspapers at the moment of the Restoration, the painstakingly detailed enumeration of the composition of various ministries (Cánovas, Jovellar, Martínez Campos), Doña Mercedes's marriage and death, with much of López de Ayala's long memoir in Congress given verbatim; the recalling of the popular song '¿Dónde vas, Alfonso Doce, dónde vas, triste de ti?' sung in the streets. At certain moments, indeed, a series of incidents—

[50] '. . . La deuda que tengo contigo de referirte lo de Sagunto aplazada queda por apremios del tiempo y del espacio, superiores a la voluntad de vuestro leal y asenderado Tito' (III: 1315).

deaths (María Cristina, Rivero, Espartero), civil disturbances, the Murcia floods—are bundled into a couple of paragraphs; these are completely superfluous without an attempt to integrate them into the action: 'peso muerto', as Montesinos calls them (*3c*, iii. 289). This procedure obviously makes for monotony and betokens a tiredness in the author, very possibly due to the necessity of dictating.

It is evidently essential to create a means of penetrating into the motivation of the founder of the Restoration system and to give an intimate view of his personality. The narrator Tito obtains this access, not through a more or less plausible—if still rather forced—personal contact with the leader, as Beramendi had with Narváez or Queen Isabella, or Santiago Ibero had with Prim, but through a highly unrealistic series of events arranged by the allegorical *Madre* Mariclío and her messengers called *Efémeras*, often by resort to telepathy or magic. From the beginning of the narrative, *la Madre* gives explicit instructions to Tito: 'Reanuda y cultiva tus antiguas amistades, y disponte a estrechar las nuevas relaciones que te salgan al paso. No desdeñes a los hombres de pro' (III: 1318). This coincides with Leona la Brava's advice: 'Arrímete a Cánovas, que es el hombre de mañana, y si no tienes medios para hacerte su amigo, yo te los proporcionaré' (1321). The degree to which implausible intervention is carried is shown when Tito is materially prevented from visiting Sagasta or Primo de Rivera by the disappearance of his court clothes, an occurrence Ido attributes to Cánovas himself! Later Tito penetrates explicitly by magic into Cánovas's thought, but it is not clear whether the meeting actually took place or is simply a figment of Tito's imagination. To record the love affair between the king and Elena Sanz he is given a magic pen which will only record the truth. A purportedly authentic record of events is given, but at the expense of sacrificing the credible narrative structure utilized in earlier *episodios*.

Tito's portrayal of Cánovas himself in action and his view of his motivation, unsatisfactorily integrated into the action though they are, have evident interest. He describes the statesman's mannerisms—'ajustándose los lentes y alzando luego la cabeza, movimientos en él muy comunes' (III: 1334)—and his *ceceo* and records his famous phrases; the statement 'que ha venido a continuar la historia de España' is quoted countless times, and the anecdote about his definition of Spaniards is also cited: 'Son españoles . . . los que no pueden ser otra cosa' (1359). The essentially passive approach we have seen in *España sin rey* and *España trágica* is continued in *Cánovas*.

On his intellectual qualities Tito comments with respect,[51] but significantly has nothing to say about what his political objectives are or whether or not he

[51] Cf. Galdós's mainly favourable attitude towards Cánovas in the contemporary newspaper articles mentioned earlier. Another sign of esteem is the fact that when he lets his fantasy loose to rewrite the history of Isabella II's reign in his 1904 article, he turns to Cánovas as the most suitable responsible guide to make of her reign 'un reinado de bienandanzas' (VI: 1418); see

has any broader vision: 'Respetaba yo a Cánovas y le admiraba por su elevado entendimiento, por su saber de historia y de política, así como por su palabra enérgica y sugestiva, esmaltada con los donaires de un ingenio sutil' (III: 1337). He is shown as seeking to bring former cantonalist revolutionaries into the Alfonsine fold and is clearly aware of the need for clemency. He is open in his criticism of Spain's 'policía detestable . . . la proyección más inútil y desmayada de nuestro matalotaje burocrático' (1335). His consummate skill in political manœuvring[52] is demonstrated in the incident of the ephemeral government of Jovellar, brought in to settle the awkward question of the electoral register, while Cánovas retains the strings of power behind the scenes. Where he goes radically wrong is in not scrupling to employ Romero Robledo's manipulative skills to fix elections—a criterion of expediency at the expense of truth or honesty—in order to create the 'montón grande de la mayoría conservadora y el montón chico de la minoría liberal dinástica'. It is, as he says, 'la espléndida mentira de la soberanía nacional', consisting of 'ficciones o decorosas pamplinas' (1350). In this view he coincides exactly—though the language is more virulent—with the activities of Villalonga in *Fortunata y Jacinta*, *Realidad*, and *Lo prohibido*, as well as with his earlier political writings of 1885.

A later meeting with Cánovas receives a more realistic, if not very convincing, explanation. Widely believed to be the author of certain articles on the king's marriage plans, actually by Segismundo, Tito is called in by Cánovas, who explains his difficulties to him. The only solution, according to the father of the Restoration, is the passage of time:

> Esta vieja nación, con sus glorias y sus tristezas, sus fuerzas y sus recuerdos, sus instituciones aristocráticas y populares y su extraordinario poder sentimental, constituye un cuerpo político de tan dura consistencia, que los hombres de Estado, cualesquiera que sean sus dotes de voluntad y entendimiento, no lo pueden alterar. (III: 1371)

His self-justification is: 'Mi deber es sofocar la tragedia nacional, conteniendo las energías étnicas dentro de la forma lírica, para que la pobre España viva mansamente hasta que lleguen días más propicios' (1371–2). It is an idea we encounter elsewhere in Galdós—Prim, for example, made the same point in *España trágica*[53]—but in the case of such a cynical and anaemic world, without ideals or vision, as Galdós portrays in 1912, it was far too complacent a criterion to satisfy Liberal reformers. In Cánovas himself it betokens a deep pessimism and a certain pusillanimity. Novelistically, it is all excessively direct.

Sagasta commands much less attention, further consideration no doubt

Chapter 7. Valle-Inclán emphasizes Cánovas's 'heavy-footed pedantry' (*4a*, Sinclair 92) in *Baza de espadas* (*4a*, 19–23).

[52] Carr (*1b*, 348) remarks that Cánovas's favourite word was *hacedero*, which he equates with Burke's 'expedient'. He was known to his contemporaries as 'el monstruo'.

[53] 'El único filósofo que puede dar obras duraderas es el Tiempo' (III: 947).

being left for the next *episodio*, which was never written.[54] Tito finds, when he visits him, that he has a true 'don de simpatía'; Galdós's former leader, for his part, is depicted as claiming that real progress is being made despite the election frauds. Yet his 'partido liberal dinástico' is clearly seen as intended, not to bring any genuine political principles into play, but to secure for its members some of the advantages of the system ('para vivir y pescar lo que se pueda'). The question is raised as to whether this was not perhaps inevitable in the circumstances, and certainly the logic of events, as documented in *Cánovas* and the novels, pointed that way. But life is not fully determined, since tendencies may be modified by imaginative leadership, such as Prim's had been. Neither Cánovas, with his ultra-cautious reliance on time as healer, nor Sagasta, depending on expediency, could offer that leadership. Regalado García comments that 'El problema de Galdós consiste en que su visión de la política era un complejo del pesimismo de Cánovas y del oportunismo de Sagasta, durante la Restauración y la Regencia' (*3d*, 514). By the time Galdós comes to write *Cánovas*, the reverse is true: neither Cánovas's pessimism nor Sagasta's opportunism satisfies him. The result is angry frustration.

Other possibilities are touched upon: one is Ruiz Zorrilla, now referred to as 'aquel gran ciudadano, rebelde y tenaz', as progressive politicians of varying persuasions—among them Castelar, Salmerón, and Pablo Nougués, as well as some *alfonsinos*—see him off at the station on his departure for exile; no doubt more would have been heard of him in the unwritten *Sagasta*. As Estébanez Calderón has pointed out (*3b*, 1982: 20–1), Galdós's earlier critical attitude towards the Radical leader has changed between 1885 and 1908. Yet it is evident that Galdós can offer little hope of a successful revolution, and he uses his greatly esteemed friend Nicolás Estévanez to express this pessimistic view to Segismundo in Paris. Estévanez is depicted as 'muy desalentado, y cree que todo intento revolucionario, ya sea zorrillista, ya sea de otro orden, quedará hecho polvo bajo el peso de esta oligarquía de tres cabezas: la femenina aristocrática, la militar masculina y la papista epicena' (**III**: 1354). All three heads of the oligarchy are shown in action in the *episodio*. Incidentally, the revolutionary uprising in which Ángel Guerra is involved around 1883 can be linked with Zorrilla's determined efforts to bring down the Alfonsine monarchy, but the leader, Brigadier Campón, is fictitious and the details of the revolt are deliberately left in obscurity.[55]

Since, as always, Galdós is fascinated by historical continuity (as Celestina says to Tito, '¡ . . . está visto que dondequiera que usted va, allí encuentra la

[54] In a letter to Teodosia Gandarias, he writes: 'El tomo que seguirá se titulará *Sagasta*, y en él retrataré de cuerpo entero al sutil, agudísimo y maravilloso D. Práxedes' (*3b*, Porter).

[55] In general terms Campón's fictitious uprising, though it took place earlier, recalls the Villacampa *pronunciamiento* of 19 Sept. 1886. See Lakhdari (*3c*). Galdós wrote an article about it on 25 Sept. 1886 (*3a*, *OI* iii. 213–20), showing no sympathy for its objectives. Subsequently he advocated the death penalty for its leaders, though he later paid tribute to the queen regent's clemency.

Historia!'), there are frequent flashbacks (Genette's 'analepses') to happenings in the past: as early as Ferdinand *el deseado*'s return in 1814 and his marriage to María Cristina, with quick references to the African war, Alcolea, Paúl y Angulo and Prim's assassination, the *mantillas blancas* demonstration, and Pavía's *coup d'état*. These have, at times but not always, some functional justification. In the main Galdós is concerned to emphasize the link between past and present developments, but quite often he cannot resist the temptation simply to pass judgements on past occurrences or personalities. This is particularly the case with Isabella II, when Tito abandons for the moment 'la mágica pluma' Mariclío had given him to offer his readers a monologue 'de su propia cosecha'. The ex-queen's defects and *tristes destinos* are once more lyrically evoked in a way which seems hardly necessary—'en tu larga vida de soberana pusiste siempre tu corazón blando sobre tu inteligencia, y abusaste irreflexivamente del poder afectivo y lo extendiste fuera de tu órbita personal, llevándolo a trastornar y corromper la vida del régimen'—as a prologue to delivering her a stinging rebuke for introducing the opera-singer Elena Sanz to her son at Vienna and the subsequent favours she bestowed on her: 'Con la tea del buen querer pegaste fuego al templo del Estado' (III: 1389).[56] Galdós takes sardonic pleasure in juxtaposing this passionate affair with the marriage of convenience contracted with María Cristina of Austria and, subsequently, the birth of Elena Sanz's illegitimate son[57] and the pregnancy of the queen.

The fortunes of another figure essentially from the past, Montpensier, achieve a temporary revival as a result of his daughter's marriage to Alfonso, but this is shattered by her death, and he remains unloved. Once more we have direct, unmediated statements: 'Era el Duque una capacidad administrativa, hombre ordenadísimo, económico, buen esposo, buen padre, y a despecho de estas apreciables dotes, nadie le quería.' His tragedy even seems to be just retribution for his public failings: 'algunos, desviando los hechos del terreno lógico al de las conjeturas supersticiosas, afirmaban que sobre don Antonio de Orleáns pesaba una maldicción: no podía ser feliz en su vida doméstica el que había sido en la pública desleal, ingrato y locamente ambicioso' (III: 1379).[58]

Throughout his work Galdós was concerned with the power of the Church and the frequent failures and very occasional successes of its spiritual leaders.[59] The treatment, in terms of structure, of the theme between the

[56] Beck notes that Isabella was a 'cross borne by Alfonso throughout his brief reign' (*1b*, 44).

[57] Since Elena is nicknamed *Doña Leonor de Guzmán*, from the operatic part she has played in Donizetti's *La Favorite*, she could potentially inaugurate a new Trastámara dynasty. She is treated with notable sympathy by Galdós.

[58] Galdós's comments on Montpensier's wealth, economic and domestic virtues, driving ambition, and ill fortune in his obituary article (10 Feb. 1890; *3a*, *OI* iv. 232–44) are so similar to the above that it is possible that he had the earlier text available when he dictated the *episodio*.

[59] For a rather uncritical account of religious content in the novels see Correa (*3c*, 1962). Ruiz Ramón (*3c*, 127–222) carries out a careful investigation of priests in Galdós, finding a dividing-line in 1891; before that few have a vocation (Nones is an exception), after that date many do; he does not, however, study the anticlericalism of the later *episodios*.

novels and the late *episodios* is very different. In the novels Church and story are bound within society in a strong metonymic pattern. In *Fortunata y Jacinta*, for example, the presentation of the Church and its institutions, severely critical as it was by implication, is carefully interwoven into the fabric of the story (*3c*, Whiston 1979). Doña Guillermina brooks no interference on merely constitutional grounds from the Protestant couple Don Horacio and Doña Malvina when they afford Mauricia help and protection: 'Don Horacio dijo que nones y que haría valer sus derechos luteranos ante el mismo Tribunal Supremo; amoscóse la otra, y doña Malvina sacó el libro de la Constitución, a lo que replicó Guillermina que ella no entendía de constituciones ni de libros de caballerías' (**V**: III. 5. iii, 364). Of course Doña Guillermina gets her way, for the governor, law or no law (and the constitution of 1876 is deliberately vague), supports her against the Protestants.

Similarly, the urban development of the north of Madrid is vitiated by the proliferation of religious constructions of the most undistinguished sort: 'Como por encanto, hemos visto levantarse en aquella zona grandes pelmazos de ladrillo, de dudoso valor arquitectónico' (**V**: II. 5. i, 227). In this environment the forbidding edifice of Las Micaelas stands, its tinny music in keeping with its mediocre architecture. As the walls of the church are completed, all glimpse of the interior is eliminated: its outer darkness matches the obscurity inside. The portrayal of the hirsute and gluttonous Nicolás Rubín as he counsels Fortunata or enters into conflict with Doña Lupe is humorous as well as being entirely negative. All of these developments we see unfolding in metonymic association before our eyes.

In *Cánovas*, by contrast, the ecclesiastical criticism is largely unmediated, unspecific, and unintegrated as well as being more frantic in tone. Segismundo, it is true, is responsible for a not unamusing parody of those ambivalent dispositions of the 1876 constitution regarding religious belief—'Todo ciudadano será *molestado* continuamente en el territorio español por sus opiniones religiosas' (**III**: 1359; my italics)—but then his opinions grow harsher. Like Tito, he foresees no hope of religious tolerance or even of the carrying out of the limited provisions of the law in the immediate future. Any progress towards justice for all will depend, we are told, on a profound change of attitude over a long period of time. Clearly, what is prophecy for Segismundo is hindsight for Galdós. It does not make for good fiction; it would have been more effective and more in accord with fictional structure to have provided some concrete example like that of the Protestant couple.

At the same time despair for the future, which clearly offers no escape for more than a generation, does not entirely eliminate the somewhat desperate Olympian hopes for an eventual breakthrough:

En el propio estado de pérfida legalidad seguiría viviendo nuestra nación año tras año, hasta que otros hombres y otras ideas nos trajeran la política de la verdad y la justicia, gobernando, no para una clase escogida de caballeros y señoras, sino para la familia total que goza y trabaja, triunfa y padece, ríe y llora en este pedazo de tierra feraz y desolada, caliente y frío, alegre y tristísimo que llamamos España. (**III**: 1359–60)

In these violent oxymorons, we are now clearly in a world of political slogans—as well as wishful thinking—rather than considered presentation or interpretation of the past.

Later in the work, in another conversation with Tito, Segismundo is even more categorically negative about the *turno pacífico* and its future results:

—Ni tú ni yo, querido Tito, podemos esperar nada del estado social y político que nos ha traído la dichosa Restauración. Los dos partidos, que se han concordado para turnar pacíficamente en el poder, son dos manadas de hombres que no aspiran más que a pastar en el presupuesto. Carecen de ideales, ningún fin elevado les mueve, no mejorarán en lo más mínimo las condiciones de vida de esta infeliz raza, pobrísima y analfabeta. Pasarán unos tras otros, dejando todo como hoy se halla, y llevarán a España a un estado de consunción que de fijo ha de acabar en muerte. No acometerán ni el problema religioso, ni el económico, ni el educativo; no harán más que burocracia pura, caciquismo, estéril trabajo de recomendaciones, favores a los amigotes, legislar sin ninguna eficacia práctica, y adelante con los farolitos. (**III**: 1396)

The passage should be compared with the description of café life from *Fortunata y Jacinta* quoted earlier. Here we have direct and passionate assertion, instead of an example of the same ideas integrated into the action of the novel.

Segismundo goes on to raise the question of whether a revolution offers a possible solution, giving first a negative answer and then asking Tito for his opinion. Tito's reply makes the same use of pessimistic hindsight:

No creo ni en los revolucionarios de nuevo cuño ni los antidiluvianos, esos que ya chillaban en los años anteriores al 68. La España que aspira a un cambio radical y violento de la política se está quedando, a mi entender, tan anémica como la otra. Han de pasar años, lustros tal vez, quizá medio siglo largo, antes que este Régimen, atacado de tuberculosis étnica, sea sustituido por otro que traiga nueva sangre y nuevos focos de lumbre mental. (**III**: 1396)

Galdós, writing more than thirty years later, in 1912, is well aware that no such change as he desired has taken place; even so he appears to take an excessively gloomy view of a period which, although it has been severely criticized, was hardly as totally inert as he claimed. Moreover, he evidently does not anticipate any radical transformation for some time after that: 'medio siglo largo' would take us, in a proleptic leap, beyond 1925, within reach, curiously enough, of the advent of the Second Republic in 1931. This may be compared with the similar forecasts made in *Los apostólicos* and *Un faccioso más*. His denunciation is all the more vehement as a result of his evident frustration.[60]

[60] For Galdós's extremely dejected state of mind at the time of writing *Cánovas* see Blanquat (*3b*, 1968).

Finally, we see how Fajardo's agreement with this prognostication causes him to adopt 'un filosofismo atrozmente cínico' and to compromise, like so many others, with the status quo: 'Por eso yo he cambiado mi rebeldía por un epicureísmo que me asegure el regalo del presente y el porvenir' (III: 1396). He becomes reconciled with his vulgar, ugly, and ambitious mother, now Countess of Casa Pampliega, who, as a result of her moneylending activities, has become extremely wealthy and, as a natural consequence, an ardent admirer of Cánovas. In order to consolidate her place in the grasping and ill-educated society of the new nobility, Doña Segismunda plunges herself into pious religious practices and arranges a marriage for Segismundo 'con una señorita honesta y de buen ver, hija única de opulento matrimonio, muy notado por su catolicismo a machamartillo y por sus conexiones con toda la gente de Iglesia' (1396), related moreover to extreme traditionalists of the Erro, Socobio, and Emperán ilk. The case of Beramendi and María Ignacia is thus repeated. When we compare Fajardo's savage cynicism with Feijoo's calculated but benevolent tolerance we become aware, not only of the difference between a character from the novels and one from the *episodios*, but of the change in Galdós's perspective in the interval from 1886 to 1912. Tito concludes pessimistically that the development of society following the norms of the Restoration, far from leading to a gradual but continuous improvement, as had been the implied—if sometimes ironic—assumption in earlier works, was in the direction of steadily increasing religious fanaticism: '"Ya veo bien clara la *lenta pero continua* evolución de nuestra bendita sociedad hacia las ollas del ultramontanismo"' (1397).

In the last chapters Galdós's obsessive hatred of the Church becomes almost hysterical, 'bordering on paranoia', as Dendle terms it (*3d*, 1980: 180).[61] Tito comments scathingly on the support given by the ignorant society ladies to the religious orders, including the munificent gift of her property by the Duchess of Pastrana to the Jesuits, for whom Galdós reserves special venom; on the deleterious effects the latter in particular have on education; and on the alarming unawareness on the part of vapid society that they are menaced by an invasion on the scale of those by the Carthaginians, Visigoths, or Arabs. In one of the last chapters Tito and Casiana capture a number of *efémeras*, who report on the swarm of religious orders of all types (more than twenty are named!) which have descended upon all corners of Spain: 'esta plaga de insectos voraces que devastará tu tierra' (III: 1404). The final denunciation of Restoration values comes in Mariclío's last speech, in which she pours scorn on 'los tiempos bobos' which, she reiterates once more, will bring 'años y lustros de atonía, de lenta parálisis, que os llevará a la

[61] It is worth noticing that Pablo Iglesias criticized *Electra* for an excessive concern with religion, which distracted attention from more essential economic issues. (*3b*, Blanquat 1966: 283).

consunción y a la muerte' (1410). The alternating parties have nothing to offer, politically, socially, educationally, or economically:

'No harán nada profundo; no creerán una nación; no remediarán la esteridad de las estepas castellanas y extremeñas; no suavizarán el malestar de las clases proletarias. Fomentarán la artillería antes que las escuelas, las pompas regias antes que las vías comerciales y los menesteres de la grande y la pequeña industria.' (III: 1410)

Once more, the worst denunciation is reserved for the Church: 'Y por último, hijo mío, verás, si vives, que acabarán por poner la enseñanza, la riqueza, el poder civil y hasta la independencia nacional, en manos de lo que llamáis vuestra Santa Madre Iglesia' (III: 1410).[62]

Finally, she advocates abandoning all inhibition in support of revolutionary and obstinately unruly behaviour, which she views as the only hope for the future:

'Alarmante es la palabra Revolución. Pero si no inventáis otra menos aterradora, no tendréis más remedio que usarla los que no queráis morir de la honda caquexia que invade el cansado cuerpo de tu Nación. Declaraos revolucionarios, díscolos si os parece mejor esta palabra, contumaces en la rebeldía. En la situación a que llegaréis andando los años, el ideal revolucionario, la actitud indómita si queréis, constituirán el único síntoma de vida.' (III: 1410)

In this passage Urey reads 'a new irony into her forecast for the future of Spain' and converts the plea into a mere matter of words: 'an ideal, an attitude, or a symptom, not to act' (*3d*, 1989: 203). On the contrary, I see it as a final manifesto of a desperate aspiration, made more frantic because it is not envisaged as an immediate possibility. On this note of despairing frustration the *episodio* comes to an abrupt end, with the bored, somnolent withdrawal of the Muse of History.

The Restoration was important to Galdós as the conclusion of the revolutionary period on which he had set such hopes. It enters into his novels as the event which brings much-appreciated relief from conflict, a period of peace and tranquillity based not on conviction but on expediency, and a vacuous sense of pretensions and complacency. It is at one with the society it exemplifies and which is explored in detail in the fictions. Its clearest consequence is the plethora of undeserved rewards it offers, together with a few examples of manifest injustice, to those who come into its all-embracing fold.

[62] Galdós was no doubt influenced here by the controversies surrounding the Liberal project, from 1906 onwards, for a *Ley de asociaciones* to place mild restrictions on clerical expansion. Galdós himself mentions its inadequacy as a crucial reason for his conversion to Republicanism (*3b*, Fuentes 52). The law, interrupted by Maura's clerically inclined government of 1907–9, and better known as the *Ley del candado*, was finally passed by Canalejas in 1910. Liberals pointed out that this was shutting the stable-door after the horse had bolted: most religious orders had become very well established in the previous twenty years. For a good account of the power of the Church in the early 20th cent. consult Ullman (*1b*, 27–47).

It serves as part of the backcloth of Isidora's tragic pursuit of chimeras and of Jacinta's disillusion. It is a brilliant metaphor for an essential aspect of Juanito Santa Cruz's domestic life.

From an *episodio* one might expect a consideration in depth of the motif forces which go to produce this new stability in society, but unfortunately *Cánovas* does not do full justice to the theme. The circumstances which gave rise to a change in technique have been addressed elsewhere. As a consequence Galdós went on to develop a different brand of fantasy by the creation of allegorical figures, first of secondary personages within a Spanish cultural context: Don Juan Urríes (Don Juan), Don Wifredo (Don Quixote), *Celestina* Tirado—and then embracing the narrator himself. The decision to abandon the technique of the involved focalized narrator—Beramendi, Santiago Ibero, Halconero—who has access to the principal historical figures, represents in himself a distinctive point of view, and moves within a circle of friends of varying persuasions, in favour of a largely allegorical structure with a fixed autonomous narrator, seems to me regrettable. The problem of breathing imaginative life into a period he now saw as essentially sterile and an increasing concern with the immediate present made his task an increasingly difficult one. His bitter hindsight, unfortunately, prevents his undertaking what he was in a unique position to do: to present in fictional form a balanced assessment of the issues which caused the men of the Restoration to act as they did, the different options they had, the reason for their decisions, etc., treating the past as present action and the future as unknowable and speculative. Something of the old qualities, and with them the apt, racy, if strident, language, remains; on occasion the genuine controversies of the time emerge and *la historia grande* and *la historia chica* intertwine. In the case of Cánovas himself, Galdós does succeed to a modest degree in capturing some insight into his motivation. On the whole, however, the new technique encourages the use of whimsy rather than constructive imagination and direct assertion rather than an interchange of opinion: of bald diegesis rather than creative mimesis. Satisfactory interlocking of fiction and history is largely inhibited, and chunks of undigested data alternate with vehement political manifestos. No doubt an increasing realization that he might not be able to fulfil the programme he had outlined in *Amadeo I* impelled Galdós to make a final political statement in the terms we have just seen.

7

The Idealistic Reconstruction of History

FOR the sake of chronological completeness, I have followed through in previous chapters Galdós's examination of contemporary history up to the Restoration in both the contemporary novels and the *episodios* of the fourth and fifth series. I now want to examine in more detail other aspects of what has every appearance of being a prolonged crisis or series of crises concerning the use of historical material.

First we should ask how far such a crisis affected the contemporary novels. Although five of these novels take place during the revolutionary period from the 1860s to 1874, and two of this group also cover the return of the Bourbon dynasty, it is noteworthy that none of the novels set during the Restoration era is closely involved in the history of the period. Peter Bly attributes this at least partly to a change in orientation towards history on the part of Galdós: 'The relative absence of any significant post-1875 coverage is due not only to the relative calm of the Restoration period, but also to the change in Galdós's use of historical matter after 1888' (*3c*, 1983: 182). I am inclined to place more emphasis on the first factor and see the change in subject as reflecting the historical context. The Restoration marked the end of the revolutionary process and the beginning of a new stability. By contrast with what had gone before, the political events of subsequent years had little distinctive resonance and offered little scope for Galdós's preferred analysis of socio-political upheaval and strife. Carlos Blanco Aguinaga explains very well that while in this period 'hechos/fechas de orden político y militar' are lacking, so that 'ya no ocurre nada de significativo más que la especulación y el pasteleo', Galdós's attitude has not become ahistorical since it remains true that 'la visión de la realidad histórica determina en Galdós la estructura misma de sus ficciones' (*3b*, 203). The essential contemporary novels were written, moreover, in the 1880s and 1890s, and culminate in *Misericordia* (1897). There has been, it is true, a lessening of historical concern in the later novels (though not as much, in my view, as Bly claims in the very brief parts II, III, and IV of his 1983 book [*3c*]), but the decisive change appears to come after *Misericordia*, when the orthodox realist pattern, with its recognizable setting and historical specificity, is interrupted. The second batch of *Episodios nacionales* begins to appear just a year later. For the moment, however, a diachronic approach continues to prevail, to the extent that Casalduero, discerning a change in attitude but not in technique, can term the fourth series anachronistic (*3b*, 1961: 160–1).

Yet the problem has always been latent in the historical narratives. From the earliest novels, Galdós had shown an insistent preoccupation with the relation between history and the fictional form he had adopted; we even saw in Chapter 1 a distinct self-reflexive tendency in *La Fontana de Oro*. Nor had he ever been entirely satisfied with the balance he was establishing between fact and fiction. By the fourth series this dissatisfaction had become intense.

Two distinct manifestations of this concern stand out. One is the change in narrative technique in 1910 from *Amadeo I* onwards, which has been discussed in the two previous chapters; I have not disguised my opinion that such a change was unfortunate and its results largely unsuccessful. Up to and including the fourth series Galdós, while very aware of the inherent difficulties of integrating historical concerns into his fiction, none the less regards historical facts as essential components of the genre to be conveyed to his readers for their own sake. Despite the marked change in Galdós's treatment of history, so evident in the final four volumes of the fifth series, I do not think, as explained in the previous chapters, that he modified his historicist approach sufficiently to embrace a fundamentally innovatory technique. Rather, he found himself led by an inappropriate desire for novelty and perhaps constrained by physical and spiritual frustration to attempt ill-conceived and disturbing variants on a well-tried formula. Tito's definition of his task, as quoted in Chapter 5, does not differ greatly from the practice of earlier *episodios*, except in the personification of Clío; the main innovation lies in the figure and actions of Tito himself.

Of the other, earlier expression of dissatisfaction I take an altogether more favourable view. It is, in my opinion, a remarkable manifestation of the rejection of standard history: the construction, within the framework of the later volumes of the fourth series, of the ideal, logical history of Spain penned by Santiuste-*Confusio* and fostered by Beramendi. The experiment does not abandon the concept of historical truth as such, but simply 'logically', poetically, juggles the facts. Not this time so radical as to interfere with the narrative structure, it consists essentially of a well-developed and significant subplot offering an alternative view, suggested to but not imposed on the reader, a justified play of fantasy (a permanent impulse in Galdós, as we have seen) which gives interesting results.

Confusio's modification of history is not entirely without precedent. Well before formulating this fully fledged alternative history, Galdós had begun to demonstrate the degree of his dissatisfaction. Essentially, these hesitations are focused on the fascinating figure of Pepe García Fajardo (Beramendi) from the very beginning of the fourth series, commenced in 1902.

The supposed author of the most sustained and ambitious first-person narration within the series, Fajardo consciously addresses his *Memorias* to posterity. From the outset he reveals a preoccupation with history and sets out to provide a record of 'los públicos acaecimientos y los privados casos' (**II**: 1413). His definition of history emphasizes continuity and the need to record the unrecorded, the *intrahistoria*, modestly but efficaciously designated as the almost imperceptible trace a snail leaves on a rock: 'los caracoles, que van soltando sobre las piedras un hilo de baba, con que imprimir su lento andar' (1428). Beramendi is conscious of this humble task, which none the less adds its mite to human progress and will have a cumulative effect in time. It corresponds to a deep-seated conviction of Galdós's—at least until the last *episodios*—that, however adverse the present and immediate future, some advance is attainable through the private efforts of well-intentioned individuals.

We have already seen in Chapter 2 Fajardo's major function as a committed reflector and commentator on Spanish politics. What needs to be emphasized here is that his literary activities are closely integrated into his political role. Thus the fact that he has a written account of his stay in Italy—his *Memorias*, which are also *Confesiones*—is skilfully used as a vehicle through which he is inserted into Madrid high society: stolen by his sister-in-law Sofía, they circulate among prominent individuals, such as Eufrasia Carrasco, who pretends to be his Italian lover Barberina[1] at a masked ball, Eugenia de Montijo, later to be Napoleon's III's empress, and Fajardo's sister, the nun Catalina. The result is that, when he is presented at court in *Narváez*, significantly by Sister Patrocinio's brother Juan Quiroga, everyone has read his missing *Memorias*, which have passed surreptitiously from La Latina convent to the palace.

Now that he is accepted, then, as a potential historian, the question arises as to what sort of history he should write. It is highly ironic that, through a rumour spread by Pepe's loving and extraordinarily naïve mother Librada, he comes to be regarded as the potential author of a *Historia del papado*, a perilous subject if ever there was one, especially for a nonconformist spirit like Fajardo's. A less serious subject is put forward at his first brief meeting with the queen, when she makes the not very welcome suggestion, typical of her sensuous and frivolous mentality, that he should write about the scandalous life of Lola Montes and her relations with Ludwig I of Bavaria. Subsequently, he is encouraged by the king-consort to undertake a history of the nineteenth century. Don Francisco's idea is of 'Una Historia imparcial, que se aparte del criterio extremado de las facciones, una relación verídica, escrita con talento, revisada por personas peritas y autorizada por la Iglesia, crea usted que sería una gran cosa. Y la publicación de esta obra no faltará quien la patrocine' (**III**: 1612). It need hardly be pointed out that such academic revision, ecclesiastical censorship, and official patronage would ensure a work of complete orthodoxy.

[1] The story is skilfully recounted by means of intertextual allusions to two famous literary affairs: Rousseau (Mme Warens) and Dante (Paolo and Francesca).

Then the possibility is raised of a history of the queen's reign and the difficulties it would entail: whether the events are too recent (the king's opinion), whether it was liable to be dull (the king's view, once more—this time highly ironic), and finally whether—Beramendi's contribution, reflecting Galdós's hindsight—it would inevitably have sad connotations as well as beautiful ones.

Shortly afterwards, Beramendi separates himself categorically from official history, which Isabella has described as 'una matrona gallardísima'. In his view, 'Ésta es la Historia oficial, académica y mentirosa. La que merece ser escrita es la del ser español, la del alma española, en la cual van confundidos Pueblo y Corona, súbditos y Reyes' (III: 1612). The queen likes this identification, and they indulge in a short dialogue, in which Isabella describes Spain as 'País grandioso el nuestro, pero empolvado'. Afterwards, Beramendi comes to doubt whether this conversation ever took place. He falls into a confused meditation ('Mi cerebro era una linterna mágica'), in which 'Isabel, vestida de manola, me decía que escribiese su historia; Lucila callaba siempre, imagen y representación del inmenso enigma' (1613). The queen is identified with the people; the symbol of the race, Lucila, remains enigmatically silent. The dividing-line between fact—even fictionalized fact—and fantasy has become more blurred than ever, and we are confronted by a further example of Fajardo's mental instability and imaginative perception. It is now time to address this issue.

Fajardo's deep sense of frustration, at once personal and political, has been centred on his search for Lucila Ansúrez, which has become a maniacal obsession. Lucila represents for him the ideal of perfection to be found in *el pueblo*, with whom she is identified: 'Amo a Lucila porque amo al pueblo; estos dos amores no son más que uno' (III: 1586). He then concludes that this *mal de Lucila* is what he calls *una efusión estética*, which requires an artistic expression, such as his *Memorias* or, possibly, a formal history. He next rectifies: it is not an *efusión estética*, but an *efusión popular* (1592), identification with the people. His meeting with the queen marks the next stage, in which Isabella and Lucila come together in his mind: the perfection he is seeking in Lucila should—but evidently does not—find its fulfilment in the queen, who in turn should, and potentially does, embody the essential qualities of *el pueblo*: 'pensé que Su Majestad y yo nos parecemos: padece *la efusión popular*' (1611). The meeting is followed by another of his frequent attacks of mental disorder. This time he refines the problem once more: 'eso que padezco, y que me ataca de vez en cuando, la *efusión* . . . ¿de qué?, la *efusión de lo ideal*, de lo que, debiendo existir, no existe' (1613). This unattained and indeed unattainable ideal further crystallizes in the idea of writing history: 'El ideal de esa historia me fascina, me atrae . . . pero ¿cómo apoderarme de él? Por eso estoy enfermo: mi mal es la perfecta conciencia de una misión, llámala aptitud, que no puedo cumplir' (1628). Thus the aesthetic (writing history), the popular (the identification of Lucila and the queen), and the ideal (history as it ought to be) converge. This mission is never completely fulfilled, of course, and Fajardo's anguished idealism goes on unabated. It causes the solicitous and practical María Ignacia[2] to burn the next instalment of the *Memorias*, after the manner of the housekeeper and the niece in *Don Quixote* (III: 12): hence the change of narration in *Los duendes de la camarilla*. When the *Memorias* do resume, in *La revolución de julio*, the effects of Merino's execution are so powerful that María Ignacia first insists that he eliminate everything dramatic, tragic, or concerned with low life (18), and finally, after the traumatic effect of the revolution and of his killing of Gracián, forbids him to continue at all (115).[3] Although this is therefore the last instalment of his memoirs, the process of idealizing history is not interrupted. While the deep dissatisfaction he expresses to Manuel Tarfe[4] remains, his unrealized aspirations are precisely those which are to be captured by Santiuste. For this reason he involves himself so closely, as employer, mentor, and friend, in the former's ideal history. *Confusio*'s mission, not Beramendi's, 'es vivir, ver tierras, pueblos y humanidad próxima y lejana; probar todas las pasiones, sufrir todos los infortunios y gustar alegrías inefables'. Beramendi, tied to his idle and ambiguous lifestyle, envies him: 'pobre, aventurero, hijo de tus obras, soberanamente' (382).

Another important indication of a new speculative direction occurs at the end of Galdós's interviews with the aged ex-queen held in 1902 and published in 1904. At this time Galdós was of course in the middle of writing the fourth series; he had just completed *Narváez* and was about to begin *Los duendes de la camarilla*. After recording his conversations with Doña Isabella, the novelist allows himself to indulge in fantasies about what might have happened during her reign: 'Reformaba yo la Historia, y hacía del reinado de Isabel, con la misma Isabel, no con otra, un reinado de bienandanzas' (VI: 1194). The essential condition for this historical revision is that she should freely marry a Prince X, one of the most suitable European princes, instead of the 'enorme desconcierto' of her marriage to Francisco de Asís, imposed upon her, as has been discussed in Chapter 2, by the diplomatic schemes of France and England. Equally essential is an able and dedicated minister,[5] and for this purpose Galdós turns to Cánovas, whom he endows in the 1840s with the

[2] It is evident that María Ignacia represents a counterbalance to Beramendi, at the same time supporting and moderating his literary and historical ambitions, and confronting his *efusiones* with pragmatic common sense.

[3] All this occurs between *Las tormentas* (1902) and *La revolución de julio* (1903–4), before the introduction of Santiuste in *O'Donnell* (1904).

[4] '¿No sabes que me revuelvo en la vulgaridad yo, poseedor de todos los bienes materiales sin haberlos ganado por mí mismo? ¿Sabes que sufro un inmenso mal, la conciencia de no haber hecho en el mundo nada bello ni grande, nada que me diferencie del común de los hombres de mi tiempo?' (III: 569).

[5] Possibly Galdós was thinking of the contemporary case of Queen Victoria, who was able to rely on Lord Melbourne's steady support at the beginning of her reign and who found a highly suitable and proper husband in Prince Albert.

mature wisdom he has acquired thirty years later: 'Este estadista ideal, que he llamado Cánovas porque los talentos y el rigor de este hombre de nuestro tiempo parécenme los más adecuados para inaugurar en aquéllos un reinado eficaz.' Other choices for prime minister in this alternative history, Prim and Sagasta, are better held in reserve for later tasks.

The choice of these three is significant. Cánovas always commanded his respect, if not his allegiance or affection.[6] His attachment to Prim, as we have seen, was enduring, and he retained much of his admiration for his old political chief Sagasta. Such a leader might have avoided what Galdós enumerates as the fundamental defects of the epoch: clerical intrigue, unrestrained self-seeking, and *caciquismo*. Evincing the considerable personal sympathy he has for the queen, he even ends up, a bit fancifully no doubt, by dreaming of her generous qualities as a 'gran revolucionaria inconsciente' being channelled towards justice and social equality.

The suggested transformation of Isabella II's reign reflects Galdós's deep concern with that troublesome era and with the personality of the queen, as well as the problems of historiography itself. The direct contact with Isabella, at a time when he was actively engaged in analysing her political acts of omission and commission, evidently stimulated him to speculate on other historical possibilities, poetically satisfying if not historically accurate. This propensity, so tentatively put forward in 1904, finds its full expression some two years later in *Confusio*'s *Historia lógico-natural*.

Let us start, however, by tracing the spiritual development of Juan Santiuste. He first appears in *O'Donnell*, where Teresa Villaescusa comes across him in incredible squalor supporting the miserably poor family of Jerónima Sánchez. From the beginning he is characterized by extreme idealism, shown in his shame at his poverty ('Mis desgracias me han inspirado el horror al aseo', **III**: 195), his unconditional adoration of Teresa, and his scrupulous honesty, when he returns a gold coin she has inadvertently thrown him. As Teresa observes, 'Juan tiene mucho talento y ve las cosas desde lo alto, desde lo muy alto' (226). As has been discussed, his simple idealism inspires a deep love in the generous Teresa, but she is not heroic enough, or she is too realistic, to grasp it.

[6] Berkowitz sums up his attitude as follows: 'he admired his political skill and unadorned eloquence. Disapproving as he was of Cánovas' extreme conservatism, he nevertheless recognized his statesmanship' (*3b*, 1948: 205). In the 1870s they had corresponded about articles to be published in the *Revista de España*. In 1880 Cánovas, as prime minister, did not support the vigorous request Galdós had made for the appointment of Benito's brother Ignacio as military governor of the Canary Islands (*3b*, Armas Ayala 1989: 318). He opposed Galdós's entry into the Royal Academy in 1883 and again in 1889 (*3b*, Berkowitz 1948: 162, 228–9). (Galdós obtained his seat, unanimously, later that year.) Cánovas did, however, attend the banquet in the novelist's honour in 1883 (*3b*, Berkowitz 170). In a letter to Galdós of 17 Jan. 1889 (*3a*, Casa-Museo) on the Academy election Castelar reports that 'Alguno de los dudosos me parece para V. segurísimo. Ya le diré quién. Pero le temo al monstruo' (i.e. Cánovas).

Connected by upbringing with a wide area of Spanish national territories,[7] Santiuste studied for three years at the university and was passionately devoted to reading. Thanks to Teresa's support, he becomes apprenticed, as a gunsmith, to Leoncio Ansúrez, and by the beginning of *Aita Tettauen* has secured, thanks to Teresa and Beramendi, a comfortable government post. In the African war he shares an exalted patriotic sentiment with Jerónimo Ansúrez, Bruno Carrasco, and the frail Vicente Halconero, to which is added the childish love of soldiers shown by young Vicente. While the excitement is too much for Halconero senior, who dies of his emotion, in Santiuste's case his patriotic exaltation takes on a bombastic expression suitable for imperialist jingoism. It accounts for his devout admiration for Castelar's oratory, and this respect in turn provides an occasion to pay incidental tribute to the patriarch of conservative Republicanism and to his staunch democratic principles.

In battle, however, Santiuste's simple heroism gradually diminishes: dead Moors, he finds, are very like dead Christians, even though their bodies are treated with less respect (III: 255). He begins to conceive doubts about the 'Dios de las batallas' (256) who is expected to favour the Spaniards; if there is a providence for Christians, it is reasonable to expect the Muslims to have one too. Finally, he is horrified by his visit to a field hospital,[8] overflowing with the dead, the maimed, and the dying (260). In his arguments with Alarcón, as we have seen, he becomes the voice of pacifist sentiments which have a distinctively Tolstoyan ring to them (*3d*, Collins; *3d*, Gilman 1981). In the midst of war Galdós is concerned to give expression, through an unstable, highly imaginative figure like Santiuste,[9] to an ideal of peace which, however unattainable in the short term, must not forgotten: 'La paz es mi sola idea, El Nasiry; la paz es mi aliento. Odio la guerra, y deseo que todos los pueblos vivan en perpetua concordia, con amplia libertad de sus costumbres y de sus religiones' (325). Santiuste also experiences the differing moral values of Christian, Muslim, and Jew, and takes a critical attitude (just as Jerónimo Ansúrez had with fashion, 239)[10] towards certain Christian characteristics—formal dress, for instance—while exalting the common humanity which unites the three races; the similarity of African Muslim and Christian Spaniard is emphasized, first by the venerable Ansúrez (232) and then in the description

[7] He was born in Havana, with a father from Burgos and a mother from Andalusia. He spent his childhood in Alicante, and then resided in Chiclana and Cadiz before moving to Madrid. Such diversity of regional experience is typical of many of Galdós's key figures: Jerónimo Ansúrez's wives, Tito, etc.

[8] Cf. Nikolai Rostov's disgust on visiting an improvised military hospital in similar circumstances in *War and Peace* (v, ch. CII).

[9] He describes himself in *Carlos VII, en la Rápita* as 'tres partes galán enamoradizo y un cuartillo de poeta' (III: 341). For Galdós unselfish sexual love is a key to understanding and compassion.

[10] It is pertinent to recall that Juanito Santa Cruz, so representative of the worst features of his class and society, is associated with fashion and the seasons (V: I, 8. i, 84).

of Alarcón as 'un perfecto agareno' (265).[11] The means by which Santiuste attains this experience is significant: it is through sexual love, raised to the level of forming part of God's purpose: 'la paz no sería buena y fecunda sin el amor, que es el aumento de las generaciones y la continuación de la obra divina. Dios no dijo "Menguad y dividíos", sino "Creced y multiplicaos"' (325). This command Santiuste carries out in his erotic relationships, real or merely aspired to, with Yohar (Jewish), Erhimo (Muslim), and Donata (Christian): he is thus an early and enthusiastic exponent of 'Make love, not war.' It is an example of Galdós's belief in the liberating effect of sexual freedom, which becomes increasingly conspicuous as the *episodios* advance. We find a reiterated questioning of the celibacy of the clergy (263, 283, 330), the most extreme case being the elderly priest Don Hilario de la Peña, who, sponsored by Queen María Victoria and finally offered a bishopric during the Republic, is determined to abolish clerical celibacy (III: 1141) and to preach the same message of *crescite et multiplicamini*.[12] Accordingly, Santiuste is assigned three different names by the three races he encounters: Yahia (John the Baptist) by the Jews, Djinn by the Muslims, and various nicknames by his Spanish friends (Profetángano, Don Bíblico), which finally resolve into the highly significant nickname *Confusio* (III: 341), first given him by the army chaplain Don Toro Godo. The name itself has a not altogether negative quality; it connotes not only 'confusion' but also 'mixing together'. Throughout this period he corresponds with his employer and patron Beramendi, whose aspirations and frustrations, as we have indicated, he largely shares.

Santiuste's transformation occurs in Chapter 7 of *Prim*, the ninth *Episodio nacional* of the fourth series, which dates from 1906. At Beramendi's dinner-table an anonymous figure is introduced who bears all the marks of being a new character; he is described as being elderly before his time, jerky in movement, wizened and emaciated, with nothing to say for himself.[13] Only after he has left does Beramendi explain that the mysterious figure is Juan Santiuste, who has been completely transformed as a result of a severe illness. Beramendi describes how his personality and interests have changed profoundly; his deep patriotism remains, but he is no longer interested in women and his real name has been finally superseded by his nickname.

De su ser anterior y del desplome de su entendimiento y de su memoria no restan más que el sentimiento patrio y una idea, una sola idea y propósito: escribir la Historia de España, no como es, sino como debiera ser, singular manía que demuestra el brote de

[11] Another example is the parallel established between El Nasiry's harem and the 'nieces' who constitute the household of Don Juan Hondón, 'arcipreste, patriarca y califa' (III: 399).

[12] There is a resemblance here with Feijoo ('El amor es la reclamación de la especie, que quiere perpetuarse', V: III. 4. v, 339). It is curious to observe two men who, like Galdós, have officially forgone matrimony and offspring advocating the idea of multiplication.

[13] Rodríguez greatly exaggerates when he speaks of 'a form of schizoid regression, almost catatonic at times [which] tends to dissociate him from the real world' (*3d*, 154–5).

un cerebro brutalmente paradójico y humorístico. Como entiendo que la ociosidad ha de perjudicarle, en vez de combatir esa manía, le estimulo para que trabaje en eso que él llama *Historia lógico-natural de los españoles de ambos mundos en el siglo XIX*... El hombre lo ha tomado con ahinco, y cuanto más trabaja, más se afianza en la fortaleza de su ser nuevo, y más aguza las dotes paradójicas y *lógico-naturales* que le han salido ahora... Cada dos o tres días despacha un capítulo, que me lee antes de ponerlo en limpio. En su estilo no se advierte ninguna extravagancia; en la narración de los hechos está lo verdaderamente anormal y graciosamente vesánico, porque *Confusio* no escribe la Historia, sino que la inventa, la compone con arreglo a lógica, dentro del principio de que los sucesos son como deben ser. (**III**: 563)

It is evident that by the summer of 1906 Galdós had decided to give a new slant to a character who had already served him for a number of significant purposes. We have seen how Teresa Villaescusa, in the same volume, changes in a similar manner after a ravaging and near fatal attack of typhoid. Such considerable changes may seem forced or inconsistent. Yet this evolution is by no means arbitrary, any more than Teresa's is: the man characterized from his first appearance by extreme idealism bordering on derangement; who has suffered heroic voluntary poverty; who has passed from blind patriotism to pacifism; who contains a strong charge of pent-up sensuality now diverted into intellectual channels; who has witnessed and appreciated the diverse perspectives of opposing races—this man possesses the appropriate agonizing range of experience—missing, to his chagrin, from Beramendi's life—to sublimate his eroticism and to feel irresistibly compelled to produce an idealistic version of Spain's history.

The main advocate of *Confusio*'s scheme, Beramendi, treats it publicly with apparent irony, but in fact lends it constant moral support and guidance. While Tarfe wonders whether the constant discussions with *Confusio* do not have an adverse effect on Beramendi's health, his wife clearly realizes that *Confusio* has great therapeutic value ('es para él el oxígeno', **III**: 570) for the marquis, out of tune with aristocratic society, frustrated by the political events of the time, and prone to periodical brainstorms. By this wifely solicitude, Beramendi is saved from the mental turmoil which overwhelms his protégé.

Confusio himself speaks of his project to Santiago Ibero in the most extravagant idealistic terms:

Mi historia no es la verdad pedestre, sino la verdad noble, la que el principio divino engendra en el seno de la Lógica humana. Yo escribo para el Universo, para los espíritus elevados en quienes mora el pensamiento total. Yo abandono el ambiente putrefacto que nos rodea; saco mis pies de este lodo de los hechos menudos, y subo, señor mío, subo hasta que mis oídos pierden el murmullo terrestre, y mis ojos el falso brillo de las mentiras barnizadas de verdades. Yo subo, señor, y arriba escribo la Historia lógica, y pinto la vida ideal. (**III**: 637–8)

Elsewhere he claims that his *History*, as well as being logical, is also valid aesthetically, in accordance with Beramendi's aspirations. It is not surprising, therefore, that it should be associated with music: 'En la vaguedad de su solitario pensamiento, relacionaba el soñador Beramendi la música de Beethoven y Mozart con la *Historia lógico-natural* del eminente compositor *Confusio*, y descubría entre uno y otro arte semejanzas notorias, que saltaba a la imaginación y al oído' (III: 587).

Evidently we cannot take at their face value these exalted pretensions to aesthetic universality and contempt for 'el ambiente putrefacto' and the 'hechos menudos' of life. Santiuste is constantly treated as somewhat deranged and even crazy,[14] and, like Maxi Rubín, he ends up, still writing his *History of Spain*, in the mental asylum of Leganés (*España trágica* III: 915). Yet the exaggeration with which his ideas are expressed should not deceive us into dismissing them as simply pointless or absurd. It seems clear to me that some of the major questions posed by contemporary Spanish history are here given, in a deliberately whimsical vein, a coherent and sustained airing which we should take seriously. Although *Confusio* in his title claims to encompass *los españoles de ambos mundos*, no attention is in fact given to the New World.

The reordering of history affects not only the whole of Isabella's life, but also the latter part of the reign of Ferdinand VII. At the French invasion of 1823—the intervention of the 'Cien Mil Hijos de San Luis'—the Cortes decide, according to *Confusio*, not just to declare the king incapable of acting, but to put him on trial (in the *episodio* there is much desultory talk of cutting off his head). They proceed to sentence him to death; after some discussion about hanging or decapitation as the most suitable method of execution, it is determined that he shall be shot.[15] Thus it is Ferdinand and not Riego, together with so many others, who pays for his political failures with his life. From the occasional references to Cromwell it is clear that the English civil war is echoed as a precedent. The parallels are not only between two successful strong men (Cromwell/Prim) but encompass an eventual failure of their causes and a subsequent restoration. Another war of independence ensues, with a reversal of the situation which applied in 1808: the supporters of the French (*afrancesados*) are now dubbed the absolutists and the constitutionalists the patriots. With the elimination of the last ten years of Ferdinand's reign, there is no place for María Cristina or the Carlist war. In fact *Confusio*, at Beramendi's prompting, later remembers Don Carlos and has him sent to the firing-squad with his brother. Later *Confusio*, in conversation with Ibero, doubts of his wisdom in killing Ferdinand, whose shadow continues to haunt the political scene.

[14] Compare the attitudes to him shown by Halconero and Ibero in *La de los tristes destinos*: 'Tuvo mucho entendimiento, y ahora está trastornado' (Halconero at III: 669); 'para mí, es más profeta que loco y más sabio que poeta' (Ibero at 670).

[15] Casalduero (*3b*, 1961: 158) notes this as an example of violent, unequivocal action against evil, comparable with Casandra's killing of Doña Juana or Bárbara's killing of Lotario (*Bárbara*, 1905).

As a consequence of the shortening of the reign, Isabella necessarily has to be born some years earlier and is turned into the daughter of Ferdinand's second wife Isabel of Braganza,[16] during whose short period as queen (1816–18) some hopes for an alleviation of Ferdinand's despotism were conceived.[17] It thus became natural for the infant queen to be brought up in Portugal under the tutelage of the Braganzas[18] during the civil war. The regency which around 1830 exercises sovereignty during her minority has an admirably broad ideological base: it consists of the Liberal Mendizábal, architect of the disentailing of the Church lands, Francisco Javier Istúriz, his erstwhile colleague and subsequent adversary,[19] and the flamboyant, popular, and idealistic Carlist leader Tomás Zumalacárregui.[20]

Eventually a Federation of Hispanic States, 'la más grandiosa fiesta de concordia, de paz y alegría que han visto las generaciones', emerges at the end of a five-year civil war, in which 'pelearon los antiguos reinos, quedando al fin condensados en las dos grandes síntesis históricas de Aragón y Castilla' (III: 575). So, to Isabella of Castile corresponds (obviously!) a Ferdinand of Aragon, nicknamed 'el Pilarón'.[21] The ideal and the real clash when *Confusio* meets Ibero, who is staying in the same boarding-house, for *Confusio* sees in Galdós's principal representative of the virtues of the race a striking resemblance to *El Pilarón*, to such an extent that he actually takes him for his created figure. Aragon is seen as contributing its distinctive *Derecho público*

[16] Isabel of Braganza had a daughter, also called Isabel, in 1817, who died in infancy (*4a*, Mesonero Romanos 169). The queen died in childbirth the next year. She was the sister of both of the pretender Don Carlos's wives, Francisco and María Teresa (the Princess of Beira).

[17] See e.g. Mesonero Romanos (*4a*, 165–71) and Butler Clarke (*1b*, 42), who comments that 'the death of his second wife, a woman of lofty views and lovable character, removed one of the few good influences about the king... The liberals founded great hopes upon her gentle influence, but it waned before her death.'

[18] The destinies of the two kingdoms were very much intertwined since Dom Miguel, the Conservative pretender to the Portuguese throne, was closely associated with Don Carlos. *Confusio*'s historical modifications therefore implied a readjustment of Portuguese history also.

[19] Mendizábal and Istúriz fought a duel in 1836, which Galdós described in *De Oñate a la Granja* as a Romantic outcome of the efforts of the newly constituted Conservatives (*moderados*) to oust Mendizábal (II: 580). As an unconditional supporter of María Cristina, Istúriz had moved to the right, and succeeded Mendizábal as prime minister that same year.

[20] Galdós's portrayal of Zumalacárregui, in the first volume of the third series, is largely sympathetic, though not perhaps to the extent of making him 'el santo mártir de la causa, su apóstol', as Avalle-Arce argues (*3d*, 373).

[21] This scheme is a remote echo of Maroto's plan, explained to Calpena as Espartero's delegate, to marry Isabel to Carlos, the son of the Carlist pretender: 'Como los Reyes Católicos, mancomunadamente, firmando juntos, pues si en aquel matrimonio se casó Aragón con Castilla, en éste se casan y conciertan dos ramas igualmente legítimas, para bien de la Nación y para establecer una paz duradera' (*Vergara* II: 1082). The example of William and Mary of Great Britain is also cited. Maroto's project was revived, to Beramendi's indignation, in a different form in 1860: 'El hijo de Montemolín se casará con la infanta Isabel, y subirá al Trono cúando cumpla veinticinco años... Isabel y Carlos reinarán juntos con igual derecho mayestático, y se titularán Segundos Reyes Católicos' (*Carlos VII, en la Rápita* III: 386). Such unachieved possibilities strengthen the verisimilitude of *Confusio*'s imagined history. In a similar vein, Galdós combines Espartero and Maroto in a character called Baldomero Maroto in *El abuelo*.

(the influence of Joaquín Costa is discernible here). The essential difference from the time of the Catholic Monarchs is that now the sovereignty of the country lies firmly in the hands of *la nación*.

The contemporary struggle depicted in *Prim* between the supporters of the Catalan general, now determined opponents of the dynasty, and those like Serrano who still uphold the queen is seen in an optimistic light. Thus, for *Confusio*, even while Prim is conspiring in Valencia, armed conflict has ceased to be an option: 'La palabra *pronunciamiento* sólo figuraba ya en el Diccionario como arcaísmo, a disposición de los pedantes' (III: 588). The fascinating narrative ends with a significant pseudo-dramatic dialogue, giving the immediacy of the present spoken word to the subject. *Confusio* and Teresa represent respectively the same encounter between an idealistic and a realistic attitude to history. For Santiuste, the San Gil insurrection is the 'campo de una batalla ideal, tan ganada por los vencedores como por los vencidos'; for Teresa, a realist and an ardent admirer of Prim's revolutionary aspirations, it is a bloody skirmish in which Ibero is at risk. Accordingly, she refutes *Confusio*'s denial of the fact of death: 'Muertos hay. Tú no has visto bien, o con tu imaginación enferma trabucas las formas reales' (652). At the very time when the savage retribution on the sergeants of San Gil described in Chapter 2 is about to be exacted, *Confusio* refuses to acknowledge the existence of vanquished or dead; he asserts that he has seen Prim among the soldiers and that 'Serrano y Prim se abrazaron.' In contrast with Teresa's 'sinful' eyes, *Confusio* claims that his vision penetrates beneath the surface: 'Veo los muertos vivos, los enemigos reconciliados, el Altar y el Trono llevados a la carpintería para que los compongan, la historia de España escrita por los orates' (653). A madman's perception of Spain, implies Galdós, could not be less real than the crazy slaughter that actually occurred.

Confusio's rectification of history is not confined to the past; as well as commenting on contemporary events as they happen, it projects his vision from the 'story time' into the future, reaching at times even beyond 'discourse time', the time of composition of the novel itself, into the real future. In *La de los tristes destinos* the theme of the eventual reconciliation between the warring factions represented by the political generals is carried forward into a time when 'los hijos de O'Donnell se abrazarán con los de Prim' (III: 670). In a mixture of prophecy and optimistic projection, *Confusio* foresees Isabella's abdication in Alfonso's favour and prophesies a long and glorious reign for Alfonso XII.

Yet reality impinges rather incongruously on the ideal history. Beramendi explains how extremely deficient Alfonso's education is, so dominated by ignorant and bigoted clerics and generals.[22] As a consequence *Confusio*,

[22] Beramendi's son Tinito, a playmate of Alfonso's, puts it plainly: 'Alfonso no sabe nada. No le enseñan más que religión y armas' (III: 693).

insisting on his obligation to produce a logical history, concedes the necessity for a revolution as 'cirugía política' (another reminiscence of Costa). Alfonso thus needs, logically, to emigrate. This exile in turn produces a vacuum, an interlude to be filled by a Cromwell-like solution, perhaps Prim, perhaps a Republic. However, for *Confusio* the revolution, far from being a new and inspiring start, is merely a cosmetic operation:

> La elegante revolución . . . es un lindo andamiaje para revocar el edificio y darle una mano de pintura exterior. Era de color algo sucio, y ahora es de un color algo limpio, pero se ensuciará en breves años . . . Luego se armará otro andamiaje . . . llámelo usted República, llámelo Monarquía restaurada. Total: revoco, raspado de la vieja costra, nuevo empaste con yeso de lo más fino, y encima pintura verde o rosa . . . (III: 780)

Such retouching will of course not last for ever, and *Confusio* foresees an eventual complete rebuilding carried out pacifically and on the same foundations, at some unspecified date in the future:

> Y el edificio, cuanto más viejo, más pintado. Pasarán años, y aquí estoy yo para derribarlo antes de que se desplome y aplaste a todos los que estamos dentro. Sobre las ruinas armaré yo el gran andamiaje *lógico-natural* para edificar de nueva planta sobre el basamento, que es del mejor granito . . . (III: 780)

With the new coat of paint furnished by the revolution, the reign of Alfonso XII can, in *Confusio*'s conception, resume and continue, happily and prosperously, until 1925.

Confusio's rewriting of history contains elements of firmness and decision, like the shooting of Ferdinand VII, which fitted the aspirations of the more militant later Galdós (*3b*, Fuentes 70). At the same time it postulates a genuine and pacific reconciliation between authentic political forces, not as a cynical deal with sinecures for everyone, as occurred with the Liberal Union or the Restoration. This compromise and reconciliation, and the long-term prospects of complete renewal, are in stark contrast with what is presented in *Cánovas*. In this *episodio* such hopes are specifically rejected by Santiago Ibero—'Cada cual obedece a sus propias revoluciones. Yo no tengo que poner los andamiajes de que habla *Confusio* para revocar un viejo caserón. Mi casa es una choza nueva y linda. En ella tengo mi trono y mi altar' (III: 781)—using the traditional terms applied to the clerically orientated monarchy. Representing an alternative view, Ibero and the transformed Teresa decide in another final dialogue, as we have seen, to go into exile, like Isabella, but in their case in search of freedom and independence from the dead weight and the prejudices of the past.

Confusio's tentative and incomplete *Historia lógico-natural* and its creator have received little support from the relatively few critics who have commented on them. Alfred Rodríguez treats Santiuste as a psychotic case history, comparable to Maxi Rubín, but does not refer to the *Historia*.[23] For Brian Dendle:

The case of Santiuste sadly illustrates the fate of those who lose touch with reality... Real history, that of Spain's sordid political manipulations, is discussed by *Confusio* (and also in part by the unstable Fajardo) as *Historia ilógica y artificial*. Rather than study Spanish life with all its imperfections, *Confusio* takes refuge in an ideal universe of the spirit... *Confusio*'s impracticality, verbalism, and blindness to the real world represent in extreme form national defects which, for Galdós, a reconstructed Spain must shun. (*3d*, 1980: 137)

Antonio Regalado García sees him in a somewhat more positive light, relating him to the goals of regeneration Galdós sought in 1906, but overlooks the visionary side of his personality:

Santiuste es, pues, como su mecenas el Marqués de Beramendi, al que complementa en su función de historiador, un intelectual saturado, aún más que aquél, de ideas imprecisas de reforma, salidas del espíritu regeneracional del 1900, que aplica retrospectivamente a la historia de España. La tendencia revisionista de la historia es una de las preocupaciones fundamentales de los pensadores españoles de principios de siglo. Santiuste, a pesar de su personalidad extremista y desequilibrada, mantiene con decoro el carácter de héroe de la novela histórica galdosiana. (*3d*, 375)

For his part, Montesinos glimpses a more specific purpose:

Hay no poco método en esta locura—por parte de Galdós, se entiende. Como éste quiere recalcar constantemente, sobre todo en sus últimos episodios, lo absurdo de la historia moderna de España, las locuras de *Confusio* pueden ser en ocasiones una especie de discreto contraste; algunos de los disparates realmente ocurridos pueden ser así reparados—tal vez lo serán de hecho—y haciéndonos ver lo que por fuerza ocurrirá, una profecía de *Confusio* subraya la torpeza del error pasado. (*3c*, iii. 130)

Claire-Nicolle Robin notes, rightly, that *Confusio* 'échappe aux classifications habituelles' but considers only his prophecies for the future, judging them to be

une manière de prouver les lois de l'histoire d'un point de vue romanesque. N'ayant besoin de rien, contrairement aux gens qui l'entourent, *Confusio* peut considérer les lois qui régissent l'enchaînement des faits sans les modifier en fonction de ses besoins personnels, comme le font ses contemporains... *Confusio* annonce tout simplement... le renversement des institutions en vigueur et le rôle de l'historien dans ce renversement: rôle à la fois d'avertisseur et de conseiller technique. (*3d*, 243–4)

Once one takes into account *Confusio*'s obviously unrealistic rectification of past history, however, the whole of the *Historia lógico-natural* takes on a much

[23] 'He is quite unable to adjust and is gradually transformed into the pathetic "Confusio", a pejorative appellation that underscores the pseudo-religious direction of his psychotic retreat from unbearable reality'; 'the most thorough of Galdós' psychopathological creations. Like Maximiliano Rubín' (*3d*, 146, 155).

more idealistic hue. It cannot in my opinion be construed as an unequivocal forecast of revolution; it offers, rather, a tenuous hope of an eventual rebuilding of the structure of Spanish political life by steady communal effort and constructive consensus.

Hans Hinterhäuser, after perceptively linking *Confusio* with the meditations upon the queen's reign published in 1904, goes on to make a largely valid comparison between the idealism displayed by Galdós and that of the Generation of '98, the former precise, concrete, rationalistic, the latter lyrical, metaphysical, irrationalist. He concludes that Galdós 'huyendo de la insuficiencia de la realidad histórica, se refugia en la utopía racionalista' (*3d*, 129–31). For my part, I find something much more positive in *Confusio*'s fantasies, however tentative, unrealistic, and optimistic (they are all these things) they may be, than mere escapism or Utopian refuge-seeking; this criticism, to my mind, applies more exactly to Tito's experiences in Graziella's grotto in *La Primera República*.

By contrast, and in accordance with her consistent emphasis on the metafictional nature of the *episodios*, *Confusio*'s formulation has for Diane Urey a special significance: 'Santiuste's history epitomizes the entire project of the *Episodios* because it demythifies history by exposing itself as another myth' (*3d*, 1989: 116). While I welcome the close attention to *Confusio* which this approach implies, it is evident that, unlike Urey, I maintain a belief in the validity of objective history and the Aristotelian distinction between history and poetry.

It is, as I see it, a fundamental mistake to treat the new *Confusio* (or his protector Beramendi) as examples or models to be avoided within Spanish society. They do not lay down for us specific lines of conduct to follow or to avoid, as Dendle assumes.[24] Nor do they set off on a regenerationist impulse, as Regalado García argues. The reader of 1906 knew very well at least the broad lines of the history of the time—the political instability of clerical intrigue and military intervention, the lack of social justice, the failure of the revolutionary hopes associated with Prim; the presence of an idealistic dreamer like Santiuste or of an ultra-conscious critic from within the system like Fajardo does not add to the realism of the presentation. Their value is not primarily documental, mimetic, symbolic, or psychological but contrastive, universalizing, even provocative. The essential clue about how to interpret *Confusio* and his *Historia* comes from Beramendi himself when he declares: 'no veo en mi protegido *Confusio* un perturbado de tantos como andan por el

[24] Dendle (*3d*, 1980: 137). Robin takes a very different and salutary view of Beramendi: 'ce n'est pas lui qui est le personnage fictif essentiel: c'est un lien entre le lecteur, la monarchie et la Révolution . . . La seule justification de ce personnage — on le voit par cette citation — est de nous donner à la fois un témoignage humain sur la personne de la reine, mais aussi de nous faire le bilan de l'ensemble du règne, sans que la pitié puisse altérer la justesse des accusations lancées contre Isabelle II' (*3d*, 231–2).

mundo; téngole por una inteligencia de fuerza irregular y ciega, que se lanza sin tino a la cacería de las verdades distantes. Yo me siento algo *Confusio*, (III: 564). 'La cacería de las verdades distantes' is an excellent definition of this inspired and wilful disregard for the *faits divers* of contemporary history in order to stretch the implied reader's imagination beyond its expected limits and guide it towards an eventual reassessment of the historical process.[25]

Thus we should take into account in the first place the strikingly fortuitous and arbitrary character of nineteenth-century Spanish political history, which caused it to be called so frequently a bad 'folletín', a 'farsa', a 'sainete', or a 'comedia' by such diverse figures, real or imaginary, as María Cristina, Narváez, Feijoo, and others. It is an attitude which is exploited with such skill by Valle-Inclán in his *visión esperpéntica* of the period. By contrast, the *Historia lógico-natural*, as its title indicates, seems much more plausible than the actual historical events.[26] At times, indeed, as in the reporting of the Mexican campaign, real occurrences are singled out as more improbable than *Confusio*'s imaginary history: 'Ved ahora el gracioso paso de Aranjuez, que, aunque parece inventado por el diablo de *Confusio*, es de incontestable realidad' (III: 566). It is clearly a case of using once more the famous Aristotelian distinction between history, which deals with what happens even if improbable, and poetry, which is concerned with what might or should happen even though it did not. As Beramendi asks Manuel Tarfe:

> Figúrate que han pasado mil años, y que los habitantes del planeta, en esa fecha remota, conocen las dos historias. ¿A cuál darán más crédito: a la de *Confusio* o a la que estarán escribiendo ahora Rico y Amat o don Antonio Flores? Yo creo que la de *Confusio* será más leída, y acabará por gozar concepto de única historia y verdadera. (III: 576)

More plausible and at the same time more satisfying, more aesthetically pleasing. As Beramendi says: 'es más divertido escribir la historia imaginada que leer la escrita. Esta suele ser embustera, y pues en ella no encuentras la verdad real, debemos procurarnos la verdad lógica y esencialmente estética' (III: 576).

What is more, the counter-history casts unexpected light on the true history and enables readers to readjust their sights with regard to historical events. It was unfortunate, for example, that Ferdinand VII's daughter (another Isabella) by Isabel de Braganza did not survive; she would presumably have been brought up in a more stable environment than her namesake was. The absolutist invasion of 1823 was just as much a foreign intervention as the

[25] Iser notes pertinently that 'If the text reproduces and confirms familiar norms, he [the reader] may remain relatively passive, whereas he is forced into intensive activity when the common ground is cut away from under him' (*2*, 1978: 84–5).

[26] Compare Lucila's quotation of her father's opinion: 'No hay cosa, por desatinada que sea, que no pueda ser verdad en este país, mayormente si es . . . contra la justicia y contra la paz de los hombres' (*Los duendes* II: 1703; *3c*, quoted in Montesinos iii. 125).

Napoleonic invasion of 1808. Logically, then, the constitutional forces could call themselves patriots and their adversaries *afrancesados*; and by extension Ferdinand VII could well be called to account as a traitor. It is perfectly possible to conceive of Isabella's receiving a more careful education—indeed almost impossible to think of her having a worse one—and the same applies to the ludicrously unsuitable marriage imposed upon her. Equally feasible and desirable was a closer collaboration with Portugal; *Confusio*'s deliberations undoubtedly reflect Prim's attempts to persuade Ferdinand of Portugal to accept the vacant throne in 1868. The ancient Spanish kingdoms and principalities might well have played a much more positive official role in nineteenth-century history, as in the modifications introduced by *Confusio*. The acute problem of centralism and regional aspirations is, unfortunately, largely overlooked in the *episodios*. Even in narratives where such questions naturally arise (the world of Mosén Juan Ruiz Hondón, for example, in *Carlos VII, en la Rápita*), little awareness is shown of these issues. Furthermore, the neat pattern of an Aragonese prince defending common law in the manner of Costa and marrying Isabella of Castile is extremely superficial and does not correspond in the least to the real problems of pressures for autonomy and decentralization led by Catalonia: a complex matter which involves complicated economic and social issues as well as political and cultural considerations.[27] Occasionally, it is true, as in *Aita Tettauen*,[28] he does show some awareness of economic issues regarding Catalonia, even if it is expressed in a somewhat elusive way in El Nasiry's chronicle; and, equally exceptionally, there is a little more penetration, as we have witnessed in Chapter 5, regarding the Basque provinces, in the otherwise rather unsatisfactory adventures of Tito in *De Cartago a Sagunto*.

From the idealistic point of view, as the arguments of *Confusio* and Beramendi determine, the revolution becomes inevitable not because of social, political, or economic factors, but in order to introduce fresh air and to give Prince Alfonso the wide education and experience he needs. Although in this case real and ideal history largely coincide, *Confusio*'s starting-point is radically different: it all happened because it ought to happen, logically and

[27] Among the plentiful bibliography see esp. Vicens Vives (*1b* 1961). For a general overview see Carr (*1b*, 538–47).

[28] In *Aita Tettauen* Galdós puts in the mouth of El Nasiry some comments which, using Catalonia as an example, ironically mark the contrast between useful industry and war: 'Acudí a ilustrar al Príncipe diciéndole que esta tropa viene de un territorio hispano que se llama *La Catalonia*, país de hombres valientes, industriosos y comerciantes; país que está todo poblado de talleres donde labran variedad de cosas útiles: papel, telas, herramientas, vidrio y loza. Como expresara extrañeza de que los *catalonios* dejaran sus telares, alfarerías y fraguas para venir a una guerra en que morirían como moscas, le respondí que allí sobra gente para todo, y que los trabajadores pacíficos no temen interrumpir su faena para ayudar a los fogosos militares, pues los pueblos de Europa saben por experiencia que después de la guerra es más fecunda la paz y mayor el bienestar de las naciones' (III: 306). A batallion of Catalan volunteers was prominent in the war (*4a*, Alarcón 1954: 971–3).

aesthetically. The Glorious Revolution of September 1868 did produce a reconciliation between Serrano and Prim, and though evidently a failure once its far-reaching expectations are taken into account, it none the less contributed, as Galdós was well aware,[29] to limited if imperfect constitutional advances which are incorporated in the Restoration. As *Confusio* appropriately termed it, it was touching up, repainting the old edifice, in the expectation that a steady progress towards complete renewal will be achieved. As for Alfonso XII, one might have expected him—though in the real history Galdós refers not infrequently to his weakly constitution (e.g. **III**: 694)—to have lived longer than twenty-eight years. It might have been more 'logical' and 'natural' for him to survive till 1925.

Most important, *Confusio* represents, as I see it, a sort of antidote against the pessimism and even despair the official history of Spain was liable to engender in early twentieth-century readers: the despair, I suggest, which plays its part in the unsatisfactory structure of the final volumes. Galdós introduces an Aristotelian concept of poetic truth into his *episodios* at a particularly crucial moment, both as regards the historical period he is describing ('story time', 1868) and in relation to the time at which he is writing ('discourse time', 1906). In Aristotle's terms, by dealing with universals rather than with singulars he endows Spanish history with a hope of eventual prosperity once the particular adversities of the nineteenth century and its many lost opportunities have been superseded. In this way he seeks to avoid the grotesque history of Spain being accepted as inevitable or as utterly predetermined. If this were so there would have been little purpose from his personal point of view in his re-entering politics the next year at the age of 64.[30] A new political start might imply a revised assessment of the past. It certainly does imply some measure, however slight, of optimism about immediate effective action. It is this ray of hope, I suggest, which *Confusio* brings to the world of the *episodio*. According to this conception, Spain's disasters and misfortunes evidently proceeded from specific historical causes, but fortuitousness and bad luck also played their part. Inasmuch as *Confusio*, even in his constant rectifications to his narrative, claimed to be unyielding in seeking a logical and natural sequence of events, he was confounding a rigorous deterministic concept of Spanish history while it was actually evolving.[31] Through *Confusio* Spain no longer appears to be condemned, in

[29] Galdós declared in an article of 1886 ('Un rey póstumo') that 'los derechos políticos se conquistaron de un modo definitivo en la revolución de 1868' (*3a*, *OI* iii. 145). For a good discussion of these problems see Goldman (*3b*, 1969; 1971; 1975).

[30] See his letter of 6 Apr. 1907 to Alfredo Vicenti, the director of *El Liberal*, collected by Fuentes (*3b*, 51–3); see also Madariaga (*3b*, 205–42).

[31] Such as Hinterhäuser discerns in Galdós's earlier philosophy of history (*3d*, 115–29). The same critic perceptively points out that 'la fe en el progreso y en el plan de la Providencia es el último recurso al que acude el autor antes de llegar a la desesperación' (118). *Confusio*'s history is in my view a reaction against such despair.

Larra's well-known image, to remain 'un país a medio hacer', to resemble 'la casta Penélope, que deshacía de noche la tela que tramaba por el día'.[32] Without explaining away or excusing the abject failure of Spain's constitutional organization, *Confusio*'s *Historia lógico-natural* sustains the remote dream of concord, reconciliation, and reform and holds out to Spaniards, in the years following the disaster of 1898, some hope that they are not inexorably doomed, Sisyphus-like, to a futile process of enduring sterility and repetition from which there is no escape. At least, that to my mind is how, for a relatively short time in the early years of the twentieth century, Galdós came to see it. Unfortunately, such tenuous hopes did not last long. His political aspirations brought more frustration than successes. The increasingly frantic tone, at times bitterly sardonic, at times despairing, at times stridently militant, we have seen in the last four *episodios* is the result.

[32] 'Ventajas de las cosas a medio hacer' (*4a*, 355b). The sentiment is echoed, in the words of José Ido del Sagrario, when the revolt of 23 Apr. 1873 takes place against the new Republic: '¡Oh España! qué haces, qué piensas, qué imaginas? *Tejes y destejes tu existencia.* Tu destino es correr tropezando y vivir muriendo' (*La Primera República* III: 1134; my italics).

Conclusion

IN his *Ideas sobre la novela* (1925)[1] Ortega proclaims the decadence or imminent demise of the novel because it is an exhausted form, 'un arsenal de posibilidades muy limitadas . . . una cantera de vientre enorme, pero vacío . . . y cuando la cantera se agota, el talento, por grande que sea, no puede hacer nada' (*2*, 1925: 389). He views it as characteristic of his *bête noire*, realism ('¡Terrible, incómoda palabra!': *2*, 1921: 142). Thus Balzac is for him (and, he claims, for his generation) 'irresistible', the world of the *Comédie humaine* has a 'carácter convencional, falso "à peu près"': what should be a 'cuadro' is a 'chafarrinón'. Zola fares even worse; following the pernicious examples of Darwin, responsible for bringing about a loss of freedom and originality, and of the physiologist Claude Bernard, Zola believes, according to Ortega, that '[el] medio es el único protagonista . . . La novela aspira a fisiología' (*2*, 1925: 390). In Spain the passing object of Ortega's scorn is Emilia Pardo Bazán; Galdós (a friend of his father's)[2] is conspicuous for his absence. In making this judgement, as in the *Meditaciones del Quijote* which preceded it, Ortega is flexing his muscles as the up-to-date literary theorist aware of current tendencies, at the same time as he is claiming leadership of a new politico-philosophical direction in opposition to Unamuno. Ortega thus becomes the Spanish champion of the rejection of nineteenth-century historical-realist values detailed in the Introduction and culminating in a textuality which, as Edward Said observed (*2*, 1983: 3–4), became the exact antithesis of history. It is precisely the critical attitude to be approached with extreme caution in any assessment of the realist novel.

The dominance of historical concepts for so long a period in the nineteenth century was bound to provoke a reaction. Hence the anti-historicist cast of most modern literary practice and theory, which, taking its lead from Nietzsche, has prejudiced a sympathetic consideration of historically orientated fiction. Historiography itself, in the influential and eloquent discourse of Hayden White, has succumbed to the charms of rhetoric. Marxist criticism has always proved an exception, but, as in the case of Lukács, inclines too much towards a narrow interpretation of socio-economic phenomena to offer a counterbalancing structural perspective of the work of fiction. Some hopeful signs exist, in the work of such critics as Foucault, Jameson, and LaCapra,

[1] Attached to *La deshumanización del arte* on its first publication in 1925; later added to *Meditaciones del Quijote*.

[2] José Ortega Munilla; see Schmidt (*3b*). For Ortega's 'silence' about Galdós consult Morón Arroyo (*3b*).

that a new reconciliation between literary and historical imperatives, assimilating the insights of both, may be emerging: this can only be good for future critical reappraisals of Galdós's work.

There is no doubt that Galdós, for the major part of his life, was fascinated by history, and that, in broad terms, he subscribed to the Hegelian or realist concept of history characteristic of the nineteenth century. He seeks, in John Lukacs's terminology, 'the *recorded past*' and participates in the 'rage for a realistic apprehension of the world', as Hayden White put it (*1a*, 1973: 45): a continuum expressed, with different emphases, by the image of the tree (organic growth) and of the stream (uninterrupted flow).

In several respects, however, we have seen Galdós deviating somewhat from the mainstream. First, he demonstrates a deeper concern than most for the minor occurrences of history, what I have called *la historia chica* (in this he appears to resemble Scott), together with an acute awareness of the cumulative undercurrent of historical development which Unamuno termed *intrahistoria*, or, in the distinctive image Galdós gives to Beramendi, 'como caracoles, que van soltando sobre las piedras un hilo de baba, con que imprimen su lento andar' (II: 1428). In addition, to continue the zoological imagery, he is, in Sir Isaiah Berlin's terms, a fox as well as a hedgehog. This is at once the sign and the consequence of his extraordinary capacity for the creation of representative types, 'embedded', in Auerbach's phrase, in historical reality and 'emplotted', as Hayden White terms it, in a fictional structure, in each case with a profound originality of its own.

Second, he is very aware, more perhaps than his contemporaries, of the precariousness of our information, which inevitably renders our grasp of history tenuous and insecure. Great imaginative creative power, as Coleridge acknowledges and Collingwood after, what George Eliot calls a 'veracious imagination', is required to fill, tentatively, the gap. This process in turn produces an 'indeterminacy' of form (*2*, Iser 1974: 283–90) conducive to reader-response interpretations, since the reader is afforded the opportunity to judge for himself such unresolved ambiguities as Fortunata's salvation or Torquemada's conversion. His work therefore possesses a distrust of traditional closure; subsequent developments after the conclusion of the novel are left unclear: How will Rosalía de Bringas fare after the revolution? Will Fortunata's son flourish in his new bourgeois environment? What are the chances of Tristana's being happy? Will Santiago and Teresa ever return to Spain? There is no question, however, of using the typical realist device of an 'epilogue' to report on the future destiny of the characters. Nothing gives us grounds for undermining, in Barthian fashion, the role of the author or, after Fish, of the text itself.

A third factor is the role of chance (Cleopatra's nose, as E. H. Carr has it), which is not eliminated from Galdós's spectrum as it is from the majority of his contemporaries'; it successfully undercuts the complete determinism implied by Collingwood and demanded (and in part practised) by Zola, while not denying the substantial effect of environment and heredity; part of his special attraction is the unresolved tension which exists between these two forces. And in the *episodios* there is an implicit appeal to what we may call the 'What if?' syndrome: What would have happened if a more suitable consort had been found for Isabella? If she had followed her popular instincts? If Espartero had resisted O'Donnell in 1856? If Progressive forces had worked together? If Prim had not been assassinated? If Pavía's coup had not taken place?

Finally, Galdós's instinctive concept of language is one which sets course once more towards tentative conclusions and ambivalence rather than towards authorial command. To express this linguistic indeterminacy he employs a type of idiom embracing several innovations, which, while not infrequent in the nineteenth century, will become much more widely prevalent in the twentieth. He employs shifts in point of view or focalization through characters acting as reflectors. Manifestations of the spoken language are highlighted as enthusiastically as Bakhtin did with *skaz*; as the latter declares, the novelist has a 'very rich verbal palette' (*2*, 1973: 166). Galdós's practice includes much direct speech, interior monologue, free indirect style, pseudo-dramatic dialogue, and Bakhtinian double-voicing; at times it approaches the stream-of-consciousness technique. He seeks to convey an immediacy based on direct 'evidential' experience. Subtle use is made of modifications of the time-scale (the various 'anachronies' as defined by Genette). Included, too, are hints of self-reflexivity, broached as early as *La Fontana de Oro* and at their most evident in *El amigo Manso*; these draw attention to the problems of narrative without disrupting the sequential structure.

Having begun his narrative work with the conventional historical novel, he was to polarize his socio-historical interests in two different directions. One of his two forms of novel, the *Episodio nacional*, has the clear mission of describing comprehensively the history of the recent past. Within the Spanish environment, this primary objective has a social importance which guarantees it an assured market. This is what the *episodios'* readers expected, obtained, and enjoyed. If in an international context such a goal is a limiting factor, as the lack of translations indicates, their continuous sale in Spain demonstrates their conspicuous success in the domestic market. Recent scholars (Dendle, Rodríguez, and particularly Diane Urey) have revealed the frequently overlooked narrative skills Galdós displays in the *episodios*, but their consistent historical purpose at the national level makes it inconceivable for me that they could be subordinated to a metafictional aim of reflecting on the role of narrative, as Urey (*3d*, 1989), following Derrida, argues.[3] For me, the

[3] This is not to say that within these conditions they do not obey narratological imperatives in their structure in ways which differ only to a limited extent from those of the novels; and it is greatly to Urey's credit to have contributed towards elucidating this important and neglected aspect of the *episodios*.

fundamental concern of the *Episodios nacionales* with the historical process as it unfolds in specific occurrences at a specific time is beyond all possible doubt.

In this historical aspect Galdós is not concerned to pass definitive judgements concerning the past, but to reflect representative contemporary attitudes which will bring out the living issues of the day. He thus provides a wide spectrum of points of view through carefully conceived characters embedded in true situations over a period of time and capable of changing with them. Centurión, Eufrasia Carrasco, Beramendi, Santiuste, Teresa Villaescusa, Santiago Ibero, and Halconero all intervene and evolve effectively in this fashion.

Writing in the second batch of *episodios* from the period after the 1898 débâcle, Galdós no longer offers a perspective which points towards easy or direct solutions to Spain's problems. The Spanish history he depicts lacks the grandeur, range, or decisive consequences of Tolstoy's depiction of the Napoleonic invasion of Russia; indeed, as has become evident, it is difficult at times to take nineteenth-century Spanish history seriously as more than, in Galdós's own terms, a *mal folletín, los tiempos bobos*, a potential *esperpento*; from another angle, it was a continual blood bath. A constant conflict unfolds in which Spaniards are shown struggling unsuccessfully to achieve a balance between progress and stability, to attain a valid compromise solution which will make for genuine political and economic reform without renewed partisanship. Attempts in this direction were made by O'Donnell and his *Unión Liberal*, by Amadeo's constitutional monarchy, and finally by the Restoration, all tending, however, and especially the latter, to sacrifice principle for expediency in the process. Moreover, violent measures are all-pervading, and make up a long and dismal list, beginning with a private act with some political implications like Merino's assassination attempt, followed by an understandable but unproductive act of revenge against a hated police chief, harsh repression of a student demonstration (*Noche de San Daniel*), a barrack insurrection and ruthless retaliation (San Gil), and a 'glorious' revolution which was ultimately a failure. After the revolution assassination removes the most far-seeing of the military leaders, Prim, a flagrant assault on parliamentary legality is carried out in Pavía's *coup d'état*, and, finally, the country faces two large-scale uprisings in the Cantonalist outbreak and the Carlist wars. Masses of lives are wasted uselessly in these savage and futile actions. The army's conduct in all this is quick in precipitate action and ineffective in government; Galdós has, as Bly has shown (*3d*, 1984; 1986), his own deflating and ambiguous portrayal of heroism and even patriotism. The corresponding role of the Church is scheming and sinister, with the result that the reactionary forces, political, military, spiritual, and economic, constitute a formidable and well-nigh unbeatable power; and power, as Foucault has elucidated, is the underlying cause of human endeavour. Clerical and oligarchical hegemony is all the more strong, as Galdós constantly emphasizes, since the Progressive forces remain incorrigibly disunited. The revolution of 1854, the Savoyan experiment, and the Federal Republic all founder on this basic failure to coalesce. Finally, forming a distinctive if unobtrusive part of his evaluation and censure of Spanish society is his sympathetic portrayal of the underprivileged status of women, a concern for their poor educational opportunities, together with a consistent if largely implicit advocacy of sexual freedom.

The general lessons to be learnt (and the *episodios* do have a didactic function) from this largely negative scenario are therefore mainly of a defensive or supportive nature: disgust at intrigue and capricious government, loathing of instinctive recourse to violence, distrust of heroic rhetoric and foreign adventures, apprehension about popular uprisings, ironic contempt for bureaucracy, despair at the neglect of education, concern for innocent victims (often women), approval of free and honest personal relations.

The *episodios* have of course had their critics. They have been accused of a conservative, bourgeois bias (*3d*, Regalado García), an accusation which neatly counteracts the traditional accusations of his being demagogic and anticlerical. The most valid objection to Galdós's historical perspective is that in some areas—economics, working-class activity, regionalism—coverage is scanty or incomplete, but these deficiencies are amply compensated for by his extremely wide-ranging political and social coverage. As a result of his determination to offer a broad, open panorama, few prejudices and preferences show through in such a way as to distort the picture. Each of the key political leaders—Narváez, O'Donnell, Prim, even Cánovas—is depicted in his own fashion as attempting to face honestly his own difficult problems, hampered in most cases by the queen's capriciousness and her clerical entourage. Only in the last four volumes does Galdós lead us more dogmatically towards an inflexible condemnation of Church and state, when—to conclude our zoological parallels—his narrative voice comes increasingly to resemble a squawking parrot.

Hinterhäuser has some other pertinent criticisms: first, that Galdós is too obvious in explaining his technique, which is true but not fundamental (*3d*, 245); and second, with more justification, that the historical–fictional mix is unbalanced and improvised, which has a tedious effect and slows down the action ('efecto molesto . . . retardador'. *3d*, 241). This is certainly true in some cases, as has been indicated; I would single out as examples *La vuelta al mundo* and parts of the last four *episodios*. As a more general statement, however, it needs to be refined. On occasion there is too conspicuous a mouthpiece—Beremendi, Halconero, for example—for his views. In other cases it is the subgenre itself rather than the narrative technique that is at fault. The dense chronicle-like structure of the *episodios* means that there tend to be too many raw facts to be accommodated, and this at times produces congestion; at decisive moments, like the 1868 revolution or Prim's murder, these events inevitably dominate and condition the course of the fictional narrative. It is inherent in the chosen form.

Where do the *episodios* stand regarding the present? From *La Fontana de Oro* onwards Galdós's historical work has a close relationship with the time of writing, since past events offer numerous lessons for and parallels to contemporary issues. It seems to me none the less an error to view the chronicling of the past primarily with reference to current concerns (*3d*, Dendle 1980; 1986). The past is not just a source of disguised symbolic references to the present; rather it is treated as worthy of analysis and meditation in and for itself. Only in this way can a valid interpretation of the past be determined and so serve to gain a better understanding of the present. In this regard the *episodios* are 'modern' historical novels, as defined, following Fleishman and Tillotson, in the Introduction, but with a more continuous, more closely interlinked historical network than the genre as a whole.

The contemporary novels have an amazing variety of scope, focus, and narrative technique. In accordance with the tenets of the realist novel, they represent 'a perennial mode of representing the world and coming to terms with it' (*4b*, Stern 32). They reflect, as Hinterhäuser rightly notes (*3d*, 247), Galdós's joy in the creation of characters. His creations are rich and substantial enough to make Balzac's personages seem crude and Dickens's uncomfortably eccentric. At the same time, many of the novels, especially the more lengthy narratives or sequences like *Fortunata y Jacinta*, *Ángel Guerra*, and the Torquemada tetralogy, have such wide horizons and leisured pace that they amply deserve to be termed, according to Henry James's select and rarefied vision, 'large, loose, baggy monsters'; but it is comforting to be in Tolstoy's company. In all of them, implicitly if not explicitly, as Blanco Aguinaga has argued (*3b*, 203), historical concerns form an intrinsic part, to no less a degree, certainly, than in the work of most European contemporaries. The historical context is particularly important in all the novels set before the Restoration and is not inconsiderable in the others. As has been established, however, these are not historical novels any more than Balzac's are, but rather, in the term used by Peter Bly, 'novels of historical imagination'.

How effective are the three types of historical allusion we have discerned in the novels? In the first type, those which reflect political occurrences as they occur, the manner in which figures like Isidora, Rosalía de Bringas, and Juanito Santa Cruz act in concert with the historical events as they take place seems to me outstanding. In this intertwining of private and public affairs Galdós's achievements seem to me fully comparable with those of other major realist novelists. I think, for instance, of Julien Sorel functioning as the Marquis de la Mole's secretary against the background of the final agony of the post-Napoleonic *ancien régime*; and of *L'Éducation sentimentale*, as the 1848 revolution, on which so many Progressive hopes rested, breaks out at the same time as Frédéric Moreau, having waited in vain for Mme Arnoux to come to him, settles for second best in Rosanette's arms.[4] Political figures, when they appear, take the relatively inconspicuous form I venture to call in the Introduction the third-hand Napoleon model of interaction with fiction. And there is no lack of examples of Galdós's deep concern for socio-economic history: the commercial developments of the first part of *Fortunata y Jacinta*, the rise of Torquemada, the tribulations of *cesantes* like Villaamil.

The second type, in which certain minor characters recapitulate past history, fulfils without doubt a specific Spanish need by bringing as rapidly as possible a full array of disparate historical events before the reader, but the accumulation of names tends to be excessively detailed and to have a localizing effect. The device is, however, carefully controlled by being the preserve of a limited number of characters—Relimpio, Estupiñá, Isabel Cordero—operating within a narrow sphere of action; it is never allowed to impinge adversely on the narration.

The most notable case of the third type (petrified occurrences from the past) is the vivid flashback, well embedded in the story, to a specific incident, the execution of the San Gil sergeants, in *Ángel Guerra*; it is more detailed and immediate, and has a more direct effect than any of the lesser incidents. Frequent references are made in particular to Prim (his *jamás* speech, his murder), and occasionally to others: Narváez, O'Donnell, Cánovas, the Empress Eugénie. These references have the effect of my fourth-hand type of Napoleon: a fleeting unelaborated association. Also falling within this group are snap physical characterizations by means of an external resemblance to a historical character, usually foreign. Such resemblances, like Mauricia with Napoleon, Bringas with Thiers, de la Caña with Cavour, Estupiñá with Rossini, are part and parcel of the realist's stock in trade; they are much favoured by Galdós, who uses them most effectively.

In Chapter 1 I noted that in his mature career Galdós did not compose novels and *Episodios* simultaneously; by the time he resumed writing *Episodios* in 1898, he was no longer writing novels, and the three that he wrote subsequently do not conform to the realist pattern. It is a clear indication that he thought of the two forms differently, and at the same time was prepared to go on using an essentially realistic technique for this type of narrative for some time after he had abandoned it in the straight novel form. Since the second batch of *Episodios nacionales* brings us near contemporary history, there is an overlap of time-scale with the novels, especially from the later stages of Isabella II's reign onwards; and by 1863 Galdós himself is an eyewitness of some events. This approximation to the present has allowed us to study those

[4] 'Mme Arnoux fails to keep the appointed rendez-vous; and so as we know, did France' (*4b*, Hemmings 1974: 178). Hemmings's comments on the meshing of private drama with the great public drama of the France of 1848 are equally pertinent to Galdós.

significant parallels, spanning about a dozen years of 'story time'—roughly 1863 to 1876—which reveal the differences between the treatment of history in the two forms.

To pinpoint this difference more clearly, let us return for a moment to the diagrams by Scholes and Spires we discussed in the Introduction (p. 27). In Scholes's graph we should place the Galdosian novel in the shaded area corresponding to the plenitude of the realist novel, alongside Stendhal, Balzac, Flaubert (and including Alas), while the *Episodio nacional* would be on the lower side of 'realism', towards the limit marked 'history', slightly beyond its more generalized subgenre, the modern historical novel. In the case of Spires's graph, Galdós's novels partake of the general uncertainty I feel about where realist novels fit in; taken collectively, they do not have a very clearly defined position within the ellipse, though individual novels might be placed nearer one or other of the surrounding headings, thus distracting from their generic similarity. Thus many (*La desheredada, La de Bringas, Fortunata y Jacinta*, the Torquemada sequence) incline towards the historical edge, while others look towards 'satire' and 'metafiction' (*El amigo Manso*) or 'metafiction' without 'satire' (*Misericordia*). The typical *episodio* would, undoubtedly, find its place on the bottom edge marked 'history', just beyond the modern historical novel but not as far as 'reportorial fiction', but the four last *episodios* would be fiercely tugged towards 'picaresque' and 'satire' in the middle/top left, an indication perhaps of the incompatibility within their generic make-up. Either scheme, Scholes's or Spires's, brings out clearly the differences between the two forms.

In the *episodios* the sort of questions posed turn generally on more or less political issues of the past such as how to reconcile progress and order, the place of clerical influence in society, the validity of military insurrection, the turbulence and frustrations of *el pueblo*, political patronage and corruption, the perennial topics of *cesantía* and usury, the rise and fall of fortunes of representative individuals. In the novels the same or similar issues arise, but in a purely societal context. The sharp and detailed portrayal of these events in the *episodios* is neatly supplemented by the effects the main upheavals have on characters in the novels: the 1868 revolution on the Bringases, Amadeo's abdication on Isidora, Alfonso's succession on Jacinta, economic and administrative change on individual *cesantes* or usurers like Villaamil and Torquemada. And the clearest impression of how the complacency and cynicism of the Restoration came about is conveyed through the wide social canvas of social life in *Fortunata y Jacinta* (opulent bourgeoisie, café job-seekers, and the representative figure of Feijoo) and the opportunistic financial intrigues of *Lo prohibido* and *La incógnita*; the underside of this apparent prosperity is shown in *Misericordia*.

The *episodios* operate, then, in an exciting world of continuous interaction between real and fictitious characters. They are perched on the edge of history, as much as other characters from the finest historically based novels: Jeanie Deans interviewed by Queen Caroline in *The Heart of Midlothian*, the steadfast Esmond confronting the Stuart pretender in *The History of Henry Esmond*, Nikolai Rostov witnessing the meeting of Napoleon and Czar Alexander and meditating on the flagrant injustice of war, in *War and Peace*. Their greater historical density and coherence mean that such *episodios* as *Prim, La de los tristes destinos,* and *España trágica* complement the schematic but imaginative references to political events found in the contemporary novels with a more thorough, balanced, on-the-scene consideration of contemporary or near-contemporary issues.

Historical 'embedding' in the novels is none the less frequent and varied. It applies particularly to representative figures with well-rounded individuality like Isidora, Rosalía, and Juanito; these are similar to figures from the *episodios*, but have more personality (except, perhaps, in Juanito's case) than is possible for even their most highly individualized counterparts like Beramendi and Teresa. The characters who most resemble those of the *episodios* (I am thinking here of Centurión, Halconero, and Segismundo Fajardo) are the less-developed figures, like Estupiñá, Villalonga, Melchor de Relimpio, or Joaquín Pez, whose main *raison d'être* is a socio-historical role; these tend also to be continually recurring characters.

Where the difference lies is in the fact that the novels also contain major figures who are much less directly conditioned by history, characters who are 'eccentric' in the literal sense: Manso, Maxi, Leré, Nazarín, Almudena, Tristana, in very different ways, come to mind. These characters are conditioned by their social environment rather than by any historical narrative sequence, though they are not in the least unrelated to or alienated from the latter. As a consequence, they are one of a kind; their place is essentially in one novel, and they do not generally reappear in another. Such a degree of independence does not occur, indeed cannot occur, in the *episodios*, where characters may be original and autonomous within a contextual frame but are all much more closely involved in contemporary issues and are carried along with them. Eufrasia Carrasco, present, like Jenara Baraona before her, at all the leading political incidents of the time, is one example who has no great depth of characterization. Teresa Villaescusa. on the other hand, has much greater individuality, but as she is constantly and intimately involved, throughout a series of volumes, with real-life figures like Sixto Cámara, O'Donnell (and the Liberal Union), and Prim, she still suffers from this essential contextual limitation.

The novel form also offers rather more scope for experimentation, both in form and in content. It caters better for the subtle use of an autobiographical structure and for the development of the dialogue novel, though neither of these forms is absent from the *episodios*. Without as definite a didactic purpose, it allows greater ambiguity of narrative tone. If the *episodios* fall rather into the first category of narration—the one dominated by its subject-matter—referred to by Cipión in *El coloquio de los perros*, the novels overlap both categories, since substance and form, inseparable in any case, are equally balanced in them.

Differences occur in both general and more specific areas. Among the first, we have seen examples in both of the queen's charitable instincts, the build-up to the Glorious Revolution (1868), *desamortización*, the search for stability, and the conciliation of interests. In all these subjects the integration seems to me uniformly skilful. There are two major detailed examples. The first is the execution of the San Gil sergeants (1866), 'emplotted' at a distance of time within the story of Ángel Guerra, and the apparently contemporary focus on the rising and executions in the *episodios*; both are equally effective. The second is Pavía's coup (1874), where the historical 'embedding' and the fictional 'emplotment' of the incident within *Fortunata y Jacinta* are masterly, but in the late *episodio* the treatment, though lively, is marred by too direct and partisan an approach. Less detailed parallels concern events of Isabella's reign (Calvo Asensio's death and the *Noche de San Daniel* of 1865, the latter better 'emplotted' than the first), the brief Savoyan episode, the *pronunciamiento* at Sagunto, Alfonso XII's entry into the capital, the pervasive clerical influence; in all these the novels surpass the late *episodios* in quality.

Some controversy arises over the question of symbolism, which it has been suggested is present in both forms. The term itself is perhaps vague and inexact when applied to Galdós's fiction: a symbol implies some degree of substitution, and this very rarely occurs in a full-blooded form in the novel, the basic structure of which lies on Jakobson's metonymic axis rather than on his metaphorical one (*2*, 76; *2*, Lodge 1990: 81). As J. P. Stern acutely observes, 'Literature of all kinds may be, and realistic literature always is, a representative but not a "standing for" in the sense of a replacement of that which it represents' (*4b*, 70). Certainly, no character represents 'Spain', 'the 1868 revolution', 'Amadeo', 'the Restoration', etc. as such; here I disagree with such distinguished major scholars as Gilman (*3c*, 1981) and Bly (*3c*, 1981), who err somewhat, in my judgement, in the direction of over-symbolizing the characters. What does occur is interweaving and parallelism, as in the cases of the *mantillas blancas* incident in *La desheredada* and the hollow Maundy Thursday ceremony in *La de Bringas*. I have no difficulty in concurring with Bly's modestly worded suggestion, 'I think that it is not inappropriate to assert that the story of Rosalía and Francisco can be construed as an allegory, or a metaphor, on the decline and fall of the House of Bourbon' (*3b*, 1981: 8). What would be inadmissible to my mind would be to claim that any incident in itself lacked literary substance except as a symbolic substitute for something else.

On the whole, symbolism is rather more plausible in the *episodios*, where a historical liaison exists *a priori*; even here, however, I find it a very restricting device. Fortunately, its use is limited. What tendency towards symbolism does occur, especially as regards the Ansúrez family or the use of names like 'Santiago Ibero', is not central or fully developed. This is not, of course, to deny that the characters have close historical affiliations. To clarify a little: if we ask 'Does Teresa (or Isidora in *La desheredada*) stand or substitute for Isabella II', I think the answer is no. If we ask 'Does she have a close parallel relationship with the queen without losing her own fictional autonomy', then the answer is yes. At their best the characters of the *episodios* have a role which is strictly analogous to, but not identical with, public figures and events, as plausible representatives of their particular time and place.

In addition, occasional parallels or contrasts with other novelists have proven fruitful. Compared with Baroja and Pereda, Galdós's description of the murder of Chico is more closely integrated into a coherent social pattern. In not dissimilar fashion, Pardo Bazán depicts her heroine Amparo in a revolutionary setting in both political and private spheres. While Oller works into an act of hero-worship some criticisms of Prim from a Catalan standpoint, Prim's ambitions and revolutionary zeal are viewed by Galdós in a largely heroic if now distant light, in sharp contrast with Valle-Inclán's highly critical satirical portrait.

With the Restoration, interest in political events diminishes in the novels, while in the *episodios* two new attitudes to contemporary history emerge. One consists of a breakdown of the old balance between story and history. A more assertive attitude to historical events emerges; politicians are lambasted directly by Mariclío or by her protégé Tito, while others, like Estévanez, are too close to Galdós himself. This forthrightness is accompanied by experiments in narrative renewal, together with the use of fantasy and clearly allegorical figures. The new technique is not as successful, in my judgement, as the well-tried one of previous works, even though the evocation of selected aspects of Restoration society, the detailed treatment of important issues, and the relation with Galdós's earlier work ensure that these compositions continue to attract attention. Second comes the fascinating idealistic reconstruction of historical events in Santiuste-*Confusio*'s *Historia*, a *tour de force* of demonstrating constructively, under a whimsical façade, what might have been. Chance is indicated once more as a determining factor.

While I have committed myself to certain general statements I consider legitimate concerning Galdós's own attitude, the broadly representative quality of his work makes any attempt to isolate Galdós's opinions more categorically, at least until his last stage, very hazardous. No political view, no single figure, not even a partial 'reflector' or 'focalizer' such as Beramendi, is offered as a model: such a tidy solution would be to rewrite history as radically as *Confusio* attempts to do. At every turn in the political maelstrom Galdós seeks to allow the diverse and heterogeneous forces at work to act and speak for themselves. He thus enables the reader to reflect on and arrive at his own conclusions about the agonized recent history of his country, which, as he well knows, will have no happy outcome within the period he is chronicling. His particular form of the historical novel reveals, it must be admitted, an absorbing concern with specific local and Spanish issues, despite occasional intrusions into foreign affairs, but this limitation is perhaps mitigated somewhat by its breadth of interpretation.

How far do my investigations tend to alter the canon? They may throw more light on one feature which has always seemed anomalous in Galdosian criticism: the wide discrepancy in the popularity of various parts of his production. Historically, three areas can be discerned in which popular taste is out of step with scholarly evaluations. First, the *episodios* have always enjoyed greater commercial success in Spain than the novels. Second, the earlier thesis novels have been and still are more popular than the maturer novels. Third, the earlier *episodios* (first and second series) are preferred to the later series. Scholarly opinion, on the contrary, puts the contemporary novels of the 1880s and 1890s well ahead,[5] followed by the early contemporary novels, especially such favourites as *Doña Perfecta* and *Marianela*. The *episodios*, despite excellent recent research, have been generally neglected and implicitly considered well below the quality of the novels. To judge by such a significant but limited marker of esteem as study in Spanish departments of the United States or United Kingdom, the *episodios* rank very low in prestige. As between the various series, there can hardly be said to be a consensus, and it is difficult to signal clear preferences, though those favoured tend to be martial texts from the first series: *Trafalgar, Zaragoza, Gerona*. In Spain and possibly other Spanish-speaking countries I suspect that the *episodios*, with the first and second series still in the lead, continue to command considerable respect, though steadily less than the novels. The result of my study, I suggest, is to add further convincing evidence to that adduced by such dedicated critics as Dendle and Urey of the high value of the later *episodios*. These should be seen, in my view, not as meditations on the novel form itself (Urey) or as novels in which history is incidental (Dendle), but as a distinctive historical form which none the less prolongs, in many ways, the fine mature narrative and psychological characteristics of the great contemporary novels. Hinterhäuser (*3d*, 231) makes the interesting point that the *episodios* are at least comparable with, and perhaps superior to, a different hybrid nineteenth-century form, the 'obras históricas "poéticas"' of Carlyle, Macaulay, Burckhardt, and Renan. In readability, viability, and immediate impact I have no doubt that the *episodios* far surpass these respected but ignored—if not forgotten—period pieces.

[5] See e.g. Hinterhäuser, who declares roundly that 'las novelas sociales, desde el punto de vista artístico, son "fundamentalmente" superiores a los *Episodios*' (*3d*, 230).

Such a practised reader of Galdós as Pedro Ortiz Armengol (*4a*, 30) has ventured to give a value-judgement of the *episodios*. It is noteworthy that of the nine he singles out as among Galdós's finest achievements four are from the fourth series and two from the third.[6] When he adds another three which he considers nearly as good, two come from the fourth and the other is the last volume of the third series. In my opinion, the *episodios* which deal with the reign of Isabella II—essentially the fourth series, plus *Bodas reales*—have a very high collective quality, the weakest being those furthest from the political action: *Aita Tettauen* and *Carlos VII, en la Rápita* and, especially, *La vuelta al mundo*. The others suffer rather less from the limitation of localized scope, since the events described are on the whole not unimportant and a non-national audience can or should relate to them. Ortiz Armengol does not highlight any of the fifth series. According to my evaluation, the first two are also very good, the remaining four less so. On this somewhat arbitrary basis for assessment we have a collection of ten works[7] which deserve much more scholarly attention than they normally attract. They show an active and highly creative author, still in his late prime—his early sixties—writing in the first decade of this century works which continue at only a somewhat less elevated level, in a form which offers both opportunity and limitation, the mastery of realist narrative of the 1880s and 1890s. My evaluation of the finest of the Galdós corpus would therefore place these ten *episodios nacionales* immediately below the great novels.

There is to my mind no question, however, of dislodging these twenty master works of the 1880s and 1890s from their pinnacle. The broad sweep of society they depict militates decisively against that. Historical reference forms a substantial part of their canvas, and on this point I hope to have added something to the insights offered by Peter Bly and others. If the historical content is particularly intense in *La desheredada, La de Bringas*, and *Fortunata y Jacinta*, it is only slightly less so in *El doctor Centeno, Tormento*, and the Torquemada series. In short, the common preoccupation with history and the strong fictional inventiveness of both novels and *episodios* bring them together, just as the diverse treatment of historical events and the exigencies of their respective genres keep them distinct in all but their shared narrative strength.

My aim in this book has been to bring together for close examination Galdós's two narrative forms, without blurring the distinctions between them—rather, to indicate more precisely these differences. It is curious that despite the conviction of many critics that there is little or no difference

[6] The nine works are: *Juan Martín, el Empecinado* (1st series), *El Grande Oriente*, *El terror de 1824* (2nd series), *Mendizábal*, *Montes de Oca* (3rd series), *Los duendes de la camarilla*, *La revolución de julio*, *Prim*, and *La de los tristes destinos* (4th series), followed by an additional three: *Bodas reales* (3rd series), *Las tormentas del 48*, and *O'Donnell* (4th series).

[7] They are, in chronological order, *Bodas reales*, *Las tormentas del 48*, *Narváez*, *Los duendes de la camarilla*, *La revolución de julio*, *O'Donnell*, *Prim*, *La de los tristes destinos*, *España sin rey*, and *España trágica*.

between the two forms, they have seldom dealt with them together; at most, more or less perfunctory cross-references have been given. To study the novels and *episodios* side by side yields unexpected points of similarity and contrast. A great deal of light, I trust, has been cast on Galdós's attitude to history in general and to the history of his country in particular, from the reign of Isabella II to the Restoration. We have discussed the use he made of historical subjects when he allowed himself to create freely in fictional form and the narrative technique he adopted; the modifications made in this technique when he conditioned himself to a rigid scheme of chronological historical presentation; how the treatment of specific comparable events differs between the two forms; and, finally, how these considerations sharpen our judgement and enhance our appreciation of the narrative work, in both the directions he chose to adopt, of the tireless innovator and experimentalist who was Benito Pérez Galdós.

APPENDICES

Appendix 1

Galdós's Published Works

Date	Novels	*Episodios nacionales*	Plays	Other (selected)
1865				Begins contributions to *Revista del Movimiento Intelectual de Europa* and *La Nación*
1868	*La Fontana de Oro* (V)			*Las Novedades*
1869				*Las Cortes*
1870	*La sombra* (IV)			'La novela en España'
1871	*El audaz* (IV)			*Revista de España*
1872				*Revista de España*
1873		*Trafalgar* (I)		
		La corte de Carlos IV (I)		
		El 19 de marzo y el 2 de mayo (I)		
		Bailén (I)		
1874		*Napoleón en Chamartín* (I)		
		Zaragoza (I)		
		Gerona (I)		
		Cádiz (I)		
		Juan Martín, el Empecinado (I)		
1875		*La batalla de los Arapiles* (I)		
		El equipaje del rey José (I)		
		Memorias de un cortesano de 1815 (I)		
1876	*Doña Perfecta* (IV)	*La segunda casaca* (I)		
		El Grande Oriente (I)		
		El 7 de julio (I)		
1877	*Gloria* (IV)	*Los cien mil hijos de San Luis* (I)		
		El terror de 1824 (I)		
1878	*Marianela* (IV)	*Un voluntario realista* (II)		
	La familia de León Roch (IV)			
1879		*Los apostólicos* (II)		*El Océano*
		Un faccioso más y [illegible]		

1881	*La desheredada* (**IV**)			
1882	*El amigo Manso* (**IV**)			
1883	*El doctor Centeno* (**IV**)			*La Prensa* (Buenos Aires) (continues until 1894)
1884	*Tormento* (**IV**)			
	La de Bringas (**IV**)			
1885	*Lo prohibido* (**IV**)			
1886–7	*Fortunata y Jacinta* (**V**)			
1888	*Miau* (**V**)			
1889	*La incógnita* (**V**)			
	Torquemada en la hoguera (**V**)			
	Realidad (**V**)			
1890–1	*Ángel Guerra* (**V**)[1]			
1892	*Tristana* (**V**)		*Realidad*	
	La loca de la casa (**V**)			
1893	*Torquemada en la cruz* (**V**)		*La loca de la casa*	
1894	*Torquemada en el Purgatorio* (**V**)		*La de San Quintín*	
			Los condenados	
1895	*Torquemada y San Pedro* (**V**)		*Voluntad*	
	Nazarín (**V**)			
	Halma (**V**)			
1896			*Doña Perfecta*	
			La fiera	
1897	*Misericordia* (**V**)			*Discursos académicos*
	El abuelo (**VI**)			
1898		*Zumalacárregui* (**II**)		
		Mendizábal (**II**)		
		De Oñate a La Granja (**II**)		
1899		*Luchana* (**II**)		
		La campaña del Maestrazgo (**II**)		
		La estafeta romántica (**II**)		
		Vergara (**II**)		

Date	Novels	*Episodios nacionales*	Plays	Other (selected)
1900		*Montes de Oca* **(II)**		
		Los ayacuchos **(II)**		
		Bodas reales **(II)**		
1901			*Electra*	
1902		*Las tormentas del 48* **(II)**	*Alma y vida*	
		Narváez **(II)**		
1903		*Los duendes de la camarilla* **(II)**	*Mariucha*	
1904		*La revolución de julio* **(III)**	*El abuelo*	*'La reina Isabel'* **(VI)**
		O'Donnell **(III)**		
1905	*Casandra* **(VI)**	*Aita Tettauen* **(III)**	*Bárbara*	
		Carlos VII, en la Rápita		
1906		*La vuelta al mundo en la 'Numancia'* **(III)**		
		Prim **(III)**		
1907		*La de los tristes destinos* **(III)**	*Zaragoza*	Political writings and speeches 1907–13
1908		*España sin rey* **(III)**	*Pedro Minio*	
1909	*El caballero encantado* **(VI)**	*España trágica* **(III)**		
1910		*Amadeo I* **(III)**	*Casandra*	
1911		*La Primera República* **(III)**		
		De Cartago a Sagunto **(III)**		
1912		*Cánovas* **(III)**		
1913			*Celia en los infiernos*	
1914			*Alceste*	
1915	*La razón de la sinrazón* **(VI)**		*Sor Simona*	
1916			*El tacaño Salomón*	*Memorias de un desmemoriado* **(VI)**
1918			*Sta. Juana de Castilla*	

[1] See the footnote to section *3a* of the Bibliography.

Appendix 2

Principal Historical Events, 1805–1920

1805	(21 Oct.) Battle of Trafalgar
1808	(17 Mar.) 'Motín de Aranjuez': fall of Godoy, abdication of Charles IV
	(2 May) Uprising against French
	Ferdinand VII confined at Valençay
	(21 July) Battle of Bailén
1812	(19 Mar.) Constitution of Cadiz
1813	(21 June) Defeat of the French at Vitoria
1814	(24 Mar.) Return of Ferdinand VII: repudiation of constitution
	(7 Apr.) Abdication of Napoleon
1815	(Mar.–June) The Hundred Days (Napoleon's return)
	(18 June) Battle of Waterloo
1820–3	'Trienio Constitucional' under Riego
1823	(7 Apr.) French invasion of the 'Cien Mil Hijos de San Luis'
	(11 June) Transfer of government to Cadiz
	(7 Nov.) Execution of Riego
1823–33	'Década ominosa'
1829	(21 Dec.) Marriage of Ferdinand VII to María Cristina
1830	July revolution in France
	(10 Oct.) Birth of Isabella
1832	(1 Oct.) Calomarde/Carlota incident
1833	(29 Sept.) Death of Ferdinand VII; regency of María Cristina
1834–40	First Carlist war
1835	(June) Death of Zumalacárregui
1835–6	Mendizábal government: *desamortización*
1836	(13 Aug.) Sergeants' mutiny at La Granja
1837	(May–Sept.) Carlists reach outskirts of Madrid
1839	(31 Aug.) 'Abrazo de Vergara' between Espartero and Maroto
1840	(12 Oct.) Abdication of regent María Cristina
1841–3	Regency of Espartero
1841	(15 Oct.) Execution of Diego de León
1843	(10 May) Galdós's birth.
	(30 July) Fall of Espartero
	(8 Nov.) Isabella II declared of age
	(28 Nov.) Disgrace of Olózaga
1844	(29 Jan.) Death of Carlota
	(3 May) Narváez ministry
1845	(23 May) Conservative constitution
1846	(3 June) Election of Pope Pius IX

	(10 Oct.) Marriage of Isabella II to Francisco de Asís and the infanta Luisa Fernanda to Montpensier
1848	'Year of Revolutions'
	(Feb.) Fall of Louis-Philippe
	Pius IX flees to Gaeta
1849	(19 Oct.) *Ministerio Relámpago*; Narváez returns to power
1851	(10 Jan.) Bravo Murillo replaces Narváez
	(2 Dec.) Louis Napoleon's *coup d'état* in France
1852	(2 Feb.) Assassination attempt on Queen Isabella by Martín Merino
1853	(19 Sept.) Sartorius in power
1854	'Revolución de julio': battle of Vicálvaro; manifesto of Manzanares; shooting of Chico
1856	(13 July) Ousting of Espartero by O'Donnell; dissolution of the Milicia Nacional. *Desamortización* by Pascual Madoz
	(12 Oct.) Fall of O'Donnell
1857	(28 Nov.) Birth of Alfonso XII
1858–63	*Ministerio largo* of O'Donnell; supremacy of the *Unión Liberal*
1858	(24 June) Opening of Canal de Lozoya
1859–60	African war; capture of Tetuan
1861	(1 Apr.) Defeat of Ortega's Carlist rising at San Carlos de la Rápita
	(18 Mar.) Annexation of Santo Domingo (evacuated 1865)
1862	(Apr.) Prim's withdrawal from Mexico
1863	(June) Death of Pedro Calvo Asensio
1865–6	War of the Pacific
1865	(10 Apr.) Riots of the *Noche de San Daniel*
	(11 June) Narváez replaced by O'Donnell
	(24 July) Recognition of kingdom of Italy
1866	(June–July) Rising and execution of the sergeants of San Gil barracks
	(10 July) Fall of O'Donnell
1867	(5 Dec.) Death of O'Donnell at Biarritz
1868	(20 Apr.) Death of Narváez; government of González Bravo
	(7 July) Exile of Unionist generals
	(Sept.) Revolution ('setembrina', 'gloriosa'): battle of Alcolea; Isabella leaves for France. Leaders Serrano and Prim
1868–70	*Interinidad*: search for constitutional monarch
1869	(20 Feb.) Speech of 'los jamases' by Prim
	(15 June) Serrano appointed regent, Prim prime minister
1870	(25 June) Abdication of Isabella
	(July–Oct.) Franco-Prussian war. Sedan: defeat and capture of Napoleon III
	(3 Nov.) Election of Amadeo as king
	(30 Dec.) Death of Prim by assassination
1871	(Mar.–May) Commune in Paris
1871–3	Reign of Amadeo I
1873	(11 Feb.) Amadeo's abdication; proclamation of First Republic; four presidents: Figueras, Pi y Margall, Salmerón, Castelar
1873–6	Second Carlist war

1873	(27 June) Concha killed at Abárzuza
	(July) Cantonalist insurrections at Cartagena etc.
1874	(3 Jan.) *Coup d'état* in parliament by Pavía; Serrano regent
	(1 Dec.) Sandhurst manifesto
	(29 Dec.) *Pronunciamiento* at Sagunto by Martínez Campos
1875	(9 Jan.) Restoration: Alfonso XII enters Spain as king
	(Jan.–Sept.) Cánovas ministry
	(Sept.–Dec.) Jovellar ministry
	(Dec.) Cánovas returns to power
1876	(June) Constitution of 1876
1878	(23 Jan.) Alfonso XII marries Mercedes, followed by her death (27 June)
	(10 Feb.) Treaty of El Zanjón in Cuba
1879	(29 Nov.) Alfonso XII marries María Cristina of Habsburg
1881	(8 Feb.) First Sagasta ministry: the *turno pacífico*
1884	(Jan.) Cánovas again prime minister
1885	(25 Nov.) Death of Alfonso XII; regency of María Cristina; Second Sagasta ministry (Galdós *diputado*)
1886	(17 May) Posthumous birth of Alfonso XIII
	(19 Sept.) Villacampa's attempted *pronunciamiento*
1897	(8 Aug.) Cánovas assassinated by anarchist Angiolillo
1898	(Apr.–July) War with the United States: *El Desastre*
	(10 Dec.) Treaty of Paris: loss of Cuba, Puerto Rico, and the Philippines
1902	(17 May) Alfonso XII comes of age
	(Dec.) Galdós's visits to ex-Queen Isabella
1904	(9 Apr.) Death of Isabella II
1905	(24 Nov.) The *Cu-cut* incident: army assault on press
1906	(20 Mar.) The *ley de jurisdicciones* extends military jurisdiction to cases of 'insulting' the nation
1907	(Apr.) Galdós re-enters politics as Republican *diputado*; leader of *Conjunción republicano-socialista*
1909	(26 July–1 Aug.) *Semana trágica* in Barcelona.
	(13 Oct.) Execution of Francisco Ferrer
	(21 Oct.) Fall of Antonio Maura; government of Segismundo Moret
1910	(23 Dec.) Canalejas brings in *Ley del Candado*
1912	(12 Nov.) Canalejas assassinated
1920	(4 Jan.) Galdós's death

Appendix 3
Family Tree of the Bourbon Dynasty in Spain

Note: Reigning monarchs are given in **bold** type; Carlist claimants in *italic*.

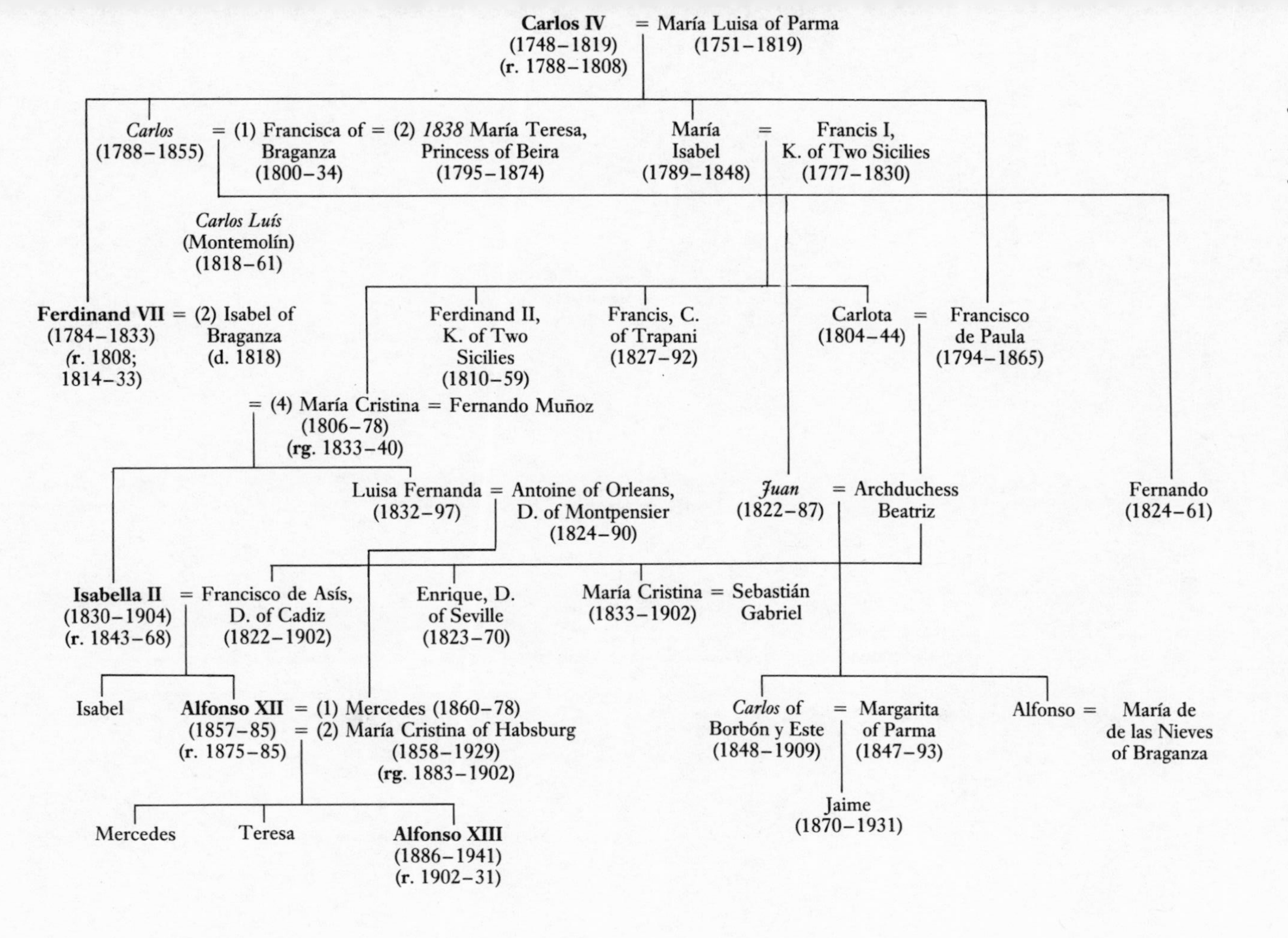
Carlos IV (1748–1819) (r. 1788–1808) = María Luisa of Parma (1751–1819)
Carlos (1788–1855) = (1) Francisca of Braganza (1800–34) = (2) *1838* María Teresa, Princess of Beira (1795–1874)
María Isabel (1789–1848) = Francis I, K. of Two Sicilies (1777–1830)
Carlos Luís (Montemolín) (1818–61)
Ferdinand VII = (2) Isabel of Braganza (d. 1818)
(1784–1833) (r. 1808; 1814–33)
Ferdinand II, K. of Two Sicilies (1810–59)
Francis, C. of Trapani (1827–92)
Carlota (1804–44) = Francisco de Paula (1794–1865)
= (4) María Cristina = Fernando Muñoz
(1806–78) (rg. 1833–40)
Luisa Fernanda (1832–97) = Antoine of Orleans, D. of Montpensier (1824–90)
Juan (1822–87) = Archduchess Beatriz
Fernando (1824–61)
Isabella II (1830–1904) (r. 1843–68) = Francisco de Asís, D. of Cadiz (1822–1902)
Enrique, D. of Seville (1823–70)
María Cristina (1833–1902) = Sebastián Gabriel
Isabel
Alfonso XII (1857–85) (r. 1875–85) = (1) Mercedes (1860–78) = (2) María Cristina of Habsburg (1858–1929) (rg. 1883–1902)
Carlos of Borbón y Este (1848–1909) = Margarita of Parma (1847–93)
Alfonso = María de de las Nieves of Braganza
Mercedes
Teresa
Alfonso XIII (1886–1941) (r. 1902–31)
Jaime (1870–1931)

Bibliography

Note. When articles are subsequently collected in book form, reference is normally given only to the latter, unless accessibility is also a factor.

1. HISTORY

(*a*) *General*

Barraclough, Geoffrey, *History in a Changing World* (Oxford UP, 1955).

Berlin, Sir Isaiah, *Historical Inevitability* (Oxford UP, 1959).

Braudel, Fernand, *On History* (University of Chicago Press, 1980).

Bulhof, Ilsa N., 'Imagination and Interpretation in History', in Leonard Schulze and Walter Wetzels (eds.), *Literature and History* (Lanham: UP of America, 1983), 3–25.

Butterfield, Sir Herbert, *The Historical Novel: An Essay* (Cambridge UP, 1924).

—— 'History as a Branch of Literature', in *History and Human Relations* (London: Collins, 1951), 225–54.

Carr, E. H., *What is History?* (New York: Knopf, 1962).

Collingwood, R. G., *The Idea of History* (Oxford UP, 1946).

Croce, Benedetto, *Theory and History of Historiography* (London: Harrap, 1921).

—— *History of Europe in the Nineteenth Century* (London: Allen and Unwin, 1934).

Gardiner, Patrick (ed.), *Theories of History* (Free Press of Glencoe, 1959).

Geyl, Peter, *The Use and Abuse of History* (New Haven: Yale UP, 1955).

—— *Debates with Historians* (Groningen: J. B. Wolters, 1955).

Huizinga, Johan, 'The Task of Cultural History' (1926), in *Men and Ideas: History, the Middle Ages, the Renaissance* (New York: Meridian, 1959), 17–76.

Kelley, Donald R., *Foundations of Modern Historical Scholarship: Language, Law and History in the French Revolution* (New York: Columbia UP, 1970).

Lukacs, John, *Historical Consciousness or the Remembered Past* (New York: Harper, 1968).

Meinecke, Friedrich, *Historism: The Rise of a New Historical Outlook* (London: Routledge and Kegan Paul, 1972). [Trans. of *Die Entstehung des Historismus* (Munich: Oldenburg, 1959).]

Popper, Karl R., *The Poverty of Historicism* (London: Routledge and Kegan Paul, 1957).

Rose, Margaret A., '"The Second Time as Farce": History as Fiction', in Leonard Schulze and Walter Wetzels (eds.), *Literature and History* (Lanham: UP of America, 1983), 27–35.

Schulze, Leonard, and Wetzels, Walter (eds.), *Literature and History* (Lanham: UP of America, 1983).

White, Hayden, *Metahistory: The Historical Imagination in Nineteenth-Century Europe* (Baltimore: Johns Hopkins UP, 1973).

—— *Tropics of Discourse: Essays in Cultural Criticism* (Baltimore: Johns Hopkins UP, 1978).

—— *The Content of the Form: Narrative Discourse and Historical Representation* (Baltimore: Johns Hopkins UP, 1987).

Woodward, E. L., *Three Studies in European Conservatism: Metternich. Guizot. The Catholic Church in the Nineteenth Century* (London: Constable, 1929).

(*b*) *Spanish*

Alonso Baquer, Miguel (ed.), *La época de la Restauración* (Catalogue of the Exhibition held in the Palacio de Velázquez, Madrid, June–Sept. 1975; Madrid: Publicación del Patronato Nacional de Museos, 1975).

The Attaché in Madrid; or, Sketches of the Court of Isabella II, trans. from the German (New York: Appleton, 1856).

Azaña, Manuel, 'Tres generaciones del Ateneo', in *La invención del Quijote y otros ensayos* (Madrid, 1934).

Ballbé, Manuel, *Orden público y militarismo en la España constitucional (1812–1983)*, 2nd edn. (Madrid: Alianza Universidad, 1985).

Ballesteros Baretta, A., *Historia de España y su influencia en la historia universal*, viii (Barcelona: Salvat, 1936).

Beck, Earl R., *A Time of Triumph and Sorrow: Spanish Politics during the Reign of Alfonso XII, 1874–1885* (Carbondale: Southern Illinois UP, 1979).

Brenan, Gerald, *The Spanish Labyrinth: An Account of the Social and Political Background of the Civil War*, 2nd edn. (Cambridge UP, 1950).

Butler Clarke, H., *Modern Spain, 1815–1898* (Cambridge UP, 1906).

Cambronero, Carlos, *Isabel II íntima* (Barcelona, 1908).

—— *Las cortes de la revolución* (Madrid, n.d.).

Carr, Raymond, *Spain, 1808–1939* (Oxford UP, 1966).

Castelar, Emilio, *Discursos parlamentarios* (3 vols.; Madrid, 1873).

—— *Correspondencia 1868–1898* (Madrid, 1908).

Chant, Roy Heman, *Spanish Tiger: The Life and Times of Ramón Cabrera* (London: Midas, 1983).

Christiansen, Eric, *The Origins of Military Power in Spain* (Oxford UP, 1967).

Comín Colomer, Eduardo, *Historia de la Primera República* (Barcelona: AHR, 1956).

Estévanez, Nicolás, *Fragmento de memorias* (Madrid, 1903).

Fernández Almagro, M., *Historia política de la España contemporánea*, i. *1868–1885* (Madrid: Alianza, 1968).

Fontana, Josep, 'Cambio económico y crisis política: Reflexiones sobre las causas de la revolución de 1868', *Cambio económico y actitudes políticas en la España del siglo XIX*, 5th edn. (Barcelona: Ariel, 1983), 97–145.

Gallenga, A., *Iberian Reminiscences: Fifteen Years' Travelling Impressions of Spain and Portugal* (2 vols.; London: Chapman and Hall, 1883).

García Delgado, José Luis (ed.), *La España de la Restauración* (Madrid: Siglo XXI, 1985).

Harvey, [Annie Jane Tennant,] *Cositas españolas; or, Every Day Life in Spain*, 2nd edn. (London: Hurst and Blackett, 1875).

Hay, John, *Castilian Days* (Boston: Osgood, 1871).

Headrick, Daniel R., *Ejército y política en España (1866–1898)* (Madrid: Tecnos, 1981).

Hennessy, C. A. M., *The Federal Republic in Spain* (Oxford UP, 1962).

Holt, Edgar, *The Carlist Wars in Spain* (Chester Springs, Pa.: Dufour, 1967).

Jover Zamora, José María, *Política, diplomacia y humanismo popular en la España del siglo XIX* (Madrid: Turner, 1976).
—— 'La época de la Restauración: Panorama político-social, 1875–1902', in Gabriel Tortella Casares, *et al.*, *Revolución burguesa, oligarquía y constitucionalismo (1834–1923)* (*Historia de España*, ed. Manuel Tuñón de Lara, 8, 2nd edn.; Barcelona: Labor, 1983), 269–406.
Jutglar Bernaus, Antonio, *Federalismo y revolución: Las ideas sociales de Pi y Margall* (Barcelona: Publicaciones de la Cátedra de Historia General de España, 1966).
Kiernan, V. G., *The Revolution of 1854 in Spanish History* (Oxford UP, 1966).
Leonardon, H., *Prim* (Ministres et Hommes d'État; Paris: Alcan, 1901).
Lida, Clara E., and Zavala, Iris M., *La revolución de 1868: Historia, pensamiento, literatura* (New York: Las Américas, 1970).
Llorca, Carmen, *Emilio Castelar, precursor de la Democracia Cristiana* (Madrid: Biblioteca Nueva, 1966).
—— *Isabel II y su tiempo*, 3rd corr. edn. (Madrid: Istmo, 1984).
Marichal, Carlos, *Spain (1834–1844): A New Society* (London: Tamesis, 1977).
Mesonero Romanos, Ramón de, *Memorias de un setentón* (Madrid: Tebas, 1975).
Nombela, Julio, *Impresiones y recuerdos* (3 vols.; Madrid: La última moda, 1910).
Ortega y Gasset, José, *Vieja y nueva política*, in *Obras*, i, 2nd corr. and ampl. edn. (Madrid: RO, 1936), 83–120.
Oyarzún, Román, *Historia del Carlismo* (Madrid: Ediciones Fe, 1939).
Parry, E. Jones, *The Spanish Marriages, 1841–1846* (London: MacMillan, 1936).
Payne, Stanley G., *Politics and the Military in Modern Spain* (Stanford UP, 1967).
Pedrol Rius, Antonio, *Quién mató a Prim* (1971), 3rd edn. (Madrid: Sociedad de Educación Atenas, 1981).
Pi y Margall, Francisco, *El reinado de Amadeo de Saboya: La República de 1873* (1874), ed. Antonio Jutglar, 2nd edn. (Madrid: Dossat, 1980).
Répide, Pedro de, *Isabel II, Reina de España* (Madrid: Espasa-Calpe, 1952).
Romanones, Count, *Salamanca: Conquistador de riqueza, gran señor* (Madrid: Espasa-Calpe, 1962).
Ruiz Salvador, Antonio, *El Ateneo científico, literario y artístico (1835–1885)* (London: Tamesis, 1971).
Santovenia, Emeterio S., *Prim, el caudillo estadista* (Madrid: Espasa-Calpe, 1933).
Scanlon, Geraldine M., *La polémica feminista en la España contemporánea* (*1865–1974*), 2nd edn. (Madrid: Akal, 1986).
Seco Serrano, Carlos, *Viñetas históricas* (Selecciones Austral; Madrid: Espasa-Calpe, 1983).
Sencourt, Robert, *The Spanish Crown, 1808–1931: An Intimate Chronicle of a Hundred Years* (New York: Scribner's, 1932).
Termes Ardévol, José, *El movimiento obrero en España: La Primera Internacional (1864–1881)* (Barcelona: Publicaciones de la Cátedra de Historia General de España, 1965).
Trend, J. B., *The Origins of Modern Spain* (Cambridge UP, 1934).
Tuñón de Lara, Manuel, *La España del siglo XIX*, ii, 12th edn. (Barcelona: Laia, 1982).
Ullman, Joan Connelly, *The Tragic Week: A Study of Anticlericalism in Spain, 1875–1912* (Cambridge, Mass.: Harvard UP, 1968).
Vicens Vives, J., *Cataluña en el siglo XIX* (Madrid: Rialp, 1961). [Trans. from *Els*

Catalans en el segle XIX (1958).]
—— (ed.), *Historia de España y América social y económica*, v. *Los siglos XIX y XX* (Barcelona: Libros Vicens-Bolsillo, 1972).

2. GENERAL LITERARY HISTORY AND THEORY

Abrams, M. H., *The Mirror and the Lamp* (Oxford UP, 1953).
Allott, Miriam, *Novelists on the Novel* (London: Routledge and Kegan Paul, 1959).
Alonso, Amado, *Ensayo sobre la novela histórica: El modernismo en 'La gloria de don Ramiro'* (1942) (Madrid: Gredos, 1984).
Alter, Robert, *Partial Magic: The Novel as Self-Conscious Genre* (Berkeley: University of California Press, 1975).
Aristotle, *Art of Poetry*, ed. W. Hamilton Fyfe (Oxford UP, 1940).
Attridge, Derek, Bennington, Geoff, and Young, Robert (eds.), *Poststructuralism and the Question of History* (Cambridge UP, 1987).
Auerbach, Erich, *Mimesis: The Representation of Reality* (Princeton UP, 1968).
Bakhtin, Mikhail, *Problems of Dostoevsky's Poetics* (Ardis, 1973); another edn., trans. Caryl Emerson (Manchester UP, 1984).
—— *The Dialogic Imagination: Four Essays* (Austin: Texas UP, 1981).
Bann, Stephen, 'The Sense of the Past: Image, Text, and Object in the Formation of Historical Consciousness in Nineteenth-Century Britain', in H. Aram Veeser (ed.), *The New Historicism* (London: Routledge, 1989), 102–15.
Barthes, Roland, 'Writing and the Novel', in *Writing Degree Zero* (1953) (New York: Hill and Wang, 1968), 29–40.
—— *S/Z* (Paris: Seuil, 1970).
—— *Image, Music, Text*, sel. and trans. Stephen Heath (London: Fontana, 1977).
—— 'The Discourse of History', trans. Stephen Bann, in E. S. Schaffer (ed.), *Comparative Criticism: A Yearbook*, iii (Cambridge UP, 1981), 3–20.
Benjamin, Walter, 'Theses on the Philosophy of History', in Hannah Arendt (ed.), *Illuminations* (New York: Harcourt, 1968), 255–66.
Benveniste, Émile, *Problèmes de linguistique générale* (Paris: Gallimard, 1966).
Berlin, Sir Isaiah, *The Hedgehog and the Fox: An Essay on Tolstoy's View of History* (London: Weidenfeld and Nicolson, 1953).
Booth, Wayne, *The Rhetoric of Fiction* (1961), 2nd expanded edn. (University of Chicago Press, 1983).
—— *A Rhetoric of Irony* (University of Chicago Press, 1974).
—— *Critical Understanding: The Powers and Limits of Pluralism* (University of Chicago Press, 1979).
Chatman, Seymour, *Story and Discourse: Narrative Structure in Fiction and Film* (Ithaca: Cornell UP, 1978).
De George, Richard and Fernande (eds.), *The Structuralists: From Marx to Lévi-Strauss* (New York: Doubleday, 1972).
de Man, Paul, *Blindness and Insight: Essays in the Rhetoric of Contemporary Criticism*, 2nd rev. edn. (Minneapolis: University of Minnesota Press, 1983).
Derrida, Jacques, *On Grammatology* (Baltimore: Johns Hopkins UP, 1974).
—— *Writing and Difference* (University of Chicago Press, 1978).

Eagleton, Terry, *Marxism and Literary Creation* (Berkeley: University of California Press, 1976).

—— *Literary Theory: An Introduction* (Oxford: Blackwell, 1983).

Eco, Umberto, *The Role of the Reader* (London: Hutchinson, 1981).

Fish, Stanley, *Is There a Text in this Class? The Authority of Interpretive Communities* (Cambridge, Mass.: Harvard UP, 1980).

Fleishman, Avrom, *The English Historical Novel: Walter Scott to Virginia Woolf* (Baltimore: Johns Hopkins UP, 1971).

Foley, Barbara, *Telling the Truth: The Theory and Practice of Documentary Fiction* (Ithaca: Cornell UP, 1986).

Forster, E. M., *Aspects of the Novel* (1927) (Harmondsworth: Penguin, 1962).

Foucault, Michel, *The Order of Things: An Archaeology of the Human Sciences* (New York: Pantheon, 1971). [Trans. from *Les Mots et les choses* (Paris: Gallimard, 1966).]

—— *The Archaeology of Knowledge* (New York: Pantheon, 1972). [Trans. from *L'Archéologie du savoir* (Paris: Gallimard, 1969).]

Friedman, Alan, *The Turn of the Novel* (Oxford UP, 1966).

Frye, Northrop, *Anatomy of Criticism: Four Essays* (Princeton UP, 1957).

Genette, Gérard, *Narrative Discourse: An Essay in Method* (Ithaca: Cornell UP, 1972).

Girard, René, *Deceit, Desire and the Novel* (Baltimore: Johns Hopkins UP, 1965). [Trans. from *Mensonge romantique et vérité romanesque* (Paris: Grasset, 1961).]

Harvey, Irene E., *Derrida and the Economy of Différance* (Bloomington: Indiana UP, 1986).

Hirsch, E. D., jun., *Validity in Interpretation* (New Haven: Yale UP, 1967).

—— *The Aims of Interpretation* (University of Chicago Press, 1976).

Hollowell, John, *Fact and Fiction: The New Journalism and the Nonfiction Novel* (Chapel Hill: University of North Carolina Press, 1977).

Iser, Wolfgang, *The Implied Reader* (Baltimore: Johns Hopkins UP, 1974).

—— *The Act of Reading: A Theory of Aesthetic Response* (Baltimore: Johns Hopkins UP, 1978).

Jakobson, Roman, 'Two Aspects of Language and Two Types of Linguistic Disturbances', in Roman Jakobson, and Morris Halle, *Fundamentals of Language* (The Hague: Mouton, 1956), 55–82.

James, Henry, *The Critical Muse: Selected Literary Criticism*, ed. Roger Gard (Harmondsworth: Penguin, 1987).

Jameson, Fredric, *The Prison-House of Language: A Critical Account of Structuralism and Russian Formalism* (Princeton UP, 1972).

—— *The Political Unconscious: Narrative as a Socially Symbolic Act* (Ithaca: Cornell UP, 1981).

Jauss, Hans Robert, *Toward an Aesthetic of Reception* (Minneapolis: Minnesota UP, 1982).

Johnson, Barbara, *The Critical Difference: Essays in the Contemporary Rhetoric of Reading* (Baltimore: Johns Hopkins UP, 1980).

Kermode, Frank, *The Sense of an Ending: Studies in the Theory of Fiction* (Oxford UP, 1966).

—— *The Genesis of Secrecy: On the Interpretation of Narrative* (Cambridge, Mass.: Harvard UP, 1979).

—— *Essays on Fiction, 1971–82* (London: Routledge and Kegan Paul, 1983).

Knapp, Steven, and Michaels, Walter Benn, 'Against Theory', in W. J. T. Mitchell, (ed.), *Against Theory: Literary Studies and the New Pragmaticism* (University of Chicago Press, 1985), 11–30.

LaCapra, Dominick, *History and Criticism* (Ithaca: Cornell UP, 1985).

—— *History, Politics and the Novel* (Ithaca: Cornell UP, 1987).

—— 'On the Line: Between History and Criticism', in *Profession 89* (New York: MLA), 4–9.

Lavers, Annette, *Roland Barthes: Structuralism and After* (London: Methuen, 1982).

Lehan, Richard, 'The Theoretical Limits of the New Historicism', *New Literary History*, 21 (1990), 533–53.

Lentricchia, Frank, *After the New Criticism* (University of Chicago Press, 1980).

Lerner, Lawrence, 'Lukacs' Theory of Realism', in *The Literary Imagination* (Sussex: Harvester Press, 1982), 137–60.

Lodge, David, *The Modes of Modern Writing: Metaphor, Metonymy and the Typology of Modern Literature* (London: Arnold, 1979).

—— *After Bakhtin: Essays on Fiction and Criticism* (London: Routledge, 1990).

Lubbock, Percy, *The Craft of Fiction* (London: Jonathan Cape, 1921).

Lukács, Georg, *The Historical Novel* (Harmondsworth: Penguin, 1969).

Matejka, Ladislev, and Pomovska, Krystyna (eds.), *Readings in Russian Poetics* (Cambridge, Mass.: MIT Press, 1971).

Miller, D. A., *Narrative and its Discontents: Problems of Closure in the Traditional Novel* (Princeton University Press, 1981).

Mitchell, W. J. T. (ed.), *On Narrative* (University of Chicago Press, 1981).

—— (ed.), *Against Theory: Literary Studies and the New Pragmaticism* (Chicago UP, 1985).

Muecke, D. C., *The Compass of Irony* (London: Methuen, 1969).

—— *Irony and the Ironic* (London: Methuen, 1982).

Norris, Christopher, *Deconstruction: Theory and Practice* (London: Methuen, 1982).

Ortega y Gasset, José, *Meditaciones del Quijote*, 2nd edn. (Madrid: Calpe, 1921).

—— *La deshumanización del arte e Ideas sobre la novela* (Madrid: RO, 1925).

Pascal, Roy, *The Dual Voice: Free Indirect Speech and its Functioning in the Nineteenth-Century European Novel* (Manchester UP, 1977).

Rimmon-Kenan, Shlomith, *Narrative Fiction: Contemporary Poetics* (London: Routledge, 1983).

Robert, Marthe, *Roman des origines et origines du roman* (Paris: Éditions de Minuit, 1963). [English trans.: *Origins of the Novel* (Bloomington: Indiana UP, 1980).]

Ryan, Michael, *Marxism and Deconstruction: A Critical Articulation* (Baltimore: Johns Hopkins UP, 1982).

Said, Edward W., *Beginnings: Intentions and Method* (New York: Basic Books, 1975).

—— *The World, the Text and the Critic* (Cambridge, Mass.: Harvard UP, 1983).

Saussure, Ferdinand de, *Course in General Linguistics* (1959), ed. Charles Bally and Albert Sechehaye (New York: McGraw-Hill, 1966). [Trans. of *Cours de Linguistique général* (Paris, 1916).]

Scholes, Robert, *Structuralism in Literature: An Introduction* (New Haven: Yale UP, 1974).

—— *Fabulation and Metafiction* (Urbana: University of Illinois Press, 1979).

—— 'Afterthoughts on Narrative', ii: [Robert Scholes,] 'Language, Narrative, and

Anti-Narrative', in W. J. T. Mitchell (ed.), *On Narrative* (Chicago: University of Chicago Press, 1981), 200–8.

—— *Textual Power, Literary Theory and the Teaching of English* (New Haven: Yale UP, 1985).

—— and Kellogg, Robert, *The Nature of Narrative* (Oxford UP, 1966).

Smart, Robert Augustin, *The Nonfiction Novel* (Lanham, Md.: UP of America, 1985).

Spilka, Mark (ed.), *Towards a Poetics of Fiction* (Bloomington: Indiana UP, 1977).

Sturrock, John (ed.), *Structuralism and Since: From Lévi-Strauss to Derrida* (Oxford UP, 1979).

Tillotson, Kathleen, *Novels of the Eighteen-Forties* (Oxford UP, 1954).

Todorov, Tzvetan, *Littérature et signification: Poétique de la prose* (Paris: Larousse, 1967).

—— *Introduction à la littérature fantastique* (Paris: Seuil, 1970). [English trans.: *The Fantastic* (Ithaca: Cornell UP, 1975).]

—— *Mikhäil Bakhtine: Le Principe dialogique* (Paris: Seuil, 1981).

Tompkins, Jane P. (ed.), *Reader-Response Criticism: From Formalism to Post-Structuralism* (Baltimore: Johns Hopkins UP, 1980).

Torgovnick, Marianna, *Closure in the Novel* (Princeton UP, 1981).

Uitti, Karl D., *Linguistics and Literary Theory* (Princeton UP, 1969).

Veeser, H. Aram (ed.), *The New Historicism* (London: Routledge, 1989).

Watt, Ian, *The Rise of the Novel* (London: Chatto and Windus, 1957).

Wright, Roger, 'Fact and Fiction in Language and Literature', *Liverpool Papers in Language and Discourse*, 3 (1990), 1–37.

3. GALDÓS

(*a*) *Works*

Hyman, Diane Beth, 'The *Fortunata y Jacinta Manuscript* of Benito Pérez Galdós' (diss. Harvard, 1972).

Menéndez y Pelayo—Pereda—Pérez Galdós, *Discursos leídos ante la Real Academia* (Madrid: Viuda e Hijos de Tello, 1897).

Pérez Galdós, Benito, Library and Archives, Casa-Museo de Galdós, Las Palmas, Gran Canaria.

—— *Los artículos políticos en la 'Revista de España', 1871–1872*, ed. Brian J. Dendle and Joseph Schraibman, Introduction by Brian J. Dendle (Lexington, Ky.: Dendle and Schraibman, 1982).

—— *El caballero encantado*, ed. Julio Rodríguez-Puértolas, 2nd edn. (Madrid: Cátedra, 1979).

—— *Las cartas desconocidas de Galdós en 'La Prensa' de Buenos Aires*, ed. William H. Shoemaker (Madrid: Cultura Hispánica/CIGC, 1973).

—— *Cartas de Pérez Galdós a Mesonero Romanos*, ed. Eulogio Varela Hervías (Madrid: Artes Gráficas Municipales, 1943).

—— '28 cartas de Galdós a Pereda', ed. Carmen Bravo Villasante, *CHA* 250–2 (1970–1), 9–51.

—— *Cuentos, teatro y censo*, ed. Federico Carlos Sainz de Robles (Madrid: Aguilar, 1977).

—— *Doña Perfecta*, ed. Rodolfo Cardona, 2nd edn. (Madrid: Cátedra, 1984).

—— *Episodios nacionales*, Edición Ilustrada (10 vols.; Madrid: La Guirnalda y Episodios Nacionales, 1882–5).
—— *Episodios nacionales: Guerra de independencia, extractada para uso de los niños* (Madrid: Sucesores de Hernando, [1909?]).
—— *Fortunata y Jacinta* (manuscript): Houghton Library, Harvard University, Cambridge, Mass.
—— *Fortunata y Jacinta*, ed. Pedro Ortiz Armengol (2 vols.; Madrid: Hernando, 1980).
—— *Fortunata y Jacinta*, ed. Francisco Caudet (2 vols.; Madrid: Cátedra, 1983).
—— *La de Bringas*, ed. Alda Blanco and Carlos Blanco Aguinaga (Madrid: Cátedra, 1983).
—— *Miau*, ed. Robert J. Weber (Barcelona: Labor, 1973).
—— *Misericordia*, con prólogo del autor (Paris: Nelson, 1913).
—— *Misericordia*, ed. Luciano García Lorenzo and Carmen Menéndez Onrubia, 2nd edn. (Madrid: Cátedra, 1982).
—— *Obras completas*, ed. Federico Carlos Sainz de Robles (Madrid: Aguilar. **I**, 12th edn. 1970; **II**, 10th edn. 1970; **III**, 9th edn. 1968; **IV**, 4th edn. 1964; **V**, 3rd edn. 1961; **VI**, lst edn. 1971).[1]
—— *Obras inéditas*, ed. Alberto Ghiraldo (4 vols.; Madrid: Renacimiento, 1923–30). [= *OI*]
[*Pérez*] *Galdós periodista* (Madrid: Banco de Crédito Industrial, 1981).
—— *Los prólogos de Galdós*, ed. William H. Shoemaker (Urbana and Mexico City: University of Illinois Press and Ediciones de Andrea, 1962).
—— *Rosalía*, ed. Alan Smith (Madrid: Cátedra, 1983).
—— *Tormento*, ed. Eamonn Rodgers (Oxford: Pergamon, 1977).
—— *Trafalgar*, ed. J. Rodríguez-Puértolas (Madrid: Cátedra, 1983).

(*b*) *Biography and general criticism*

Alas, Leopoldo [Clarín], *Benito Pérez Galdós: Estudio crítico-biográfico* (Madrid, 1889).
Alonso, Amado, 'Lo español y lo universal en la obra de Galdós', in *Materia y forma en poesía*, 3rd edn. (Madrid: Gredos, 1977), 201–21.
Amor y Vázquez, José, 'Galdós, Valle-Inclán, esperpento', *Act I* (1977), 189–200.
Anderson, Farris, 'Madrid, los balcones y la historia: Mesonero Romanos y Galdós', *CHA* 464 (1989), 63–75.
Andreu, Alicia G., *Galdós y la literatura popular* (Madrid: Sociedad General Española de Librería, 1982).
Arencibia, Yolanda, 'La guerra y la patria en el pensamiento de Galdós', *Boletín Millares Carlo*, 5 (1987), 195–205.
Armas Ayala, Alfonso, 'Galdós y sus contemporáneos', *AG* suppl. (1976), 7–19.
—— *Galdós: Lectura de una vida* (Santa Cruz de Tenerife: Servicio de Publicaciones de la Caja General de Ahorros de Canarias, 1989).
Ayala, Francisco, *La novela: Galdós y Unamuno* (Barcelona: Seix Barral, 1974).
—— *Galdós en su tiempo* (Santander: CIGC, 1978).

[1] Volume **VI**, subtitled *Novelas. Miscelánea*, repeats the novels from *Ángel Guerra* to *Misericordia* and then includes the titles from *El abuelo* onwards. My references are always to volume V when there is duplication; volume **VI** is used only for the later novels and *Miscelánea*.

Ballantyne, Margaret A., 'Indice de la *Revista de España* bajo la dirección de Galdós', *Hisp.* 73 (1990), 332–44.
Baquero Goyanes, Mariano, 'Perspectivismo irónico en Galdós', *CHA* 250–2 (1970–1), 143–60.
Berkowitz, H. Chonon, *Pérez Galdós: Spanish Liberal Crusader* (Madison: University of Wisconsin Press, 1948).
—— *La biblioteca de Benito Pérez Galdós* (Las Palmas: El Museo Canario, 1951).
Beverley, John, '"Seeing History": Reflections on Galdós' *El abuelo*', *AG* 10 (1975), 55–9.
Beyrie, Jacques, *Galdós et son Mythe* (3 vols.; Lille and Paris: Champion, 1980).
Blanco Aguinaga, Carlos, 'Silencios y cambios de rumbo: Sobre la determinación histórica de las ficciones de Galdós', in *GyH* (1988), 187–206.
Blanquat, Josette, 'Au temps d'*Electra*', *BH* 68 (1966), 253–308.
—— 'Documentos galdosianos: 1912', *AG* 3 (1968), 143–50.
Bly, Peter A., 'Galdós, the Madrid Royal Palace and the September 1868 Revolution', *RCEH* 5 (1980), 1–17.
—— *Vision and the Visual Arts in Galdós: A Study of the Novels and Newspaper Articles* (Liverpool: Francis Cairns, 1986).
—— 'La comitiva borbónica en la obra galdosiana: hacia una tipología', in A. David Kossoff *et al.* (eds.), *Act 8 AIH (Brown University, 1983)* (Madrid: Istmo, 1986), i. 255–62.
—— 'Introducción', *GyH* (1988), 13–32.
Boo, Matilde L., 'La perspectiva de Galdós en el asesinato del obispo Martínez Izquierdo', *AG* 12 (1977), 141–5.
Botrel, Jean-François, 'Benito Pérez Galdós, ¿Escritor nacional?', *Act I* (1977), 60–79.
—— 'Le succès d'édition des œuvres de Benito Pérez Galdós (1 & 2)', *Anales de literatura española* (Alicante), 3 (1984), 119–57; ibid. 4 (1985), 29–66.
Bravo-Villasante, Carmen, *Galdós visto por sí mismo* (Madrid: Novelas y cuentas, 1970).
Cabañas, Pablo, 'Moratín en la obra de Galdós', *Act 2 AIH* (Nijmegen: Instituto de la Universidad de Nimega, 1967), 217–26.
Cardona, Rodolfo, 'Don Benito el Prudente', *AG* suppl.: Los Angeles Symposium (1976), 127–52.
Cartas a Galdós, ed. Soledad Ortega (Madrid: RO, 1964).
Casalduero, Joaquín, *Vida y obra de Galdós (1843–1920)*, 2nd ampl. edn. (Madrid: Gredos, 1961).
—— 'El tren como símbolo: El progreso, la clase social, la cibernética de Galdós', *AG* 5 (1970), 15–22.
Cazottes, Gisèle, 'Un jugement de Galdós sur Bécquer', *BH* 77 (1975), 140–53.
Chamberlin, V. A., 'Galdós' Sephardic Types', *Symp.* 17 (1963), 85–100.
—— 'Galdós and the *Movimiento Pro-Sefardita*', *AG* 16 (1981), 91–7.
Clavería, Carlos, 'Sobre la veta fantástica en la obra de Galdós', *Atlante* (London) 1/2 (Apr. 1953), 78–86; ibid. 1/3 (July 1953), 267–74.
—— 'El pensamiento histórico de Galdós', *RNC* 121–2 (1957), 170–7.
Dendle, Brian J., 'Galdós and the Death of Prim', *AG* 4 (1969), 63–71.
—— 'Isidora, the *mantillas blancas* and the Attempted Assassination of Alfonso XII', *AG* 17 (1982), 51–4.
—— 'Galdós in *El año político*', *AG* 19 (1984), 87–107.

—— 'Galdós and Sol y Ortega', *HR* 53 (1985), 437–47.

Estébanez Calderón, Demetrio, 'Evolución política de Galdós y su repercusión en la obra literaria', *AG* 17 (1982), 7–23.

—— 'El lenguaje político de Galdós: "Revolución" y "restauración" en *Fortunata y Jacinta* y en los *Episodios* de la última serie', *BBMP* 61 (1985), 259–83.

Faus Sevilla, Pilar, *La sociedad española del siglo XIX en la obra de Pérez Galdós* (Valencia: Nachar, 1972).

Finkenthal, Sydney, 'Galdós en 1913', in A. M. Gordon and E. Rugg (eds), *Act 6 AIH* (University of Toronto, 1980), 245–47.

Fox, E. Inman, 'Galdós' *Electra*: A Detailed Study of its Historical Significance and the Polemic between Martínez Ruiz and Maeztu', *AG* 1 (1966), 131–44.

Fuentes, Víctor, *Galdós demócrata y republicano (escritos y discursos 1907–1913)* (Santa Cruz de Tenerife: CIGC/Universidad de La Laguna, 1982).

Goldman, Peter B., 'Galdós and the Politics of Reconciliation', *AG* 4 (1969), 73–87.

—— 'Historical Perspective and Political Bias: Comments on Recent Galdós Criticism', *AG* 6 (1971), 114–24.

—— 'Galdós and the Nineteenth-Century Novel: The Need for an Interdisciplinary Approach', *AG* 10 (1975), 5–18.

Guimerá Peraza, Marcos, *Maura y Galdós* (Las Palmas: CIGC, 1967).

—— 'El pleito de Galdós (1896–1899)', *Act I* (1977), 80–105.

Gullón, Germán, 'La historización de la "nueva" crítica: El caso de Galdós', in A. David Kossoff *et al.* (eds.), *Act 8 AIH (Brown University, 1983)* (Madrid: Istmo, 1986), i. 661–9.

Hoar, Leo J., jun., *Benito Pérez Galdós y la Revista del Movimiento Intelectual de Europa* (Madrid: Insula, 1968).

Lambert, A. F., 'Galdós and Concha-Ruth Morell', *AG* 8 (1973), 33–49.

Letamendía, Emily, 'Pérez Galdós and *El Océano: 1879–1880*', *AG* 10 (1975), 83–9.

Llorens, Vicente, *Aspectos sociales de la literatura española* (Madrid: Castalia, 1974).

Madariaga, Benito, *Pérez Galdós: Biografía santanderina* (Santander: Instituto Cultural de Cantabria, 1979).

Miller, Stephen, *El mundo de Galdós* (Santander: Sociedad Menéndez Pelayo, 1983).

Morón Arroyo, Ciriaco, 'Galdós y Ortega y Gasset: Historia de un silencio', *AG* 1 (1966), 143–50.

Nuez, Sebastián de la, 'Correspondencia epistolar entre Maura y Galdós (1989–1914)', *Anuario de estudios atlánticos*, 20 (1974), 613–68.

—— 'Las últimas novelas de Galdós a través de un epistolario amoroso', *Cent.* (1989), 205–16.

Palomo Vázquez, Pilar, 'Galdós y Mesonero (una vez más: costumbrismo y novela)', *Cent.* (1989), 217–38.

Pattison, Walter, *Benito Pérez Galdós* (Boston: Twayne, 1975).

Percival, Anthony, 'Galdós y lo autobiográfico: Notas sobre Memorias de un desmemoriado', in Giuseppe Bellini (ed.), *Act 7 AIH* (Rome: Bulzoni, 1982), 807–15.

—— *Galdós and his Critics* (University of Toronto Press, 1985).

Pérez Vidal, J., 'Galdós y la noche de San Daniel', *RHM* 17 (1951), 94–110.

Porter, Phoebe A., 'La correspondencia de Benito Pérez Galdós con Teodosia Gandarias' (Unpublished article).

Ricard, Robert, 'El asesinato del obispo Martínez Izquierdo (1886) y el clero

madrileño de la época de Galdós', *AG* 1 (1966), 125–9.
—— 'Cartas de Ricardo Ruiz Orsatti a Galdós acerca de Marruecos (1901–1910)', *AG* 3 (1968), 99–117.
Rodgers, Eamonn, 'Teoría literaria y filosofía de la historia en el primer Galdós', *GyH* (1988), 35–47.
Sackett, Theodore A., *Pérez Galdós: An Annotated Bibliography* (Albuquerque: University of New Mexico Press, 1968).
Schmidt, Ruth A., 'José Ortega Munilla: Friend, Critic and Disciple of Galdós', *AG* 6 (1971), 107–11.
Schyfter, Sara E., *The Jew in the Novels of Benito Pérez Galdós* (London: Tamesis, 1978).
Shoemaker, William H., 'Una amistad literaria: La correspondencia epistolar entre Galdós y Narciso Oller', *BRABL* 30 (1963–4), 247–306.
—— 'Galdós' Letters to Gerardo', *AG* 19 (1984), 151–9.
Smith, Alan, 'Galdós y Flaubert', *AG* 18 (1983), 25–37.
Smith, Paul Julian, 'Galdós, Valera, Lacan', in *The Body Hispanic: Gender and Sexuality in Spanish and Spanish-American Literature* (Oxford UP, 1989), 69–104.
Smith, V. A. and Varey, J. E., '*Esperpeuto*: Some Early Usages in the Novels of Galdós,' *GS I* (1970), 195–204.
Utt, Roger L., 'Galdós' Early Journalism in Madrid and the *Las Novedades* (Dis)Connection', *AG* 19 (1984), 71–85.
Varey, J. E., 'Galdós in the Light of Recent Criticism', *GS I* (1970), 1–35.
Woodbridge, Hensley C., 'Benito Pérez Galdós: A Selected Annotated Bibliography', *Hisp.* 53 (1970), 899–971.
Ynduráin, Francisco, *Galdós entre la novela y el folletín* (Madrid: Taurus, 1970).

(*c*) *Novels*

Aldaraca, Bridget, 'The Revolution of 1868 and the Rebellion of Rosalía Bringas', *AG* 18 (1983), 49–60.
Anderson, Farris, *Espacio urbano y novela: Madrid en 'Fortunata y Jacinta'* (Madrid: Porrúa, 1985).
Andreu, Alicia G., 'El folletín como intertexto en *Tormento*', *AG* 17 (1982), 55–61.
—— *Modelos dialógicos en la narrativa de Benito Pérez Galdós* (Amsterdam and Philadelphia: John Benjamins and Purdue University Monographs in Romance Languages, 1989).
Aparici Llanas, María Pilar, *Las novelas de tesis de Benito Pérez Galdós* (Barcelona: CSIC, 1982).
Bacarisse, Salvador, 'The Realism of Galdós: Some Reflections on Language and the Perception of Reality', *BHS* 42 (1965), 239–50.
Bieder, Maryellen, 'La comunicación narrativa en *El amigo Manso* de Benito Pérez Galdós', in A. David Kossoff *et al.* (eds.), *Act 8 AIH (Brown University, 1983)* (Madrid: Istmo, 1986), i. 243–53.
Blanco, Alda, 'Dinero, relaciones sociales y significación en *Lo prohibido*', *AG* 18 (1983), 61–73.
Blanco Aguinaga, Carlos, *La historia y el texto literario: Tres novelas de Galdós* (Madrid: Nuestra Cultura, 1978).
Bly, Peter A., 'Egotism and Charity in *Marianela*', *AG* 7 (1972), 49–66.
—— 'Fortunata and No 11, Cava de San Miguel', *Hispanófila*, 59 (1977), 31–48.

—— 'From Disorder to Order: The Pattern of *Arreglar* References in Galdós' *Tormento* and *La de Bringas*', *Neo.* 62 (1978), 392–405.

—— 'Sex, Egotism and Social Regeneration in Galdós' *El caballero encantado*', *Hisp.* 62 (1979), 20–9.

—— *Pérez Galdós: La de Bringas* (CG; 1981).

—— *Galdós' Novel of the Historical Imagination: A Study of the Contemporary Novels* (Liverpool: Francis Cairns, 1983).

Boring, Phyllis Zatlin, 'The Streets of Madrid as a Structuring Device in *Fortunata y Jacinta*', *AG* 13 (1978), 13–22.

Bosch, Rafael, 'Galdós y la teoría de la novela', *AG* 2 (1967), 169–84.

Braun, Lucille V., 'Galdós' Recreation of Ernestina Manuel de Villegas as Guillermina Pacheco', *HR* 38 (1970), 32–55.

—— 'The Novelistic Function of Mauricia la Dura in Galdós' *Fortunata y Jacinta*', *Symp.* 31 (1977), 277–89.

Bravo-Villasante, Carmen, 'El naturalismo de Galdós y el mundo de *La desheredada*', *CHA* 77 (1969), 479–86.

Cardona, Rodolfo, 'Nuevos enfoques críticos con referencia a la obra de Galdós', *CHA* 250–2 (1970–1), 58–72.

Chamberlin, Vernon, 'The *muletilla*: An Important Facet of Galdós' Characterization Technique', *HR* 29 (1961), 296–309.

Charnon Deutsch, Lou, 'Inhabited Space in B. P. Galdós' *Tormento*', *AG* 10 (1975), 35–43.

Correa, Gustavo, *El simbolismo religioso en las novelas de Pérez Galdós* (Madrid: Gredos, 1962).

—— *Realidad, ficción y símbolo en las novelas de Pérez Galdós*, 2nd edn. (Madrid: Gredos, 1977).

Dendle, Brian J., 'On the Supposed "Naturalism" of Galdós: *La desheredada*', in *Papers on Romance Literary Relations* (Penn State University, 1980), 12–21.

—— 'Isidora, the *mantillas blancas* and the Attempted Assassination of Alfonso XII', *AG* 17 (1982), 51–4.

Dérozier, Albert, 'El *pueblo* de Pérez Galdós en *La Fontana de Oro*', *CHA* 250–2 (1970–1), 285–311.

Elizalde, Ignacio, *Pérez Galdós y su novelística* (Bilbao: Universidad de Deusto, 1981).

Engler, Kay, 'Notes on the Narrative Structure of *Fortunata y Jacinta*', *Symp.* 24 (1970), 111–27; repr. in EC: *FyJ* 235–53.

—— *The Structure of Realism: The* novelas contemporáneas *of Benito Pérez Galdós* (North Carolina Studies in the Romance Languages and Literatures; Chapel Hill: University of North Carolina, 1977).

Eoff, Sherman H., *The Novels of Pérez Galdós: The Concept of Life as Dynamic Process* (St Louis: Washington UP, 1954).

—— *The Modern Spanish Novel* (New York UP, 1961).

Fernández-Cifuentes, Luís, 'Entre Gobseck y Torquemada', *AG* 17 (1982), 71–84.

Folley, T., 'Clothes and the Man: An Aspect of Benito Pérez Galdós' Method of Literary Characterization', *BHS* 49 (1972), 30–9.

Fontanella, Lee, '*Doña Perfecta* as Historiographic Lesson', *AG* 11 (1976), 59–69.

Gilman, Stephen, *Galdós and the Art of the European Novel: 1867–1887* (Princeton UP, 1981).

Gimeno Casalduero, Joaquín, 'Una novela y dos desenlaces', *Atenea*, 32/88 (1955), 6–8.

—— 'Los dos enlaces de *La Fontana de Oro*', *AG* suppl. (1976), 55–69.

—— 'La caracterización plástica del personaje en la obra de Pérez Galdós: Del tipo al individuo', *AG* 7 (1972), 19–25.

Gold, Hazel, 'Francisco's Folly: Picturing Reality in Galdós' *La de Bringas*', *HR* 54 (1986), 47–66.

—— 'Looking for the Doctor in the House: Critical Expectations and Novelistic Structure in Galdós' *El doctor Centeno*', *PQ* 68 (1989), 219–40.

Goldman, Peter B., 'Feijoo and Mr Singer: Notes on the *aburguesamiento* of Fortunata', *REH-PR* 9 (1982), 105–13.

—— (ed.), *Conflicting Realities: Four Readings of a Chapter by Pérez Galdós* (*'Fortunata y Jacinta', Part III, Chapter IV*) (London: Tamesis, 1984).

—— '"Cada peldaño tenía su historia": Conciencia histórica y conciencia social en *Fortunata y Jacinta*', *GyH* (1988), 145–65.

Gordon, Michael, 'The Medical Background of Galdós' *La desheredada*', *AG* 7 (1972), 67–77.

—— '"Lo que falta a un enfermo le sobra a otro": Galdós' Conception of Humanity in *La desheredada*', *AG* 12 (1977), 29–37.

Gullón, Agnes Moncy, 'The Bird Motif and the Introductory Motif: Structure in *Fortunata y Jacinta*', *AG* 9 (1974), 51–75.

Gullón, Germán, 'Unidad de *El doctor Centeno*', *CHA* 250–2 (1970–1), 579–85.

Gullón, Ricardo, *Galdós, novelista moderno* (1966), 3rd edn. (Madrid: Gredos, 1973).

—— *Técnicas de Galdós*, 2nd edn. (Madrid: Taurus, 1980).

Hafter, Munroe Z., 'The Hero in Galdós' *La Fontana de Oro*', *MP* 57 (1959), 36–43.

—— 'Ironic Reprise in Galdós' Novels', *PMLA* 86 (1961), 233–9.

Hall, H. B., 'Torquemada: The Man and his Language', *GS I* (1970), 136–63.

Hoar, Leo J., jun., '"Dos de Mayo de 1808. Dos de Septiembre de 1870," por Benito Pérez Galdós, un cuento extraviado y el posible prototipo de sus *Episodios nacionales*', *CHA* 250–2 (1970–1), 312–39.

Jones, C. A., 'Galdós's *Marianela* and the Approach to Reality', *MLR* 56 (1961), 515–19.

Kronik, John W., '*El amigo Manso* and the Game of Fictive Antonomy', *AG* 12 (1977), 71–94.

—— 'Narraciones interiores en *Fortunata y Jacinta*', in José Amor y Vázquez and A. David Kossoff (eds.), *Homenaje a Juan López-Morillas. De Cadalso a Aleixandre: Estudios sobre literatura e historia intelectual española* (Madrid: Castalia, 1982), 275–91.

—— 'Feijoo and the Fabrication of Fortunata', in Goldman (ed.), *Conflicting Realities* (1984), 39–72.

Krow-Lucal, Martha, '*Un faccioso más* y León Roch: Fin y nuevo comienzo', *Act II* (1978), i. 170–80.

Labanyi, J. M., 'The Political Significance of *La desheredada*', *AG* 14 (1979), 51–7.

Lakhdari, Sadi, 'Le pronunciamiento de Villacampa dans *Ángel Guerra* de Pérez Galdós', *Ibérica* (Cahiers ibériques et ibéro-américains de l'Université de Paris-Sorbonne) (1985), 47–65.

Lambert, A. F., 'Galdós and the Anti-bureaucratic Tradition', *BHS* 53 (1976), 35–49.

Lida, Denah, 'De Almudena y su lenguaje', *NRFH* 15 (1961), 297–308.
—— 'Galdós y sus santas modernas', *AG* 10 (1975), 19–31.
—— 'Galdós y el 68: Dos perspectivas', in Alva V. Ebersole (ed.), *Perspectivas de la novela* (Valencia: Albatros and Hispanófila, 1979), 37–48.
López-Landy, Ricardo, *El espacio novelesco en la obra de Galdós* (Madrid: Ediciones de Cultura Hispánica del Centro de Cooperación Iberoamericana, 1979).
López-Morillas, Juan, 'Historia y novela en el Galdós primerizo: En torno a *La Fontana de Oro*', in *Hacia el 98: Literatura, sociedad, ideología* (Barcelona: Ariel, 1972), 43–77.
Macklin, J. J., 'Benito Pérez Galdós: *Fortunata y Jacinta 1886–87*', in David A. Williams (ed.), *The Monster in the Mirror: Studies in Nineteenth-Century Realism* (Oxford and Hull: Oxford UP, 1978), 179–203.
Miller, Stephen, '*La de Bringas* as *Bildungsroman*: A Feminist Reading', *RQ* 34 (1987), 189–97.
Montesinos, José F., *Galdós* (3 vols.; Madrid: Castalia, 1968–73).
Nimetz, Michael, *Humor in Galdós: A Study of the 'Novelas contemporáneas'* (New Haven: Yale UP, 1968).
Ortiz Armengol, Pedro, *Relojes y tiempo en 'Fortunata y Jacinta'* (Santa Cruz de Tenerife: CIGC, 1978).
—— *De cómo llegó a Inglaterra —y a quién, y a dónde— el primer ejemplar de 'Fortunata y Jacinta', enviado por su autor* (London: [private printing,] 1981).
—— *Apuntaciones para 'Fortunata y Jacinta'* (Madrid: Editorial Universidad Complutense, 1987).
Parker, A. A., 'Villaamil: Tragic Hero or Comic Failure?' *AG* 4 (1969), 13–23.
Pattison, Walter, '*La Fontana de Oro*: Its Early History', *AG* 15 (1980), 5–9.
Petit, Marie-Claire, *Galdós et 'La Fontana de Oro': Genèse de l'œuvre d'un romancier et les sources balzaciennes de 'Fortunata y Jacinta'* (Paris: Ediciones Hispano-Americanas, 1972).
—— *Les Personnages féminins dans les romans de Benito Pérez Galdós* (Lyons: Université, 1972).
Ramsden, Herbert, 'The Question of Responsibility in Galdós' *Miau*', *AG* 6 (1971), 63–78.
Raphaël, S., 'Un extraño viaje de novios', *AG* 3 (1968), 35–49.
Ribbans, Geoffrey, 'Contemporary History in the Structure and Characterization of *Fortunata y Jacinta*', *GS I* (1970), 90–113; repr. in EC: *FyJ* 47–70.
—— *B. Pérez Galdós: 'Fortunata y Jacinta'* (CG; 1977).
—— 'La figura Villaamil en Miau', *Act I* (1977), 397–413.
—— 'El carácter de Mauricia "la dura" en la estructura de *Fortunata y Jacinta*', *Act 5 AIH* (Bordeaux, 1977), ii. 713–21.
—— *Reality Plain or Fancy? Some Reflections on Galdós' Concept of Realism* (Liverpool UP, 1986).
—— 'Feijoo: Policeman, Inventor, Egotist, Failure?', *AG* 22 (1987), 71–87.
—— '*Doña Perfecta*: Yet Another Ending', *MLN* 105 (1990), 203–25.
Ricard, Robert, *Galdós et ses romans* (Paris: PUF, 1961).
—— *Aspects de Galdós* (Paris: PUF, 1963).
Risley, William R., 'Setting in the Galdós Novel', *HR* 46 (1978), 23–40.
Robin, Claire-Nicolle, *Le Naturalisme dans 'La desheredada' de Pérez Galdós* (Paris: Les

Belles Lettres, 1976).
Rodgers, Eamonn, *B. Pérez Galdós: 'Miau'* (CG; 1978).
—— *From Enlightenment to Realism: The Novels of Galdós, 1870–1887* (Dublin: [private printing,] 1987).
Rodríguez, Alfred, *Estudios sobre la novela de Galdós* (Madrid: Porrúa Turanzas, 1967).
Rodríguez, Rodney T., 'Las máscaras del engaño en *Tormento*', in A. David Kossoff *et al.* (eds.), *Act 8 AIH (Brown University, 1983)* (Madrid: Istmo, 1986), ii. 517–24.
Rodríguez-Puértolas, Julio, 'Galdós y *El caballero encantado*', *AG* 7 (1972), 117–32.
—— *Galdós: Burguesía y revolución* (Madrid: Turner, 1975).
Round, Nicholas G., 'Rosalía Bringas' Children', *AG* 6 (1971), 43–50.
—— 'Time and Torquemada: Three Notes on Galdosian Chronology', *AG* 6 (1971), 79–97.
—— 'Villaamil's Three Lives', *BHS* 63 (1986), 19–32.
Ruiz Ramón, Francisco, *Tres personajes galdosianos* (Madrid: RO, 1964).
Ruiz Salvador, Antonio, 'La función del trasfondo histórico en *La desheredada*', *AG* 1 (1966), 53–62.
Sánchez, Roberto G., 'The Function of Dates and Deadlines in Galdós' *La de Bringas*', *HR* 46 (1978), 299–311.
Scanlon, Geraldine, '*El doctor Centeno*: A Study in Obsolescent Values', *BHS* 55 (1978), 245–53.
—— *Pérez Galdós, 'Marianela'* (CG; 1988).
—— and Jones, R. O., '*Miau*: Prelude to a Reassessment', *AG* 6 (1971), 53–62.
Schraibman, Joseph, *Dreams in the Novels of Galdós* (New York: Hispanic Institute, 1960).
—— 'Galdós y Unamuno', in Germán Bleiberg and E. Inman Fox (eds.), *Spanish Thought and Letters in the Twentieth Century (Centenary of Miguel de Unamuno)* (Nashville: Vanderbilt UP, 1966), 451–82.
Shoemaker, W. H., 'Galdós' Literary Creativity: D. José Ido del Sagrario', *HR* 19 (1951), 204–37. [Repr. in *Estudios sobre Galdós* (Madrid: Castalia, 1970), 85–122.]
—— 'Galdós' Classical Scene in *La de Bringas*', *HR* 27 (1959), 423–34. [Repr. in *Estudios sobre Galdós*, 145–58.]
—— *The Novelistic Art of Galdós* (2 vols.; Valencia: Albatros and Hispanófila, 1980).
Sinnigen, J. H., 'Individual, Class, and Society in *Fortunata y Jacinta*', *GS II* (1974), 49–68; repr. in EC: *FyJ* 71–93.
—— 'Ideología, reflejo y estructuras literarias en Galdós: El ejemplo *de Tormento*', in A. M. Gordon and E. Rugg (eds.), *Act 6 AIH* (University of Toronto, 1980), 711–13.
—— 'Galdós's *Tormento*: Political Partisanship/Literary Structures', *AG* 15 (1980), 73–82.
Smieja, Florian, 'An Alternative Ending to *La Fontana de Oro*', *MLR* 41 (1966), 426–33.
Smith, Gilbert, 'Galdós, *Tristana*, and Letters from Concha-Ruth Morell', *AG* 10 (1975), 91–120.
Sobejano, Gonzalo, 'Muerte del solitario (Benito Pérez Galdós: *Fortunata y Jacinta*, 4.ª, II, 6)', in Andrés Amorós *et al.*, *El comentario de textos*, iii. *La novela realista* (Madrid: Castalia, 1979, 203–54).
Suárez, Manuel, 'Torquemada y Gobseck', *Act II*, ii. (1980), 369–82.

Tarrío, Ángel, *Lectura semiológica de 'Fortunata y Jacinta'* (Las Palmas: CIGC, 1982).
Terry, Arthur, '*Lo prohibido*: Unreliable Narrator and Untruthful Narrative', *GS I*, (1970), 62–89.
Turkovich, Ludmilla B., 'Tolstoj and Galdós: Affinities and Coincidences Reviewed', in Victor Terras (ed.), *American Contributions to the Eighth International Congress of Slavists (Zagreb and Ljubljana, September 3–9, 1978)*, xxi. *Literature* (Columbus, Ohio: Slavish Publishers, 1983), 707–33.
Turner, Harriet S., 'Family Ties and Tyrannies: A Reassessment of Jacinta', *HR* 51 (1983), 1–22.
Urbina, Eduardo, 'Mesías y redentores: Contraste estructural y motivo temático en *Fortunata y Jacinta*', *BH* 83 (1981), 379–98.
Urey, Diane F., *Galdós and the Irony of Language* (Cambridge UP, 1982).
—— 'Problems in Defining Objects of Critical Analysis in the *Torquemada* Novels of Galdós', *KRQ* 31 (1984), 189–96.
Varey, J. E., 'Francisco Bringas, nuestro buen Thiers', *AG* 1 (1966), 63–9.
Weber, Robert J. (ed.), *The* Miau *Manuscript of Benito Pérez Galdós: A Critical Study* (Berkeley: University of California Press, 1964).
Whiston, James, 'The Materialism of Life: Religion in *Fortunata y Jacinta*', *AG* 14 (1979), 65–81.
—— 'Determinism and Freedom in *Fortunata y Jacinta*', *BHS* 57 (1980), 113–27.
Willem, Linda M., 'The Narrative Premise of the Dual Ending to Galdós's *La Fontana de Oro*', *RN* 28 (1987), 51–6.
Wright, Chad, 'The Representational Qualities of Isidora Rufete's House and her Son Riquín in Benito Pérez Galdós's Novel *La desheredada*', *RF* 83 (1971), 230–45.
—— 'Imagery of Light and Darkness in *La de Bringas*', *AG* 13 (1978), 5–12.
Zahareas, Anthony, 'The Tragic Sense in *Fortunata y Jacinta*', *Symp.* 19 (1965), 38–49. [Spanish trans.: 'El sentido de la tragedia en *Fortunata y Jacinta*', *AG* 3 (1968), 25–34.]
Zlotchew, Clark M., 'The Genial Inquisitor of *El audaz*', *AG* 20/1 (1985), 29–34.

(*d*) Episodios nacionales

Note. In general, the bibliography is confined to studies directly referring to *Bodas reales* and the 4th and 5th series or issues relevant to them.

Avalle-Arce, Juan Bautista, '*Zumalacárregui*', *CHA* 250–2 (1970–1), 356–73.
Behiels, Lieve, 'La literatura italiana en *Las tormentas del 48* de Benito Pérez Galdós', in A. Vilanova (ed.), *Act 10 AIH* (Barcelona, 1989) (Barcelona: Universitat, 1992), ii. 1193–1201.
Benítez, Rubén, 'Jenara de Baraona, narradora galdosiana', *HR* 53 (1985), 307–27.
Bly, Peter A., 'For Self or Country? Conflicting Lessons in the First Series of the *Episodios nacionales*?', *KRQ* 31 (1984), 117–24.
—— 'On Heroes: Galdós and the Idea of Military Leadership', *RCEH* 10 (1986), 339–51.
Bush, Peter, 'The Craftsmanship and Literary Values of the Third Series of *Episodios nacionales*', *AG* 16 (1981), 33–56.
Cardona, Rodolfo, 'Apostillas a los *Episodios nacionales de Benito Pérez Galdós* de Hans Hinterhäuser', *AG* 3 (1968), 119–42.

—— '*Mendizábal: Grandes esperanzas*', *GyH* (1988), 99–112.

Carr, Raymond, 'A New View of Galdós', *AG* 3 (1968), 185–9.

Casalduero, Joaquín, 'Historia y novela', *CHA* 250–2 (1970–1), 135–42.

Collins, Vera, 'Tolstoy and Galdós' Santiuste: Their Ideology on War and their Spiritual Conversion', *Hisp.* 53 (1970), 836–41.

Dendle, Brian J., 'The First Cordero: *Elia* and the *Episodios nacionales*', *AG* 7 (1972), 103–5.

—— *Galdós: The Mature Thought* (Lexington, Ky.: UP of Kentucky, 1980).

—— *The Early Historical Novels* (Columbia: University of Missouri Press, 1986).

Enguídanos, Miguel, 'Mariclío, musa galdosiana', *PSA* 63, (June 1961), 235–49; repr. in EC: *G* 427–36.

García Barrón, Carlos, 'Fuentes históricas y literarias de *La vuelta al mundo en la "Numancia"*', *AG* 18 (1983), 111–24.

Gilman, Stephen, 'Judíos, moros y cristianos en las historias de don Benito y de don Américo', in *Homenaje a Sánchez Barbudo* (Madison: University of Wisconsin, 1981), 25–36.

—— 'The Fifth Series of *Episodios nacionales*: Memories of Remembering', *BHS* 63 (1986), 47–52.

Gullón, Ricardo, 'La historia como materia novelable', *AG* 5 (1970), 23–37; repr. in EC: *G* 403–26.

—— '*Episodios nacionales*: Problemas de estructura', *LD* 4 (1974), 33–59.

Hinterhäuser, Hans, *Los 'Episodios nacionales' de Benito Pérez Galdós* (Madrid: Gredos, 1963).

Journeau, Brigitte, 'Histoire et mythe dans *Las tormentas del 48* de Pérez Galdós', in C. Dumas (ed.), *Les Mythes et leur expression au XIXe siècle dans le monde hispanique et ibéro-américain*, (Lille: Press Universitaire de Lille, 1988), 137–53.

—— 'Histoire et création dans *Narváez* de Pérez Galdós', in *Hommage à Claude Dumas: Histoire et création* (Lille: PUF, 1990), 67–75.

Jover Zamora, J. M., 'El fusilamiento de los sargentos de San Gil (1866) en el relato de Pérez Galdós: Los dos primeros capítulos de *La de los tristes destinos*', in *El comentario de textos*, ii. *De Galdós a García Márquez* (Madrid: Castalia, 1974), 15–110; repr. in *Política, diplomacia y humanismo popular en la España del siglo XIX* (Madrid: Turner, 1976), 365–427.

Lécuyer, M. C., and Serrano Carlos, *La Guerre d'Afrique et ses répercussions en Espagne 1859–1904* (Paris: PUF, 1975). [On *Aita Tettauen*, pp. 293–356.]

Letamendía, Emily, 'Pérez Galdós and *El Océano*: 1879–1880', *AG* 10 (1975), 83–9.

Lida, Clara E., 'Galdós y los *Episodios nacionales*: Una historia del liberalismo español', *AG* 13 (1968), 61–77.

Martínez Ruiz, Juan, 'Ficción y realidad judeoespañola en el *Aita Tettauen* de Benito Pérez Galdós', *RFE* 59 (1977), 145–82.

Navarro González, Alberto, '*Los Episodios nacionales* extractados por Galdós', *Act I* (1977), 164–76.

Pabón, Jesús, 'El espadón en la novela', *CHA* 265–7 (1972), 220–33.

Regalado García, Antonio, *Benito Pérez Galdós y la novela histórica española, 1868–1912* (Madrid: Insula, 1966).

Ricard, Robert, 'Notes sur le genèse de l'*Aita Tettauen* de Galdós', *BH 37* (1935), 475–7.

—— 'Mito, sueño, historia y realidad en *Prim*', *CHA* 250–2 (1970–1), 340–55.
—— 'Pour un cinquantenaire: Structure et inspiration de *Carlos VII, en la Rápita* (1905)', *BH* 57 (1955), 70–3.
Robin, Claire-Nicolle, '*La de los tristes destinos*: Un roman historique tardif de Benito Pérez Galdós', in *Recherches sur le roman historique en Europe: XVIII–XIX siècles*, i (Besançon and Paris: Les belles lettres, 1977), 211–53.
Rodríguez, Alfred, *An Introduction to the 'Episodios nacionales' of Galdós* (New York: Las Américas, 1967).
Schraibman, José, 'Pedro Antonio de Alarcón y Galdós: Dos visiones de la guerra de Africa (1859–60)', *LT* (Nueva Época), 1 (1987), 539–47.
Seco Serrano, Carlos, 'Los *Episodios nacionales* como fuente histórica', *CHA* 250–2 (1970–1), 256–84.
Shoemaker, William H., 'Galdós' *La de los tristes destinos* and its Shakespearian Connections', *MLN* 71 (1956), 114–19; repr. in *Estudios sobre Galdós* (Madrid: Castalia, 1970), 139–44.
Smith, Alan, 'El epílogo a la primera edición de *La batalla de los Arapiles*', *AG* 17 (1982), 105–8.
Torres Nebrera, Gregorio, '*Aita Tettauen*: Texto y contexto de un episodio nacional', *Cent.* (1989), 385–407.
Urey, Diane Faye, 'Isabel II and Historical Truth in the Fourth Series of Galdós' *Episodios nacionales*', *MLN* 98 (1983), 189–207.
—— 'Words for Things: The Discourse of History in Galdós' *O'Donnell*', *BHS* 63 (1986), 33–46.
—— 'La revisión como proceso textual en los *Episodios nacionales*: El caso de *Bodas reales*', in *GyH* (1988), 113–30.
—— *The Novel Histories of Galdós* (Princeton UP, 1989). [Includes revised versions of 1983, 1986, 1988.]
Varela, Antonio, 'Galdós's Last *Episodios nacionales*', *Hisp.* 70 (1987), 31–9.

4. OTHER LITERARY AND CRITICAL WORKS

(*a*) *Hispanic*

Alarcón, Pedro Antonio de, *Obras completas*, 2nd edn. (Madrid: Faz, 1954).
—— *Obras olvidadas*, ed. Cyrus de Coster (Madrid: Studia Humanitas and Porrúa, 1984).
Alas, Leopoldo ['Clarín'], *La Regenta*, ed. Gonzalo Sobejano (2 vols.; Madrid: Castalia, 1981).
Azorín, *Castilla*, ed. Juan Manuel Rozas (Barcelona: Labor, 1973).
Baroja, Pío, *Obras completas* (8 vols.; Madrid: Biblioteca Nueva, 1948).
Bauer, Beth Wietelmann, 'Innovación y apertura: La novela realista del siglo XIX ante el problema del desenlace', *HR* 59 (1991), 187–203.
Bécquer, Gustavo Adolfo, *Libro de los gorriones*, ed. María del Pilar Palomo (Madrid: Cupsa, 1977).
Beser, Sergio, *Leopoldo Alas, crítico literario* (Madrid: Gredos, 1968).
Borges, Jorge Luis, *Ficciones* (Buenos Aires: Emecé, 1956).
Carrasquer, Francisco, *'Imán' y la novela histórica de Sender* (London: Tamesis, 1970).
Cervantes, Miguel de, *Novelas ejemplares*, ii, ed. Harry Sieber, 6th edn. (Madrid: Cátedra, 1984).

Coloma, Luis, *Pequeñeces*, ed. Rubén Benítez (Madrid: Cátedra, 1975).
Dendle, Brian J., *The Spanish Novel of Religious Thesis 1876–1936* (Princeton UP and Madrid: Castalia, 1968).
Earle, Peter G., 'Unamuno: *Historia* and *intrahistoria*', in Germán Bleiberg and E. Inman Fox (eds.) *Spanish Thought and Letters in the Twentieth Century* (*Centenary of Miguel de Unamuno*) (Nashville: Vanderbilt UP, 1966), 179–86.
Ebersole, Alva V. (ed.), *Perspectivas de la novela* (Valencia: Albatros and Hispanófila, 1979).
Extramiana, José, *La guerra de los vascos en el 98: Unamuno—Valle Inclán—Baroja* (San Sebastián: Haranburu, 1983).
Ferreras, Juan Ignacio, *Catálogo de novelas y novelistas españoles del siglo XIX* (Madrid: Cátedra, 1979).
Fletcher, Madeleine de Gogorza, *The Spanish Historical Novel 1870–1970* (London: Tamesis, 1974).
Gilabert, Joan, *Narcís Oller: Estudio comparativo con la novela castellana del siglo XIX* (Barcelona: Marte, 1977).
González Arias, Francisca, '*La tribuna* de Emilia Pardo Bazán como novela histórica', *AG* 19 (1984), 133–40.
Gullón, Germán, *El narrador en la novela del siglo XIX* (Madrid: Taurus, 1976).
Henn, David, 'Aspectos políticos de *La tribuna* de Emilia Pardo Bazán', in Jesús Riuz-Veintimilla (ed.), *Estudios dedicados a James Leslie Brooks* (Barcelona: Puvill, 1984), 77–90.
Hurtado, J., and González Palencia, A., *Historia de la literatura española*, 2nd edn. (Madrid, 1925).
Jara, René, and Vidal, Hernán (eds.), *Testimonio y literatura* (Institute for the Studies of Ideologies and Literature, 3; Minneapolis, 1986) [René Jara, 'Prólogo', 1–6; Miguel Barnet, 'La novela-testimonio', 280–302.]
Laín Entralgo, Pedro, *España como problema* (2 vols.; Madrid: Aguilar, 1954).
Larra, Mariano José de, *Obras*, ed. C. Seco Serrano, i (BAE 128; Madrid, 1960).
Longhurst, Carlos, *Las novelas históricas de Pío Baroja* (Madrid: Guadarrama, 1974).
López-Morillas, Juan, 'La Revolución de Septiembre y la novela española', *Hacia el 98: Literatura, sociedad, ideología* (Barcelona: Ariel, 1972), 9–41.
Machado, Antonio, *Soledades. Galerías. Otros poemas*, ed. Geoffrey Ribbans, 7th edn. (Madrid: Cátedra, 1990).
Menéndez Pelayo, Unamuno, Palacio Valdés, *Epistolario a Clarín*, ed. Adolfo Alas (Madrid: Escorial, 1941).
Mesonero Romanos, Ramón de, *Escenas matritenses*, ed. Ángeles Cardona de Gilbert and Francisca Sallés de Martínez, 2nd edn. (Barcelona: Bruguera, 1970).
Miller, Stephen, 'Introduction to the Illustrated Fiction of the Generation of 1868', *RQ* 35 (1988), 281–7.
Oleza, Juan, *La novela del XIX* (Barcelona: Laia, 1984).
Oller, Narcís, *Obres completes* (Barcelona: Selecta, 1948).
Ortiz Armengol, Pedro, *Aviraneta y diez más* (Madrid: Prensa Española, 1970).
Pardo Bazán, Emilia, *La tribuna*, ed. Benito Varela Jácome (Madrid: Cátedra, 1975).
Pattison, Walter, *El naturalismo español: Historia externa de un movimiento literario* (Madrid: Gredos, 1969).
Peers, E. Allison, *Spain: A Companion to Spanish Studies* (London: Methuen, 1929).
Pereda, José María de, *Pedro Sánchez* (1883), 2nd edn. (2 vols.; Clásicos Castellanos;

Madrid: Espasa-Calpe, 1968).

Ribbans, Geoffrey, 'La evolución de la novelística unamuniana: *Amor y pedagogía y Niebla*', in *Niebla y soledad* (Madrid: Gredos, 1971), 83–107.

Román Gutiérrez, Isabel, *Persona y forma: Una historia interna de la española del siglo XIX*, ii. *La novela realista* (Seville: Alfar, 1988).

Rutherford, John, *Leopoldo Alas: La regenta* (1984) (CG; 1974).

Sáez de Melgar, Faustina, *La cruz del olivar*, ed. Alicia Andreu, *AG* suppl. (1980).

Scanlon, Geraldine, 'Class and Gender in Pardo Bazán's *La tribuna*', *BHS* 67 (1990), 137–50.

Schiavo, Leda, *Historia y novela en Valle-Inclán: Para leer 'El ruedo ibérico'* (Madrid: Castalia, 1980).

Sender, Ramón J., *Míster Witt en el Cantón* (1936), ed. José María Jover (Madrid: Castalia, 1987).

—— *Réquiem por un campesino español*, 2nd edn. (Buenos Aires: Proyección, 1966); ed. Patricia McDermott (Manchester UP, 1991). [1st edn. under title *Mosén Millán* (Mexico: Aquelarre, 1953.]

Sinclair, Alison, *Valle-Inclán's 'Ruedo ibérico': A Popular View of Revolution* (London: Tamesis, 1977).

Spires, Robert C., *Beyond the Metafictional Mode: Directions in the Modern Spanish Novel* (Lexington, Ky.: UP of Kentucky, 1984).

Ulmer, Gregory, *The Legend of Herostratus: Existential Envy in Rousseau and Unamuno* (Gainsville: UP of Florida, 1977).

Unamuno, Miguel de, *Obras completas*, ed. M. García Blanco (9 vols.; Madrid: Escelicer, 1968–71).

Valle-Inclán, Ramón María, *Luces de bohemia* (1921), ed. A. Zamora Vicente, 24th edn. (Madrid: Espasa-Calpe Austral, 1990).

—— *La corte de los milagros* (1927) (Madrid: Castalia, 1973).

—— *Viva mi dueño* (1928) (Madrid: Espasa-Calpe Austral, 1961).

—— *Baza de espadas* (1932), 2nd edn. (Madrid: Espasa-Calpe Austral, 1971).

Varela, José Luís, *Larra y España* (Madrid: Espasa-Calpe, 1983).

Varela Jácome, Benito, *Estructuras novelísticas del siglo XIX* (Barcelona: Aubí, 1974).

(*b*) *Other literatures*

Balzac, Honoré de, *Œuvres complètes*, ed. Marcel Bruteron and Henri Longnon, i (Paris: Conard, 1931).

Bellos, David, *Balzac: La Cousine Bette* (Critical Guides to French Texts; London: Grant and Cutler, 1980).

Brombert, Victor, *Stendhal: Fiction and the Themes of Freedom* (New York: Random, 1968).

Carlyle, Thomas, 'Heroes, Hero-worship and the Heroic in History', in *The Works*, v (Centenary Edition, London: Chapman and Hall, 1901).

Christian, R. F., *Tolstoy's 'War and Peace'* (Oxford UP, 1962).

Coleridge, Samuel Taylor, *Biographia Literaria* (Everymans' Library; London: Dent and New York: Dutton, 1906).

Flaubert, Gustave, *Madame Bovary: Mœurs de province* (1857), ed. Claudine Gothot-Mersch (Paris: Garnier, 1971).

—— *Salammbô* (1862), ed. René Dumesnil (2 vols.; Paris: Société Les Belles Lettres, 1944).
—— *L'Éducation sentimentale: Histoire d'un jeune homme* (1869), ed. P. M. Weatherill (Paris: Garnier, 1984).
—— *Correspondance*, ed. Jeran Bruneau (5 vols.; Paris: Conard, 1910).
—— *Bouvard et Pécuchet* (1881) (Paris: Conard, 1923).
Gledson, John, *The Deceptive Realism of Machado de Assís: A Dissenting Interpretation of 'Don Casmurro'* (Liverpool: Francis Cairns, 1984).
Green, Anne, *Flaubert and the Historical Novel: 'Salammbô' Reassessed* (Cambridge UP, 1982).
Hardy, Florence Emily, *The Life of Thomas Hardy, 1840–1928* (London: Macmillan, 1962).
Hardy, Thomas, *Tess of the D'Urbervilles*, ed. Juliet Grindle and Simon Gatrell (Oxford UP, 1983).
Hemmings, F. W. J., *An Interpretation of 'La Comédie humaine'* (New York: Random, 1967).
—— (ed.), *The Age of Realism* (Harmondsworth: Penguin, 1974).
Levin, Harry, *The Gates of Horn: A Study of Five French Realists* (Oxford UP, 1966).
Lewes, G. H., 'Historical Romance', *Westminster Review, 45* (Mar. 1846), 34–55.
Manzoni, Alessandro, *I promessi sposi* (1827), ed. Giovanni Titta Rosa (Milan: Mursia, 1963).
—— *Del romanzo storico . . . Opere varie* (Milan, 1850).
—— *On the Historical Novel*, trans. with an Introduction by Sandra Bermann (Lincoln: University of Nebraska Press, 1984).
Mérimée, Prosper, *Chronique du règne de Charles IX* (1829) (Paris: Nelson, 1944).
Nietzsche, Friedrich, 'On the Uses and Disavantages of History for Life', in *Untimely Meditations*, Introduction and trans. by J. P. Stern (Cambridge UP, 1983).
Proust, Marcel, A *la recherche du temps perdu*, ed. Pierre Charac and André Ferré (Paris: Pléiade, 1954). [English trans.: *Remembrance of Things Past*, trans. C. K. Scott-Moncrieff (and Andreas Mayor) (2 vols.; New York: Random, 1961).]
Pugh, Anthony, *Balzac's Recurring Characters* (Toronto UP, 1974).
Richards, I. A., *Coleridge on Imagination* (Bloomington: Indiana UP, 1960).
Roberts, David D., *Benedetto Croce and the Uses of Historicism* (Berkeley: University of California Press, 1987).
Schopenhauer, Arthur, *Essays and Aphorisms* (Harmondsworth: Penguin, 1970).
Schor, Naomi, *Breaking the Chain: Women, Theory and French Realist Fiction* (New York: Columbia UP, 1985).
Scott, Sir Walter, *Waverley* (1814) (WLE 1; Boston: Sanborn, Carter, Bazin, n.d.).
—— *Rob Roy* (1818) (WLE 4; Boston: Sanborn, Carter, Bazin, n.d.).
—— *The Heart of Midlothian* (1818) (WLE 6; Boston: Sanborn, Carter, Bazin, n.d.).
—— *Ivanhoe* (1819) (WLE 8; Boston: Sanborn, Carter, Bazin, n.d.).
—— *The Journal*, ed. W. E. K. Anderson (Oxford UP, 1972).
Shattuck, Roger, *Proust's Binoculars: A Study of Memory, Time and Recognition in 'A la Recherche du Temps Perdu'* (Princeton UP, 1983).
Stendhal, *Romans et nouvelles*, i, ed. Henri Martineau (Paris: Bibliothèque de la Pléiade, 1966).
Stern, J. P., *On Realism* (London: Routledge and Kegan Paul, 1973).

Taine, Hippolyte, *Histoire de la littérature anglaise* (1863), 15th edn. (5 vols.; Paris: Hachette, 1921). [English trans.: *History of English literature* (5 vols.; Edinburgh: Edmonston and Douglas, 1873).]

Thackeray, William Makepeace, *Vanity Fair* (1847) (HFCE; London: Macmillan, 1911).

—— *The History of Henry Esmond, Esq.* (1852) (HFCE; London: Macmillan, 1911).

Tolstoy, *War and Peace* (1862–9) (3 vols; Everymans' Library; London: Dent and New York: Dutton, 1911).

Ullman, Stephen, *Style in the French Novel* (Cambridge UP, 1957).

Vigny, Alfred de, *Cinq-Mars* (1826), ed. Annie Picherot (Paris: Gallimard, 1980).

Williams, David A.(ed.), *The Monster in the Mirror: Studies in Nineteenth-Century Realism* (Oxford and Hull: Oxford UP, 1978).

Woolf, Virginia, *The Waves* (New York: Harcourt, 1931).

Zola, Émile, 'Le roman expérimental', in *Œuvres critiques*, i (*Œuvres complètes*, 10, ed. Henri Mitterand; Paris: Cercle du Livre Précieux, 1968).

INDEX

Only those fictional characters with significant historical associations are included; they are indicated with an asterisk.